135

Mirror Neuron Systems

For other titles published in this series, go to
http://www.springer.com/series/7626

Jaime A. Pineda
Editor

Mirror Neuron Systems

The Role of Mirroring Processes in Social Cognition

Editor
Jaime A. Pineda
University of California
San Diego, La Jolla, CA
USA
pineda@cogsci.ucsd.edu

ISBN: 978-1-934115-34-3 e-ISBN: 978-1-59745-479-7
DOI: 10.1007/978-1-59745-479-7

Library of Congress Control Number: 2008937030

Printed on acid-free paper

springer.com

Preface

The aim of this book is to bring together social scientists, cognitive scientists, psychologists, neuroscientists, neuropsychologists and others to promote a dialogue about the variety of processes involved in social cognition, as well as the relevance of mirroring neural systems to those processes. Social cognition is a broad discipline that encompasses many issues not yet adequately addressed by neurobiologists. Yet, it is a strong belief that framing these issues in terms of the neural basis of social cognition, especially within an evolutionary perspective, can be a very fruitful strategy. This book includes some of the leading thinkers in the nascent field of mirroring processes and reflects the authors' attempts to till common ground from a variety of perspectives. The book raises contrary views and addresses some of the most vexing yet core questions in the field – providing the basis for extended discussion among interested readers and laying down guidelines for future research.

It has been argued that interaction with members of one's own social group enhances cognitive development in primates and especially humans (Barrett & Henzi, 2005). Byrne and Whiten (1988), Donald (1991), and others have speculated that abilities such as cooperation, deception, and *imitation* led to increasingly complex social interactions among primates resulting in a tremendous expansion of the cerebral cortex. The evolutionary significance of an imitation capability in primates is matched by its ontological consequences. Prior to the discovery of mirror neurons, a widely accepted account for what Baron-Cohen has called "mindreading" involved "simulation," according to which the fundamental cognitive process lying at the basis of this human ability was the use of one's own cognitive resources in order to simulate another's mind. As Pierre Jacob has argued, the discovery of mirror neurons in the macaque brain by Giacomo Rizzolatti and his group in Parma Italy (di Pellegrino, Fadiga, Fogassi, Gallese, & Rizzolatti, 1992) raised the intriguing possibility that "mirroring" may constitute instances of this type of mental simulation. Such a neurobiological account expands the explanatory scope and empirical base for simulation as a necessary component of social behavior.

The discovery of a mirror neuron system (MNS) has created an immense amount of interest because of the fact that such a system may be foundational to mindreading. This would go a long way in providing a rational basis for the study of social cognition. Most would agree that the human ability to represent one's own as well as other's perceptions, emotions, intentions, desires, and beliefs is necessary for human social behavior. While general cognitive processes of perception, attention, memory, and language are important in our social competence, there is enough available evidence to suggest the existence of specific processes unique to social interaction (Blakemore, Winston, & Frith, 2004). Recent brain imagery studies in humans, for example, have revealed that the insula is the same brain structure that is active when a person experiences an emotion such as disgust, as well as when the same person perceives another person experiencing disgust. This supports the notion of a "mirroring" theory of social cognition, according to which the basis of human social cognition would be provided by a variety of mirroring brain mechanisms.

Originally discovered in macaque monkeys, mirror neurons in ventral premotor area (area F5) possess the intriguing property of being activated by both the performance as well as the observation of specific motor actions. Such a property was quickly seen as a likely candidate for coding actions requiring both an ego and non-egocentric representation, such as necessary for imitation as well as for understanding actions. Furthermore, brain imaging studies have shown that activation of the motor system is functionally specific, that is, premotor and motor cortices, as well as parietal cortex are activated in a somatotopic way based on the modality and muscle groups being observed (Buccino et al., 2001; Blakemore et al., 2004). However, the view that mirror neurons are involved in imitation created the interesting paradox that while macaque monkeys possess such cells in premotor areas, the overwhelming evidence suggests they cannot easily imitate. Clearly then, mirror neurons must either play different functional roles in the two species or they must be involved in a unique common role. Recent evidence has shown that non-human primates can actually reason about other minds, even to the extent of using the intentional structure of behavior to distinguish identical actions (Lyons, Santos, & Keil, 2006). This is consistent with a "predictive" rather than "imitative" function for mirror neurons. That is, these cells are exquisitely sensitive to the goal of the observed action rather than the lower-level kinematic features of that action. This would suggest that such specialized cells may be more involved in inferring intentionality rather than in imitating actions. Evidence consistent with this explanation comes from monkey studies at the single cell level (Umiltà et al., 2001; Fogassi et al., 2005) and from human studies using fMRI (Iacoboni et al., 2005). Furthermore, the existence of "logically-related cells" as compared to "congruent" cells in the inferior frontal gyrus and their relative ratio in human and non-human primate brains provides a potential explanation for why monkeys don't imitate but can reason about intentionality and why humans are superb at both.

The question of how it might be that mirror neurons, simulation, emulation, and other such related neural mechanisms underlie the type of social cognition principles that motivate the field is a timely one. This book emphasizes important issues relevant to mirroring activity that are foundational to understanding the link to social cognition. *What is the ontogeny and phylogeny of mirroring processes?* Gallese proposes a neurophysiological hypothesis – the "neural exploitation hypothesis" – to explain how aspects of social cognition are mediated by brain mechanisms originally evolved for sensory-motor integration. Keysers and Gazzola describe a neurophysiological mechanism involving Hebbian learning that can account for mirror neuron's shared representations. Champoux et al. address the critical issue of the developmental trajectory of mirror neurons and argue that motor behaviors in utero may serve as building blocks for the appearance of the MNS. *Is mirroring necessary and sufficient to account for simulation and human social behavior?* Oberman and Ramachandran expand the definition of mirroring as a remapping from one domain into another and hence expand the explanatory power of this process. Similarly, Pineda et al. extend the definition for mirroring-like processes that occur at all levels of information processing in the central nervous system producing a gradient of faculties that vary in complexity from stimulus enhancement, response facilitation, emotional contagion, mimicry, simulation, and emulation to imitation, empathy, and theory of mind. Rochat and Passos-Ferreira argue that imitation and mirroring processes are necessary but not sufficient conditions for children to develop human sociality since it does not allow for the co-construction of meanings with others.

Are there different types of mirroring activity? Gallagher is critical of the idea of mirror neuron activation as simulation and proposes a non-mirror neuron based alternative. Semin and Cacioppo view mirroring as a monitoring synchronization process that is only a supportive component of social cognition but that can occur in at least three different forms of co-action. *Is mirroring replicative or predictive?* Iacoboni argues that mirroring allows for sharing of mental states between individuals that basically solves the problem of other minds. Furthermore, he proposes that such a mechanism may be the basis of a secular morality built upon biological predispositions. *Do mirror neurons provide the adaptive advantage for understanding the mental states of others in an effortless and automatic way?* In their chapter, Chong and Mattingley argue that MNS is open to top-down processes such as cognitive strategy, learned associations and selective attention. Guimmarra and Bradshaw describe how synesthesia for pain is most likely a consequence of disinhibited activation, or central sensitization, of an adaptive system for the empathic perception of pain in another. Winkielman et al. apply embodied cognition explanations to the understanding of emotionally significant information. In his chapter, Goldman argues for a connection between mirror process and mindreading with the caveat that not all mental attribution can be mirror-based. *Is the complexity of behavioral imitation matched by complexity in the cellular properties of mirror neurons?* Lyons theorizes that the neural basis of imitation is hampered by the oversimplification of imitation's complex cognitive reality. *Is mirror neuron system dysfunction the basis for social dysfunction in autism?* The chapter by Bernier and Dawson provides support for a dysfunctional mirror neuron system in both children and adults with autism. In contrast, Southgate et al. argues for an alternative explanation for the deficits in imitation seen in ASD individuals. They argue that a lack of sensitivity to cues that help identify "what" to imitate may be what is dysfunctional and not necessarily the mirror neurons themselves.

In summary, this book offers a new and exciting discussion of issues relevant to the neural basis of social cognition.

La Jolla CA Jaime A. Pineda

References

Barrett, L., & Henzi, P. (2005). The social nature of primate cognition. *Proceedings of Biological Science, 22*, 272(1575), 1865–75.

Blakemore, S.J., Winston, J., & Frith, U. (2004). Social cognitive neuroscience: Where are we heading? *Trends in Cognitive Science,* May; *8*(5), 216–22.

Byrne, R.W., & Whiten, A. (1988). *Machiavellian intelligence: Social expertise and the evolution of intellect in monkeys, apes, and humans.* Oxford: Oxford University Press.

Buccino, G., Binkofski, F., Fink, G.R., Fadiga, L., Fogassi, L., Gallese, V., et al. (2001). Action observation activates premotor and parietal areas in a somatotopic manner: An fMRI study. *European Journal of Neuroscience, 13*(2), 400–404.

di Pellegrino, G., Fadiga, L., Fogassi, L., Gallese, V., & Rizzolatti, G. (1992). Understanding motor events: A neurophysiological study. *Experimental Brain Research, 91*, 176–80.

Donald, (1991). *Origins of the modern mind: Three stages in the evolution of culture and cognition.* Cambridge, MA: Harvard University Press.

Fogassi, L., Ferrar, P.F., Gesierich, B., Rozzi, S., Chersi, F., & Rizzolatti, G. (2005). Parietal lobe: From action organization to intention understanding. *Science, 302*, 662–667.

Iacoboni, M., Molnar-Szakacs, I., Gallese, V., Buccino, G., Mazziotta, J., & Rizzolatti, G. (2005). Grasping the intentions of others with one's own mirror neuron system. *PLoS Biology, 3*, 529–35.

Lyons, D. E., Santos, L. R., & Keil, F. C. (2006). Reflections of other minds: How primate social cognition can inform the function of mirror neurons. *Current Opinion in Neurobiology, 16*(2):230–4.

Umiltà, M. A., Kohler, E., Gallese, V., Fogassi, L., Fadiga, L., Keysers, C., et al. (2001). "I know what you are doing": A neurophysiological study. *Neuron, 32*, 91–101.

Contents

Part IV Relationship to Cognitive Processes

Part V Disorders of Mirroring

Part VI Alternative Views

Contributors

Raphael Bernier, PhD
Department of Psychiatry and Behavioral Sciences, University of Washington, Seattle, WA, USA

John L. Bradshaw, PhD
Experimental Neuropsychology Research Unit, School of Psychology, Psychiatry and Psychological Medicine, Monash University, Clayton, Victoria, Australia

John T. Cacioppo, PhD
Department of Psychology, University of Chicago, Chicago, IL, USA, Cacioppo@uchicago.edu

François Champoux, MSc, MScS(A)
Psychology Department and CHU Sainte-Justine, University of Montreal, Montreal, Quebec, Canada

Trevor T-J Chong, PhD
School of Behavioural Science, University of Melbourne, Parkville, Victoria, Australia

Roy Cox, MS
Department of Cognitive Science, University of California, San Diego, CA, USA; Biomedical Sciences Program, University of Amsterdam, Amsterdam, The Netherlands

Gergely Csibra, PhD
School of Psychology, Birkbeck College, University of London, London, UK, g.csibra@bbk.ac.uk

Geraldine Dawson, PhD
Department of Psychiatry and Behavioral Sciences, University of Washington, Seattle, WA, USA

Christine Désy, BSc
Psychology Department and CHU Sainte-Justine, University of Montreal, Montreal, Quebec, Canada

Shaun Gallagher, PhD
Department of Philosophy and Cognitive Sciences, University of Central Florida, Orlando, FL, USA, gallaghr@mail.ucf.edu

Vittorio Gallese, MD
Department of Neuroscience, Section of Physiology, University of Parma, 43100 Parma, Italy, vittorio.gallese@unipr.it

Valeria Gazzola, PhD
BCN Neuro-Imaging-Centre, University Medical Center Groningen, University of Groningen, Groningen, The Netherlands, v.gazzola@med.umcg.nl

György Gergely, PhD
Department of Developmental Research of the Institute for Psychological Research, Hungarian Academy of Sciences, Budapest, Hungary, gergelyg@mtapi.hu

Melita J. Giummarra, BA
Experimental Neuropsychology Research Unit, School of Psychology, Psychiatry and Psychological Medicine, Monash University, Clayton, Victoria, Australia; National Ageing Research Institute, Parkville, Victoria, Australia, melita.giummarra@med.monash.edu.au

Alvin I. Goldman, PhD
Department of Philosophy, Center for Cognitive Science, Rutgers, The State University of New Jersey, New Brunswick/Piscataway, NJ, USA

Marco Iacoboni, MD, PhD
Ahmanson-Lovelace Brain Mapping Center, Departments of Psychiatry and Biobehavioral Sciences, Semel Institute for Neuroscience and Human Behavior, Brain Research Institute, David Geffen School of Medicine at UCLA, Los Angeles, CA, USA

Christian Keysers, PhD
BCN Neuro-Imaging-Centre, University Medical Center Groningen, University of Groningen, Groningen, The Netherlands, c.keysers@med.umcg.nl

Jean-François Lepage, BSc
Psychology Department and CHU Sainte-Justine, University of Montreal, Montreal, Quebec, Canada

Mélissa Lortie, BSc
Psychology Department and CHU Sainte-Justine, University of Montreal, Montreal, Quebec, Canada

Derek E. Lyons, PhD
Department of Psychology, Yale University, New Haven, CT, USA, derek.lyons@yale.edu

Jason B. Mattingley, PhD
Cognitive Neuroscience Laboratory, Department of Psychology and Queensland Brain Institute, University of Queensland, St. Lucia, Queensland, Australia

Adrienne Moore, MA
Department of Cognitive Science, University of California, San Diego, CA, USA

Paula M. Niedenthal, PhD
Centre National de la Recherche Scientifique (CNRS) and University of Clermont-Ferrand, France

Lindsay M. Oberman, PhD
Department of Neurology, Harvard University, Boston, MA, USA; Department of Psychology, University of California, San Diego, CA, USA

Claudia Passos-Ferreira, PhD
UERJ, State University of Rio de Janeiro, Rio de Janeiro, Brazil

Jaime A. Pineda, PhD
Department of Cognitive Science and Neurosciences Program, University of California, San Diego, CA, USA, pineda@cogsci.ucsd.edu

V.S. Ramachandran, MD, PhD
Departments of Neuroscience and Psychology, University of California, San Diego, CA, USA

Philippe Rochat, PhD
Department of Psychology, Emory University, Atlanta, GA, USA

Gün R. Semin, PhD
Department of Social Psychology, Free University Amsterdam, Amsterdam, The Netherlands, GR.Semin@psy.vu.nl

Victoria Southgate, PhD
School of Psychology, Birkbeck College, University of London, London, UK, v.southgate@bbk.ac.uk

Hugo Théoret, PhD
Psychology Department and CHU Sainte-Justine, University of Montreal, Montreal, Quebec, Canada, hugo.theoret@umontreal.ca

Piotr Winkielman, PhD
Department of Psychology, University of California, San Diego, CA, USA, pwinkiel@ucsd.edu

Part I
What Is Imitation?

Unifying Social Cognition

Christian Keysers and Valeria Gazzola

Abstract Humans have an almost uncanny capacity: they can simply observe other conspecifics and get deep intuitive insights into their minds. Since the discovery of mirror neurons, this capacity seems a little less mysterious. Here, we will review evidence that suggests that when we witness the actions, sensations, and emotions of other individuals, we activate our premotor, parietal, somatosensory, and emotional structures as if we were performing similar actions or experiencing similar emotions and sensations. These activations are stronger in more empathic individuals. These activations are stronger in more empathic individuals. This suggests that empathy relies, at least in part, on our brain's spontaneous transformation of what we see in what we would have felt.

Keywords Mirror System · Social Cognition · Emotions · Actions · Sensations · Empathy · Theory of Mind

As humans, we have an almost uncanny capacity: we can look at other people and guess what is going on in their minds. If we look at someone having a phone call, and see his face decompose, what we perceive is his sadness and inner turmoil. The way we describe such event reflects a mentalistic stance: "I think John just got some bad news. He looked so sad during this phone call . . .". You would not say: "John picked up the phone, and his zigomatic muscles swiftly relaxed, causing the corner of his mouth to turn down." All that really meets the eyes of observers are the behaviors of the people around them: their inner lives are neural states that are hidden from the observers' sight. The obvious question thus is how our brain perceives the mental states of others despite their inaccessibility. In this chapter, we would like to propose that an important element for our understanding of the inner states of others lies in the fact that while we witness the actions, emotions, and sensations of other individuals, our brain spontaneously

---

C. Keysers
BCN Neuro-Imaging-Centre, University Medical Center Groningen,
University of Groningen, Groningen, The Netherlands
e-mail: c.keysers@med.umcg.nl

J.A. Pineda (ed.), *Mirror Neuron Systems*, DOI: 10.1007/978-1-59745-479-7_1,

activates representations of our own actions, emotions, and sensations that make us literally bath in a state that resembles the inner state of the people we observe. Understanding what goes on in their minds is then facilitated by the fact that we feel the similar state inside us. Additional processes that are still relatively poorly understood are then required to interpret these simulated inner states, in a way that may resemble the way in which we interpret our own inner states. Furthermore, we will venture an explanation of how our brain could learn to associate the state of other individuals with similar states inside of ourselves: by learning Hebbian associations between our own visible behaviors and inner states.

For this aim, we will first review evidence for the fact that we activate our own motor representations while perceiving the actions of others. Second, we will review evidence for the fact that we activate our own somatosensory and nociceptive representations while viewing the somatosensory and nociceptive experiences of others. Finally, we will review evidence for the fact that we activate representations of our own emotional states and emotional facial expressions while witnessing the emotions of other individuals. We will then address the problem of how we interpret these simulated inner states and how the link between visible behaviors and hidden inner states could be established. These proposals are further developments of earlier models (Gallese, Keysers, & Rizzolatti, 2004; Keysers & Gazzola, 2006).

1 Shared Representations for Actions

1.1 Primates

In primates, three brain areas have been particularly associated with the perception of the actions of other individuals: the superior temporal sulcus (STS), area PF of the inferior parietal lobule, and area F5 of the ventral premotor cortex (see Fig. 1 left). Two of these areas, PF and F5, have been shown to contain neurons called 'mirror neurons.' Mirror neurons (MN) are a subclass of visuo-motor neurons that have a peculiar combination of response properties. First, they respond during the execution of particular actions, even if the monkey cannot see himself perform the action, for instance, because its eyes are closed. Second, they respond during the observation of another individual that performs similar actions. The third area, the STS, has so far not been shown to respond during motor execution without visual feedback and has therefore not been shown to contain mirror neurons. We will briefly report the key properties of these three areas.

1.1.1 STS

The STS receives dense visual input from both the ventral and dorsal visual stream, and contains neurons responding to visual movement in general (Hietanen & Perrett, 1993, 1996), and biological movements in particular (Puce & Perrett, 2003). Different cells in this area show selectivity for particular biological

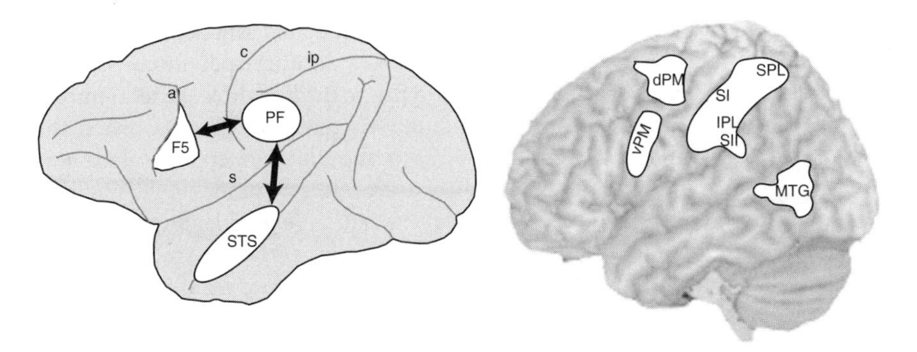

Fig. 1 *Left*: Lateral view of the macaque brain with the location of F5, PF, and STS together with their anatomical connections (*arrows*). The following sulci are shown: a = arcuate, c = central, ip = intraparietal, s = sylvian sulcus. *Right*: Corresponding view of the human brain with the brain areas involved both in action observation and execution, adapted from Gazzola et al. (2007a)

actions: certain cells are selectively responding to faces (Keysers, Xiao, Foldiak, & Perrett, 2001; Foldiak, Xiao, Keysers, Edwards, & Perrett, 2004), including particular facial expressions, others to the sight of grasping or walking, or even the presence of other individuals in hidden locations (Baker, Keysers, Jellema, Wicker, & Perrett, 2001). This combination of selectivity could serve to describe the behavior of other individuals. A small number of studies have investigated the response of this area while the monkey itself executes actions (Hietanen & Perrett, 1993, 1996; Keysers & Perrett, 2004). In studies of Hietanen et al. (1993, 1996), half of the neurons responding to movements in a particular direction also responded if the movement was caused by the monkey's own actions. The other half greatly reduced their response if the movement was produced by the monkey itself. This reduction demonstrates the fact that information about the monkey's own actions reach the STS. It remains unclear whether this latter information is motor or proprioceptive in nature. In a pilot experiment (briefly reported in (Keysers & Perrett, 2004)), we tested cells in the STS while the monkey executed actions with its eyes closed and we failed to find cells that responded in these conditions. Information about the monkey's own actions has thus been shown to inhibit responses in the STS but so far not to cause increases in firing rate as would be expected for true mirror neurons.

The area STS has reciprocal connections with area PF, and PF, as we will see below, also contains neurons that respond both during the observation of the actions of other individuals and the execution of similar actions (Seltzer & Pandya, 1978, 1994).

1.1.2 PF

Area PF contains a variety of neurons responding during the execution of hand and mouth actions, including grasping and placing (Fogassi et al., 2005).

Interestingly some of these neurons are MN, which also augment their firing rate while the monkey is immobile but observes another individual perform similar actions (Fogassi et al., 2005). A good proportion of these parietal mirror neurons respond during a particular action (e.g., grasping), both when the monkey uses its right or left hand or even its mouth to grasp. The action of grasping is thus represented independently of the effector used, suggesting that the unit of representation in this area is truly an action (e.g., grasping) and not a particular motor plan to do so. Other parietal mirror neurons are more tied to a particular effector and may thus represent a lower or more detailed level (i.e., closer to a particular motor plan) representations of particular actions. Using single cell recordings, it is difficult to quantify precisely what proportion of neurons in area PF show mirror properties or what proportion of neurons show effector independence, because experimenters interested in mirror neurons will generally record more intensively from locations where mirror neurons were recorded, leading to a systematic overestimation of the proportion of mirror neurons. Interestingly, the response of some of the mirror neurons in PF depends on the sequence of actions a particular action is embedded in: some grasping neurons, for instance, respond more strongly when the monkey grasps to place the object in a bowl than if the same object is grasped to be eaten (Fogassi et al., 2005). This suggests that representations can extend beyond single actions into action sequences. Area PF is reciprocally connected both with STS (Seltzer & Pandya, 1978, 1994) and with area F5 of the ventral premotor cortex (Matelli, Camarda, Glickstein, & Rizzolatti, 1986).

1.1.3 F5

Area F5 is the best studied of the brain areas involved in the observation of actions (Gallese, Fadiga, Fogassi, & Rizzolatti, 1996; Kohler et al., 2002; Keysers et al., 2003; Fujii, Hihara, & Iriki, in press). About 10–20% of the recorded neurons in F5 respond both during the execution and the observation of actions. This subset of F5 neurons vary in their selectivity (Gallese et al., 1996; Kohler et al., 2002; Keysers et al., 2003; Fujii et al., in press), with a label of strictly or broadly congruent MN describing these differences (Gallese et al., 1996). Strictly congruent MN have a range of effective executed and observed actions that is extremely similar, while broadly congruent MN have both similarities and differences between effective executed and observed actions. A prototypical strictly congruent MN, for instance, only responds during the execution of a precision grasp with the hand, but not during the execution of the same action with the mouth or the execution of a whole hand prehension. The same neuron will also respond more strongly to the sight of precision grasping with the hand compared to whole hand prehension or grasping with the mouth. A broadly congruent mirror neuron may, on the other hand, show difference between visual and execution preferences. It may respond only during the execution of precision grasping but both to the observation of grasping with hand or mouth. Strictly congruent MN have received most attention because they are the most striking evidence of the idea that a particular observed action is mapped onto motor representations of the exact same action. Broadly congruent MNs though are about twice as frequent as strictly congruent mirror neurons, and they have the potentially important property of activating in the brain of the observer a number of alternative ways for achieving the observed actions. While observing someone grasping a straw with his mouth, for instance, through broadly congruent MN, the observer would activate both his hand and mouth motor programs to achieve the same goal. Combining these hand and mouth motor programs, the broadly congruent MNs transform the observation of a particular way of achieving a goal into an effector independent representation of how this goal could be achieved. We will come back to the functional importance of such effector independent representations later.

Some neurons in F5 respond during the execution of actions and the sound of similar actions (Kohler et al., 2002; Keysers et al., 2003). Some of these auditory-motor neurons also respond to the vision of similar actions and have been termed audio-visual MN (Kohler et al., 2002; Keysers et al., 2003). Typical examples are neurons responding while the monkey rips a sheet of paper apart, and while viewing or listening to another individual perform the same action. Responses to another action, such as opening the shell of a peanut, would be lower in all cases. These audio-visual MN are also effector independent. The sound of an action such as paper-ripping gives very little information about the way in which the paper is being ripped: whether you rip it with your feet or with your hands, or with one hand while holding the other side of the paper with your teeth will produce the same sound. The response of the neuron to the sound of the action is thus inherently effector independent.

Interestingly, there is a roughly somatotopical organization within area F5, with dorsal sectors showing a prevalence of hand actions and ventral aspects a prevalence of mouth actions. In intermediate aspects, many neurons respond similarly to hand-and-mouth actions, and neurons with preference for one effector are often found side by side with neurons with preference for other effectors within this intermediate sector. As we will see later, this somatotopy is a useful tool for human neuroimaging, but is increasingly reinterpreted as an organization along goals more than along effectors.

1.2 Humans

A vast amount of studies have examined the presence of mirror neurons for actions in humans. Reviewing all these studies would go beyond the scope of this review and can be found elsewhere (e.g., Keysers & Perrett, 2004; Keysers & Gazzola, 2006; Rizzolatti & Craighero, 2004). After a very brief description of the gist of this literature, we will concentrate on novel developments within the human mirror system literature.

1.2.1 The Core Human Mirror System

Within the neuroimaging literature in humans, two areas are considered to form the core of the human mirror neuron system. These include the ventral premotor cortex, in particular Brodmann Area (BA) 6/44, and the rostral inferior parietal lobule (see Fig. 1 right). These areas are considered to be the homologues of area F5 and PF where mirror neurons have been found in the monkey. Both of these areas are consistently activated both while viewing the actions of other individuals and while participants execute similar actions (Grafton, Arbib, Fadiga, & Rizzolatti, 1996; Decety et al., 1997; Iacoboni et al., 1999; Buccino et al., 2001; Iacoboni et al., 2001; Grezes, Armony, Rowe, & Passingham, 2003; Buccino et al., 2004a, 2004b; Wheaton, Thompson, Syngeniotis, Abbott, & Puce, 2004; Calvo-Merino, Glaser, Grezes, Passingham, & Haggard, 2005; Iacoboni et al., 2005; Pelphrey, Morris, Michelich, Allison, & McCarthy, 2005; Gazzola, Aziz-Zadeh, & Keysers, 2006; Hamilton & Grafton, 2006; Gazzola et al., 2007a; Lahav, Saltzman, & Schlaug, 2007).

In addition, the middle temporal gyrus (MTG) and adjacent STS are often found to show augmented Blood Oxygen Level Dependent (BOLD) responses both during action execution and action observation (Iacoboni et al., 2001; Grezes et al., 2003; Gazzola et al., 2006; Gazzola et al., 2007a). While during the sound or vision of actions, this augmentation is fully compatible with single cell recordings in the STS of monkeys, the augmentation of BOLD responses during action execution without visual feedback (Iacoboni et al., 2001; Gazzola et al., 2006, 2007a) is in contrast to the reduction of firing rate found in the monkey during the monkey's own actions (Hietanen & Perrett, 1993, 1996). Two explanations have been offered to account for this finding. (1) The BOLD augmentation during action execution reflects the existence of mirror neurons in the STS that represents the expected visual and auditory consequences of the agent's own actions even if the agent cannot see him/herself (Iacoboni et al., 2001). After vainly searching for six months in the monkey's STS for cells showing such mirror like properties in the STS, other authors favor the idea that (2) the BOLD signal augmentation might instead reflect the augmented blood flow required to perform the metabolically demanding task of inhibiting neurons in the STS that would otherwise respond to the sight or sound of ones own actions (Keysers & Perrett, 2004; Gazzola et al., 2006, 2007a). Explanation 2 is in accord with existing primate data (Hietanen & Perrett, 1993, 1996) and the fact that augmentations of the BOLD signal can occur without changes in the firing rate of the neurons because of the metabolic cost of (inhibitory) synapses (Logothetis, 2003).

The core mirror system is thus considered to be composed of the premotor cortex and inferior parietal lobule, while the MTG/STS is considered to be a very close partner of the mirror system.

Neuroimaging techniques of course can only identify the voxels, within these areas, that are activated in both observation and execution, but cannot demonstrate that single neurons within these voxels respond in both cases. Attempts to show that mirror *neurons* (as opposed to mirror *voxels*) indeed exist in humans

thus have to derive from other methods. Transcranial magnetic stimu
in particular, although it is applied at the level of entire brain regions,
encouraging indirect evidence for the presence of mirror neurons. TM
primary motor cortex can evoke so-called motor evoked potentials (MEP), that
is, muscle activity that can be measured using Eteclromyographic (EMG) electrodes. A number of investigators have been able to show that observing an action that involves similar muscle movements facilitates these MEP, providing evidence for the fact that in humans also the vision of the actions of other individuals modulates neurons involved in executing similar actions (Gangitano, Mottaghy, & Pascual-Leone, 2001; Fadiga, Craighero, Buccino, & Rizzolatti, 2002; Aziz-Zadeh, Iacoboni, Zaidel, Wilson, & Mazziotta, 2004; Montagna, Cerri, Borroni, & Baldissera, 2005). In addition to TMS experiments, purely behavioral studies have provided evidence for the convergence of observed and executed actions: observing certain actions can interfere with the execution of incompatible actions (Kilner, Paulignan, & Blakemore, 2003; Chaminade, Franklin, Oztop, & Cheng, 2005; Press, Bird, Flach, & Heyes, 2005). In the experiment of Kilner and colleagues, for instance, subjects had to move their hands up and down along an imaginary straight vertical line. Interestingly, simultaneously seeing another individual move its hands from left to right and back (an incompatible movement) led the observer to deviate from the straight vertical line (Kilner et al., 2003), demonstrating again the convergence of seen and executed actions in the human brain. Although these methods provide evidence for the existence of a convergence of visual, auditory, and motor information in the human brain, they provide little information about where in the brain this convergence first occurs. In theory, the monkey's motor command could activate spinal motor neurons along a path that differs entirely from the visual information, and the facilitation and interference could occur entirely at spinal level.

Finally, in principle, singe cell recordings could occasionally be performed in human subjects undergoing surgery for the treatment of epilepsy. These recordings could provide direct evidence for mirror neurons in humans. The location of these recordings in though dictated by the suspected foci of epileptic activity, which are rare in premotor or posterior parietal cortex. Therefore, to our knowledge, such recordings have so far not led to direct evidence for MN in humans.

1.2.2 Shared Voxels and Shared Circuits

Neuroimaging shows that certain voxels are activated both during motor execution and observation, but neuroimaging cannot show that these shared voxels contain mirror neurons. They could simply contain interleaved populations of purely motor and purely visual neurons. Other techniques provide evidence for sensory motor convergence, but cannot indicate where in the brain this convergence occurs. At present, although the existence of MN in monkeys in premotor and parietal cortex makes it likely that humans have MN in premotor and parietal voxels that are active both during action execution and observation, it is advisable to use the term shared voxels (sVx) or shared circuits instead of mirror voxels or mirror neuron system when referring to human neuroimaging studies.

1.2.3 Audiovisual Shared Circuits

Often, we can recognize the action of another individual even if we can only hear it. A number of studies suggest that humans possess an auditory mirror neuron system (Fadiga et al., 2002; Aziz-Zadeh et al., 2004; Pizzamiglio et al., 2005; Bangert et al., 2006; Gazzola et al., 2006; Lahav et al., 2007) that resembles the one found in monkeys (Kohler et al., 2002; Keysers et al., 2003). Gazzola et al. (2006), for instance, showed that the sound of actions activates premotor and parietal voxels also activated during the execution of similar actions. Many of the same voxels also responded to the sight of similar actions. Interestingly, the voxels that were activated by the sound, vision, and execution of actions were more numerous and more strongly activated in the left hemisphere, raising the intriguing possibility that left lateralization of spoken language in humans could be due to the fact that language builds upon the abstract representation of actions that are implemented in a left lateralized multimodal mirror neuron system (Kohler et al., 2002; Keysers et al., 2003; Aziz-Zadeh et al., 2004; Gazzola et al., 2006). The left hemisphere dominance of the human auditory mirror system finds support from a study (Aziz-Zadeh et al., 2004) showing that MEPs are facilitated more by the sound of actions when TMS is applied to the left compared to the right hemisphere.

1.2.4 Congruent Selectivity

If mirror activity is to be informative about what other individuals do, the vision/sound of a particular action 'A' needs to activate motor programs in the observer that subserve the execution of similar actions more than those subserving the execution of a dissimilar action 'B.' Vice versa, the vision/sound of action B needs to activate motor programs for B more than for A. This feature has been demonstrated in monkeys at the level of single neurons (Gallese et al., 1996; Keysers et al., 2003). In humans, TMS experiments show that observation of particular actions or phases of actions selectively facilitate TMS-evoked motor potentials involved in these (phases of) actions (Gangitano et al., 2001; Fadiga et al., 2002; Aziz-Zadeh et al., 2004; Montagna et al., 2005). Neuroimaging experiments have reported a certain somatotopy of activations during the vision of actions similar to those reported in the monkey, with hand actions causing more dorsal activations than mouth actions (Buccino et al., 2001; Wheaton et al., 2004), but these experiments did not measure brain activity during action execution to ensure the congruence between the visual and motor preferences of the voxels. Recently, Gazzola et al. tested subjects while performing and listening to hand and mouth actions, and provided the first direct neuroimaging evidence for congruent mapping in humans: a dorsal premotor cluster responded more to hand than mouth actions during execution and listening, and a ventral premotor cluster more to the sound and execution of mouth actions compared to hand actions. In addition to these locations that discriminated hand and mouth actions, many sVx in the premotor, parietal, and temporal cortex responded similarly to actions of the hand and mouth. In analogy to the combination of strictly and broadly congruent MN in the monkey, the premotor cortex thus appears to transform the sound of particular actions into a combination of effector specific and effector unspecific motor programs.

1.2.5 Goal Matching

As mentioned in Section 1.1.3, most MNs are broadly congruent. Some broadly congruent neurons match the sight of grasping with the mouth or hand onto motor programs for grasping with the hand, and others onto motor programs for grasping with the mouth. One might thus speculate that the primate MNS matches actions not only in terms of details (precise effector, velocity of movement etc.) but also, and probably even more so, in terms of goal,[1] that is, what is achieved by the action (e.g., grasping), independently of how it is achieved. The predominance of goals in the mirror system of the monkey becomes even clearer for sounds of actions, which contain in themselves no information about the effector that was used to do the action: a crunching sound, for instance, can be the result of stepping on a potato chip with your foot, or crunching it with your fist or with your teeth.

A number of recent fMRI experiments have investigated the importance of goals in the human MNS. Gazzola et al. (2007a) have compared the vision of a robotic claw performing a variety of actions with that of a human hand doing similar actions. The actions were well known, every day actions such as taking and swirling a glass of whine, scooping soup out of a bowl etc. Interestingly, motor areas activated when subjects performed these actions within the scanner were recruited to the same extent by seeing a robot or a human perform these actions. This finding supports the idea that the MNS predominantly matches goals because the details of the action differed greatly between the robotic and

[1] The term 'goals' will be used throughout this chapter with the meaning of 'what is being achieved.' In that sense, the goal of grasping a glass is to take possession of this object in order to use it in some way. In this very pragmatic sense, a goal does not apply higher psychological concepts such as a sense of agency etc. It is a useful concept though, as it expresses what is common to performing a certain action (e.g., grasping) independently of how it is done (with your mouth, foot, hand, or a tool). Naively, one might wonder if such a concept is not mentalistic, detached from neuroscientific evidence. This though seems not to be the case: signing a sheet of paper with your foot or hand (Rijntjes et al., 1999), for instance, relies on the same location of the premotor cortex, suggesting that the premotor cortex indeed represents goals, which can then be flexibly mapped onto the most suited effector depending on environmental contingencies.

human versions: Human actions had natural biological kinematics and the familiar aspect of a human hand which can be easily matched on the details of the observer's own hand actions using strictly congruent MN. In contrast, the robotic actions had very artificial motion parameters and a claw that looked very different from a hand. The details of these robotic actions cannot be mapped on the details of the hand actions of the observer using strictly congruent MN. Nevertheless, the activations in the MNS did not differ significantly, suggesting that broadly congruent MN somehow matched the goals of the observed actions onto motor programs with similar goals independently of the discrepancies in effector and movement details (see (Gazzola et al., 2007b) for a discussion of why previous studies failed to find mirror activations to robotic actions). This finding finds support from another experiment in which two subjects born without hands and arms watched hand actions (Gazzola et al., submitted) and were compared with 16 control subjects watching the same actions. Both in the aplasic individuals and in controls, the vision of hand actions activated robustly a network of brain areas generally associated with the human MNS, and there were no significant differences between the groups. Both of these experiments raise the question of what motor programs are activated when subjects view actions that deviate from their own embodiment (e.g., robotic actions for control subjects and hand actions for aplasic individuals). The same subjects were also asked to perform actions in the scanner using their mouth and feet and for the controls, using their hands. Comparing the amplitude of activations in these conditions, the authors identified regions that respond similarly during the execution of actions with all effectors ('effector independent') and those that were activated more during the use of a particular effector (effector specific areas). It turned out that about half of the visual activations were in effector independent areas, while the other half fell within the effector specific regions where the observer would use to perform the observed actions (hand for the controls and mainly foot for the aplasics) despite the fact that this effector was not the one used by the observed individual (robotic claw? hand and hand? foot). This indicates that the MNS appears to map the goal of the observed action on a combination of effector independent motor programs and those effector dependent motor programs that the observer would use to achieve the goal independently of whether this is the effector used by the demonstrator or not.

These findings shed new light on two issues. First, although they are compatible with the idea of the mirror system matching observed actions onto corresponding motor programs ('direct matching hypothesis'), they indicate what 'corresponding motor program' really means: not only actions with directly corresponding details but also, and maybe mainly, motor programs with corresponding goal. Second, they shed new light on somatotopy in the premotor cortex: if different effector activates the same region, the dorso-ventral organization of the premotor cortex is likely to reflect an organization in terms of goals, with goals that primarily relate to the mouth represented ventrally, and goals primarily relating to the hand more dorsally. These goals are then usually achieved using the dominant effector but can be implemented using different effector (Rijntjes et al., 1999; Gazzola et al., 2006, 2007b)

This interpretation also helps understand two behavioral observations that would seem puzzling if the MNS would match the details of an action and not its goals. First, from early on, children tend to imitate the goals of observed actions and not the details of the actions (Bekkering, Wohlschlager, & Gattis, 2000; Gergely, Bekkering, & Kiraly, 2002; Wohlschlager, Gattis, & Bekkering, 2003; Williamson & Markman, 2006). Second, monkeys have mirror neurons but they do not imitate the details of observed actions. On the other hand, they do learn to achieve goals by observing other individuals perform these goals (Subiaul, Cantlon, Holloway, & Terrace, 2004).

A number of other studies confirm the importance of goals within the MNS. Hamilton et al. (2006) elegantly showed that presenting participants two actions in a row that have the same goal but different means causes a more severe decrease in the BOLD signal (i.e., selective adaptation) than showing two actions with the same means but different goals. Given that selective adaptation is though to indicate that the same neurons were used in the processing of the consecutive stimuli, this finding suggests that even if two actions have separate means but similar goals, they activate the same neurons within the MNS. Finally, Fogassi et al. (2005) and Iacoboni et al. (2005) showed that the vision of a particular action (e.g., grasping a glass) causes different responses in the MNS if the context indicate that the action serves a different goal (e.g., grasping to eat or drink vs. grasping to place).

1.2.6 MNS and Empathy

If the MNS contributes to an intuitive understanding of what other people do, one would expect that people that are more empathic would activate their MNS more strongly. To address this question, Gazzola et al. (2006) measured how empathic 16 subjects were using the interpersonal reactivity index (IRI) of Davis (1983) before scanning them while they executed or listened to various actions. While the six subjects with highest scores in the perspective taking subscale of the IRI showed very strong activations during the sound of actions in the premotor and parietal regions active during the execution of similar actions, the 6 least empathic individuals almost entirely failed to show significant activations in these regions (Fig. 2). The amplitude of activations in the MNS while listening to the sound of actions directly correlated with how empathic individuals were in premotor, parietal, and somatosensory areas, with r-values in excess of 0.6, suggesting that ~40% of the inter-individuals differences in empathy can be accounted for by differences in the putative MNS. While only correlational in nature, this is the first finding that differences in empathy may depend on differences in the MNS. As we will see below, a similar correlation appears also to exist during the perception of facial expressions. In addition, lesions in the premotor cortex have been shown to impair the recognition of the gestures of other individuals, suggesting that this region maybe

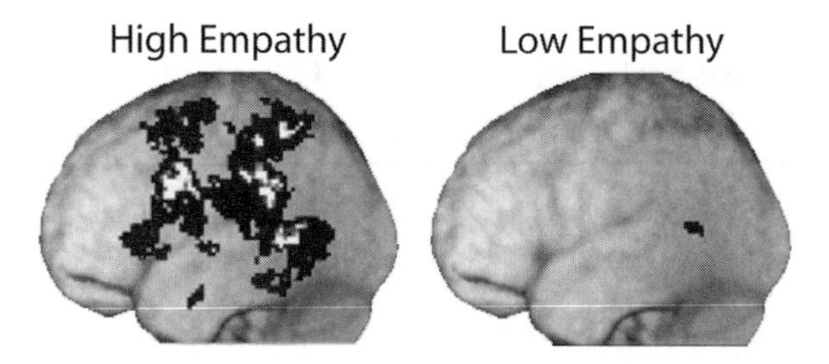

Fig. 2 *Left*, voxels involved both during the sound of actions and the execution of similar actions in participants ranking high in empathy according to self report. *Right* same for participants ranking low in empathy. Adapted from Gazzola et al. (2006)

necessary for a normal understanding of the actions of other individuals (Saygin, Wilson, Dronkers, & Bates, 2004), a finding confirmed by rTMS 'virtual lesions' (Heiser, Iacoboni, Maeda, Marcus, & Mazziotta, 2003). Together this data supports the idea that the MNS may indeed contribute to our intuitive understanding of the actions of other people.

1.2.7 Are Shared Circuits Limited to the Premotor and Parietal Lobe?

While in the monkey only areas F5 (Gallese et al., 1996; Umilta et al., 2001; Kohler et al., 2002; Keysers et al., 2003), PF (Fogassi et al., 2005) and the anterior bank of the intraparietal sulcus (Fujii et al., in press) have been thoroughly examined for the presence of mirror neurons, human neuroimaging has generated data for the entire brain. Of particular interest, a number of studies have examined the same subjects during action execution and action observation (Iacoboni et al., 1999; Grezes et al., 2003; Buccino et al., 2004b; Gazzola et al., 2006, 2007a). In these experiments, it is possible to define which voxels of the brain were involved both during the observation and the execution of actions. The existence of such shared voxels cannot demonstrate the presence of MN but can identify candidate areas that could contain such neurons. Next to the classic mirror regions, these experiments have identified a number of brain areas that contain shared voxels (Fig. 1 right). In particular, the somatosensory cortices (primary and secondary) often contain voxels active both during action execution and observation/listening (Nishitani & Hari, 2000; Grezes et al., 2003; Raos, Evangeliou, & Savaki, 2004; Gazzola et al., 2006, 2007a; Gazzola & Keysers, submitted), a finding we will come back to in the next chapter. A number of additional areas show sVx in a number of experiments: the dorsal premotor cortex (Grezes et al., 2003; Calvo-Merino et al., 2005; Calvo-Merino, Grezes, Glaser, Passingham, & Haggard, 2006; Gazzola et al., 2006; Molnar-Szakacs, Kaplan, Greenfield, & Iacoboni, 2006; Gazzola et al., 2007a; Gazzola et al., submitted), the superior parietal lobule (Gazzola et al., 2006, 2007a, submitted), MTG (Iacoboni et al., 2001; Gazzola et al., 2006, 2007a, submitted) and the cerebellum (Grezes et al., 2003; Gazzola et al., 2006, 2007a).

1.2.8 Single Subject Analysis Using Unsmoothed Data Instead of Conventional Group Analysis Reveal the Consistency of Shared Voxels Within and Outside the Premotor and Posterior Parietal Cortex

In conventional fMRI studies the interpretation of shared voxels in terms of MN is particularly difficult because shared voxel at the level of a random effect group analysis can occur without any of the single subjects showing shared voxels. There are two reasons for this problem. First, data in most fMRI studies is smoothed using filters of 6–10 mm. A voxel at the border between two unimodal territories can then appear to be shared simply because the smoothing makes the properties of unimodal territories artificially blur into each other. Second, data is usually analyzed in terms of group analyses of normalized data. This approach is valid if after normalization all subjects have the same functional brain areas in exactly the same location. Unfortunately, the border of cytoarchitectonic (Amunts et al., 1999; Grefkes, Geyer, Schormann, Roland, & Zilles, 2001) and functional (Amiez, Kostopoulos, Champod, & Petrides, 2006) areas is very variable from subject to subject and a voxels at the margin between two functional areas (e.g., Primary somatosensory cortex (SI) and Superior-parietal-lobule (SPL)) within a group analysis will thus contain voxels truly belonging to SI in some subjects and truly belonging to the SPL in other subjects. If SI would, for instance, only be activated while performing an action (due to the somatosensory and kinesthetic consequences of performing the action) and the SPL would only be activated during the observation of actions, at the group level, voxels at the border between SPL and SI would appear to be activated during both action execution and action observation, because the *t*-tests would indicate that the average activation is significantly above zero both during action execution (due to the proportion of subjects that had their SI in this voxel) and action observation (due to the proportion of subjects that had their SPL in this voxel). Both these problems can be overcome by analyzing the data of each subject separately (thereby avoiding the problem of mixing subjects) and by using unsmoothed data. Doing so, Gazzola et al. (submitted) confirmed that shared voxels can be reliably defined in all subjects in premotor and inferior parietal locations, but that these classic areas constitute only about 20% of the shared voxels in the brain during action observation and execution. The other 80% were mainly in somatosensory cortices, the MTG/STS, the superior parietal lobule, the cerebellum, and the middle cingulate cortex. All these areas are known to integrate visual and motor/somatosensory information, and could very well contain mirror neurons. Single cell recordings in these areas will constitute an important step in the next decade to understand the spatial extent of the MNS.

1.3 Conclusions

Monkeys have mirror neurons that respond during the vision, the sound, and the execution of certain actions. The most frequent type of mirror neurons are broadly congruent mirror neurons that map observed actions onto motor programs with similar goals, generalizing beyond the precise effector used by the demonstrator. These neurons have so far been found in the monkey ventral premotor and parietal cortex. Humans appear to have a similar system that matches the goals of observed or heard actions onto matching motor programs in the observer. This system has the capacity to bridge differences in embodiment between observer and agent. It provides a neural correlate for the tendency during imitation to achieve the observed goal using whatever means make most sense for the observer. Neuroimaging though suggest that the mirror system may extend beyond the premotor cortex and inferior parietal lobule, encompassing potentially the mid temporal gyrus, the superior parietal lobule, somatosensory cortices, and the cerebellum. Activity in the putative human mirror system is stronger in more empathic individuals, suggesting a link between these activations and our capacity to share the goals of other individuals.

2 Sensations

2.1 Touch

If shared circuits maybe essential for our understanding of the actions of others, how about their sensations? If we see a spider crawl on James Bond's chest in *Dr. No*, we literally shiver, as if the spider crawled on our own skin. What brain mechanisms might be responsible for this spontaneous sharing of the sensations of others? May shared circuits exist for the sensations of touch or pain?

A number of investigations now suggest this to be the case. First, Keysers et al. (2004) showed participants movies of legs being brushed. In control movies, the same legs were approached by an object, but never touched. In the second part of the experiment, the same participants were touched on their legs. Within the secondary somatosensory cortex (SII or cytoarchitectonic area OP1), shared voxels for touch were found: they responded more during the vision of touch compared to the control movies and more during the experience of touch than the baseline. Again, neuroimaging cannot show that the same neurons within SII indeed respond during the experience and the observation of touch, but this finding suggests that SII could be part of a circuitry that transforms the vision of the sensations of other individuals into what it would feel like to be touched in similar ways. Using single subject analysis, they confirmed the presence of these voxels in the SII of most of their single subjects, supporting the idea that the overlap between the vision of touch and the experience of touch was not the result of group analysis alone.

Later, Blakemore, Bristow, Bird, Frith, and Ward (2005) were intrigued by subject ('C') reporting that when she sees someone else being touched on the face, she literally feels the touch on her own skin. They scanned C and a group of normal controls while touching them on their faces and necks. In a following session, they showed video clips of someone else being touched on the same locations. As in our study, the experience of touch activated primary and secondary somatosensory cortices. During observation, they found SI and SII activation. In C, these activations were significantly stronger, potentially explaining why she literally felt observed touch on her own skin.

It therefore appears as though seeing someone else being touched activated a somatosensory representation of touch in the observers, as if they had been touched themselves. This finding is particularly important as it demonstrates that the concept of shared circuits put forward for actions appears to be applicable to a very different system: that of touch.

Interestingly, somatosensory activations are also common to the observation and execution of actions (Grezes et al., 2003; Gazzola et al., 2006, 2007a). This suggests that during action observation, activations of somatosensory cortices may reflect a transformation of the observed kinesthetic and somatosensory information into a representation of what it would have felt like to perform a similar action.

2.2 *Pain*

Painful stimulation of the skin and the observation of a similar stimulation applied to others also appear to share a common circuitry. First Hutchison et al. recorded from single neurons in the anterior cingulate cortex (ACC) of surgical patients and found a neuron that responded both when pinpricking the patient and when the surgeon was pinpricking himself (Hutchison, Davis, Lozano, Tasker, & Dostrovsky, 1999), demonstrating the existence of (at least one) mirror neuron for pain. The first fMRI investigation of this phenomenon by Singer et al. (2004) examined brain activity of female participants in four conditions. In the condition PainSelf, subjects viewed a symbolic cue on the screen and felt a mildly painful electroshock on their own hand. In the condition NoPainSelf, a different cue was shown and a milder electrostimulation cause a non-painful sensation in the hand. In the conditions PainPartner and NoPain-Partner, cues indicated that the partner in life of the participant received a painful or non-painful electrostimulation of the skin. Singer et al. found that the contrast PainSelf-NoPainSelf and the contrast PainPartner-NoPainPartner activated overlapping regions in the anterior cingulate cortex and the anterior insula. A replication of that study with male and female participants (Singer et al., 2006) indicated that while the anterior insula was shared between pain experience and observation in all groups of subjects, the ACC shared activation was specific to female subjects. Singer et al. were the first to correlate brain activity in shared voxels with empathy scores of their participants, and found that brain activity was stronger in more empathic individuals, strengthening the idea that activating one's own pain areas while witnessing the pain of others indeed may reflect the neural correlate of empathy for pain.

Singer et al. also manipulated the relationship between the participant being scanned and the actors receiving the electroshock (Singer et al., 2006). She contrasted two actors, one that had previously played in an unfair manner with the participant in sequential prisoner's dilemma game, and one that had played in a fair manner. She found that for female participants, shared voxels were similarly activated whether the participants knew that a fair or unfair actor was being shocked. For males, only knowing that a fair actor was being electro-shocked activated pain regions of the anterior insula.

A number of other studies have confirmed that the sight of pain activates regions of the brain associated with the processing of one's own pain, including somatosensory, insular, and anterior cingulate cortices (Morrison, Lloyd, di Pellegrino, & Roberts, 2004; Avenanti, Bueti, Galati, & Aglioti, 2005; Botvinick et al., 2005; Jackson, Meltzoff, & Decety, 2005; Avenanti, Paluello, Bufalari, & Aglioti, 2006). Many of these experiments have used more naturalistic stimuli, in which participants could directly see a painful event happening to someone else (e.g., a needle entering the skin of an actor). The fact that both the symbolic cues of Singer et al. and more naturalistic stimuli activate the pain matrix indicates that the brain can activate a simulation of pain using a variety of stimuli.

For the case of pain, some issues though remain unclear. Singer et al. and Jackson et al. failed to find activations in somatosensory cortices while subjects empathized with the pain of others, while other investigators (Bufalari, Aprile, Avenanti, Di Russo, & Aglioti, 2007) find that even the primary somatosensory cortex is activated during the vision of pain in others in agreement with the aforementioned findings on touch (Keysers et al., 2004; Blakemore et al., 2005). Singer et al. may have failed to find somatosensory areas because they subtracted a non-painful tactile sensation. Jackson et al. may have failed to find it because of the lack of motion in their stimuli. These differences though will require further clarification. Second, the anterior insula is the most robustly reported region of overlap between seen and experienced pain. What though is actually represented in the insula? Very similar locations of the insula become activated during the sight and experience of disgust and pleasure (Wicker et al., 2003; Jabbi, Swart, & Keysers, 2007) (see below), raising the question of whether pain in particular or emotionally laden bodily sensations in general are represented in the anterior insula. Finally, unlike the case of actions (Gazzola & Keysers, submitted) and touch (Keysers et al., 2004), attempts to demonstrate shared voxels using unsmoothed data and single subject analysis have failed to reliably find shared voxels in the ACC (Morrison & Downing, 2007).

3 Emotions

The capacity to understand and share the emotions of other individuals is possibly what most of us associate with the word empathy. The word 'emotion' comes from the Latin verb *emovere* which means something that moves out, and this is exactly what emotions are from a social neuroscientific point of view: an inner state of another individual that comes out through his movements and can be perceived by another individual. There are always two aspects to our perception of the emotions of others: the bodily movements and behaviors that signal the emotion and the inner state of the other individual that can be deduced from these behaviors.

To take a dramatic example, if on a bus trip through the mountains we witness someone looking pale, suddenly retching, and filling a paper bag with clumps of undigested food, we clearly witness an action but cannot help but feel the deep state of physical nausea of that unfortunate motion-sick travel companion. A number of experiments now show that while we witness such emotions, we activate (1) insular voxels that would be activate if we would experience similar emotional states and (2) premotor and parietal sensory-motor voxels that would be active if we would perform the actions that signaled the emotion (i.e., facial and bodily movements).

3.1 Sharing the Emotional State and Shared Voxels in the Insula/Frontal Operculum

As seen above, the perception of pain in other individuals activates voxels in the anterior insula that are also recruited while we experience pain ourselves. A series of imaging studies by Phillips and collaborators (Phillips et al., 1997, 1998) suggested that the anterior insula is also implicated in the perception of the disgusted facial expressions of others. The same area has independently been implicated in the experience of disgust (Small et al., 2003). It therefore appears as though the insula, in addition to pain, also provides a shared circuit for the experience and the perception of disgust.

This hypothesis was directly tested in two fMRI studies. Wicker et al. (2003) measured brain activity while subjects viewed short movie clips of actors sniffing the content of a glass and reacting with a pleased, neutral, or disgusted facial expression. Thereafter, they exposed the subjects to pleasant or disgusting odorants through an anesthesia mask in the scanner. Both during observation and experience, the emotion of disgust was intensely perceived and experienced while the emotion of pleasure was less intense in both cases, and serve more as a control condition as a truly emotional condition. Results indicate that voxels in the anterior insula were activated both by experience of disgust and the observation of disgusted facial expressions of others (Wicker et al., 2003). These voxels were not significantly activated by the pleasant odorants or the vision of the pleased facial expressions of others, but this may have been related to the less intense experience and perception of pleasure. The location of the voxels involved in the experience of disgust and in the observation of disgust in this fMRI experiment overlap in location with the insular damage of a patient reporting a reduced experience of disgust and a deficient capacity to recognize disgust in others (Calder, Keane, Manes, Antoun, & Young, 2000; Adolphs, Tranel, & Damasio, 2003), suggesting that this circuit is indeed necessary for our understanding of disgust in others.

Jabbi et al. replicated the experiment of Wicker et al. using tastes instead of smells to induce disgust and used more intense expressions of pleasure in their movies (Jabbi et al., 2007). In addition they also measured how empathic their individuals reported to be using the IRI scale (Davis, 1983). During the observation of both pleased and disgusted facial expressions, subjects activated voxels in their insula more strongly if they reported being more empathic. The same voxels were also activated by the intense sensation of gustatory dislike induced by tasting quinine. This result indicates that the anterior insula is thus not liked to the experience of olfaction in particular, but the strong visceral sensations associated with smells or tastes more generally. In addition, the correlation between activations while subjects viewed pleased facial expressions and their empathy scores indicate that the role of the insula in empathy is not limited to negative emotions such as disgust or pain, but includes also positive emotions.

The evidence for shared voxels for various emotions in the insula raises the question of what aspect of these emotions is represented in that particular cortical location. Penfield and Faulk (1955) stimulated similar locations of the anterior insula in epileptic patients undergoing exploratory surgery. They found that subjects reported a variety of visceral sensations, including unpleasant sensations in their digestive system. This finding supports the idea that insular activity during the observation of the facial expressions of others does not reflect abstract cognitive representations of these emotions but rather more embodied visceral sensations that might be an integral part of experiencing these emotions. In addition, insular stimulation did not cause overt facial movements as one would expect if the facial expression of disgust would be represented in this area.

Interestingly, just as we showed for the shared circuits for actions, the insula also appears to receive auditory information about the disgusted emotional state of others. Adolphs and colleagues (2003) showed that their patient B with extensive insular lesions was unable to recognize disgust, even if it was acted out with distinctive sounds of disgust, such as retching and vocal prosody.

One of the challenges for the next years will be to investigate the insula using singe cell recordings in primates during the observation of facial expressions in order to examine more precisely the nature of the neural representations triggered during the experience and observation of emotions. So far, such recordings have only been performed during the experience of tastes and smells. Cells have been found with opposite preference (e.g., quinine and sucrose) with similar frequency within the volume of a single fMRI voxel, suggesting that the apparent indiscriminative BOLD response to the sight of pleased, disgusted, and painful facial expressions in the insula may actually result from highly selective representations at the single cell level that are blurred at the level of fMRI voxels (Verhagen, Kadohisa, & Rolls, 2004).

3.2 Sharing the Facial Expression of Other Individuals in the Motor MNS

In the experiment described above, investigators mapped regions involved in experiencing emotions and observing those of others (Wicker et al., 2003; Jabbi et al., 2007). A complementary approach is that to map the regions involved in producing deliberate facial expressions and to investigate if these facial motor circuits are involved in perceiving the facial expressions of others. Carr, Iacoboni, Dubeau, Mazziotta, and Lenzi (2003) piloted this approach by asking their subjects to view or imitate facial expressions seen in photographs of emotional facial expressions. They found that a network comprising the ventral premotor cortex and the MTG/STS, similar to the classical MNS for goal directed actions, was involved both during the observation and the imitation of facial expressions, suggesting that the brain of the observer transformed the facial expressions of the observer into a motor representation of similar expressions. In addition, the insula and amygdalae were activated during the imitation and observation of facial expressions, suggesting that the emotional states fitting the facial expressions were also activated. A problem with the approach of Carr et al. is the fact that subjects had interleaved blocks in which they simply viewed facial expressions and blocks were they had to imitate facial expressions. In this context, the premotor activation they measured during observation could have simply reflected a task-specific preparation to imitate the same facial expression in subsequent trials. In addition, Carr mixed different emotions in their blocks (disgust, fear, happiness etc.), making it impossible to determine whether any of the areas involved discriminate different emotions.

More recently, van der Gaag, Minderaa, & Keysers (2007, in press) have addressed these questions in three experiments conducted on the same subjects. They first scanned their subjects in a passive viewing task in which participants had to observe short movies of actors displaying emotional and non-emotional facial expressions. Participants then received new instructions: if, for instance, they saw two happy faces one after another, they had to press the button 'same emotion,' if they saw a happy followed by a sad face, the button 'different emotion,' and were scanned again. At the end, subjects received new instructions again and had to imitate the facial expressions they saw. These three experiments indicated that even during passive viewing, a network of temporal, posterior parietal, somatosensory, premotor, insular cortices, and amydgalae were activated for all facial expressions, suggesting that the brain spontaneously transforms the facial expressions of other individuals into a sensory-motor simulation of what it would feel like to make that facial expression and experience that emotion. The discrimination and imitation task though augmented the responses in these brain areas, showing that although spontaneously activated, these circuits can be modulated by attention and task demands. When comparing different facial expressions with each other, it became apparent that despite the dramatic differences in the emotions shown in the movie clips, brain activity was generally rather similar. The amygdalae, for instance, responded to the sight of fearful facial expressions, in agreement with previous reports, but did so as strongly as to neutral, disgusted, and happy facial expressions in all three tasks (van der Gaag et al., 2007). Also the insula, which responded significantly to the vision of disgusted facial expressions in accord with previous findings (Wicker et al., 2003; Jabbi et al., 2007) did not do so more than to happy or fearful faces (van der Gaag et al., in press). It will thus remain for future studies to determine if particular emotions can be discriminated from the *pattern* of brain activity in shared voxels induced during the observation of facial expressions.

3.3 Facial Expressions and Empathy and Autism

Jabbi et al. (2007) showed that activations in the anterior insula during the observation of the facial expressions of others are stronger in more empathic individuals. Pfeifer et al. found a similar correlation within both the premotor and insular cortex while children observed or imitated the facial expression of others (Pfeifer, J. H., Iacoboni, M., Mazziotta, J. C., Dapretto, M. (2005). Mirror Neuron System Activity in Children and Its Relation to Empathy and Interpersonal Competence. Program No. 660.24. 2005 Abstract Viewer/Itinerary Planner. Washington, DC: Society for Neuroscience. Online) and further showed that children with more intense activations in these regions function better in social situations according to their teachers. Interestingly Dapretto et al. (2006) found that autistic children activate their premotor and insular cortex less than typically developing individuals during the observation and imitation of facial expressions. Together, these findings support the idea that activations in premotor, somatosensory, and insular regions during the observation of the facial expressions of others are linked to empathy and are compatible with the idea that simulating both the facial expressions and emotional states of other individuals may underpin our intuitive understanding of the emotions of other individuals.

While fMRI cannot indicate whether the somatosensory and premotor activations we find to occur during the vision of facial expressions are indeed necessary for understanding the emotions of other individuals, lesion studies can address this question. Adolphs, Damasio, Tranel, Cooper, and Damasio (2000) examined the capacity to label facial expressions in a large number of patients with acquired brain lesions. They noticed that patients with lesions in somatosensory and premotor areas reliably showed deficits in labeling facial expressions, supporting that these areas are indeed essential for correctly labeling the facial expressions of others. Impairment in labeling could be due to a complete lack of understanding of what goes on in other individuals or to impairments in the process that transforms our intuitive gut feeling of what goes on in other individuals into words.

3.4 Summary

Empirical evidence now supports the idea that in analogy to actions and sensations, also the emotions of other individuals are transformed into a representation of the observer's own emotions. This process recruits premotor regions that are also involved in producing the same facial expressions, somatosensory areas that are also involved in experiencing similar facial expressions and the insula that is involved in experiencing emotional states such as pain and nausea. Without these brain regions, subjects report difficulties in labeling the facial expressions of other individuals. Activations in these regions are stronger in more empathic and less autistic individuals.

4 Shared Circuits for Actions, Sensations, and Emotions

Subsuming the above evidence, it appears that in three systems – actions, sensations, and emotions – certain brain areas are involved both in first person experience (*I* do, *I* feel) and third person perspective (knowing what *he* does or *he* feels). These areas or circuits, that we call shared circuits, involve the premotor, somatosensory, posterior parietal, temporal cortex, and cerebellum for the case of actions; somatosensory cortices for touch, the premotor, somatosensory, posterior parietal, insular and temporal cortex for emotions and the somatosensory cortices, insula, and ACC for pain. In all these cases, observing what other people do or feel appears to be transformed into an inner representation of what we would do or feel in similar situations – as if we would be in the skin of the person we observe. The idea of shared circuits, initially put forward for actions (Gallese and Goldman, 1998) therefore appears much broader (Gallese et al., 2004; Keysers & Gazzola, 2006).

In light of this evidence, a tentative dynamic picture of the brain processes involved in social simulation appears to emerge. First, social situations appear to be processed by the superior temporal sulcus to a high degree of sensory sophistication, including multimodal audio-visual representations of complex actions. These representations privilege the third person perspective, with reduced responses if the origin of the stimulus is endogenous (Hietanen & Perrett, 1993, 1996). Through the recruitment of shared circuits, the brain then adds specific first person elements to this description. If an action is seen, the parietal, premotor areas and potentially the cerebellum adds an inner representation of the motor programs, the observer would use to achieve the observed goals to the sensory third person description. If somatosensory events are witnessed, be it with (Hari et al., 1998; Grezes et al., 2003; Raos et al., 2004; Gazzola et al., 2006, 2007a; van der Gaag et al., in press; Gazzola & Keysers, submitted; Gazzola et al., submitted) or without an action (Keysers et al., 2004; Blakemore et al., 2005), the somatosensory cortices add an inner representation of what it would file like to perform a similar action or be touched in a similar way. If pain is witnessed, the somatosensory (Bufalari et al., 2007), anterior insular (Morrison et al., 2004; Singer et al., 2004; Botvinick et al., 2005; Jackson et al., 2005; Singer et al., 2006; Bufalari et al., 2007), and anterior cingulate (Hutchison et al., 1999; Morrison et al., 2004; Singer et al., 2004; Botvinick et al., 2005; Jackson et al., 2005; Bufalari et al., 2007) cortices add a sense of pain. If facial expressions of emotions are witnessed, premotor and somatosensory cortices appear to add a sense of what it would feel like to make a similar facial expression (Adolphs et al., 2000; Carr et al., 2003; Dapretto et al., 2006; van der Gaag et al., in press) and the insula adds a gut felling for what the other individual is feeling (Calder et al., 2000; Adolphs et al., 2003; Carr et al., 2003; Wicker et al., 2003; Jabbi et al., 2007).

As a result of these neural activities, the observer does not only see or hear what movements the other individual is performing. The observer shares a substantial amount of neural activity with the observed individual, including activity relating to motor, somatosensory, and emotional/visceral states. These shared neural activities reflect what would be going on in the observer's brain if he/she would have done the observed actions or been exposed to the observed stimulations. We thus appear to observe other individuals through the potentially deforming mirror of our own sensations and actions. This will provide valid insights into other individuals only to the extent to which the observed individuals are similar to us. By simulating the brain activity of other individuals using our own brain activity, the brain implicitly makes the assumption that other individuals are similar to us. This assumption becomes particularly clear in cases where the observer and observed differ in embodiment: typically developed individuals activate their premotor, somatosensory, and parietal hand representations while observing robots although robots do not have a premotor cortex. This shared activity thus clearly does not reflect the neural state of the robot, but that which the observer would have while performing similar actions.

Neuroimaging studies which examine the neural activity while a group of subjects perform particular actions or experience particular somatosensory stimulation show how similar the basic pattern of activation is across different human beings. This observation suggests that the shared circuits of our brain do not make a fundamental mistake by simulating the brain activity of other individuals by assuming implicitly that the neural states of the people they observe are similar to their own.

We do not normally confuse our own actions/sensations/emotions and those of other individuals because although shared circuits react in similar ways to our own experiences and those of others, many other areas clearly discriminate between these two cases. Our own actions include strong M1 activation and weaker STS activations, while those of others inhibit M1 but strongly activate the STS. When we are touched, our SI is strongly active, while it is much less active while we witness touch occurring to others. In contrast, patient C who is literally confused about who is being touched shows abnormally strong SI activity during the sight of touch (Blakemore et al., 2005). Even the parietal and premotor regions that contain mirror neurons contain many neurons that only respond either during the monkey's own actions or those of another individual (Gallese et al., 1996; Kohler et al., 2002; Keysers et al., 2003; Fogassi et al., 2005; Fujii et al., in press). In this context, the distinction between self and other is quite simple, but remains essential for a social cognition based on shared circuits to work (Gallese and Goldman, 1998; Decety and Sommerville, 2003). Some authors now search for brain areas that explicitly differentiate self from other. Both the right inferior parietal lobule and the posterior cingulate gyrus have been implicated in this function (Decety and Sommerville, 2003 and Vogt and Laureys, 2005 for reviews).

The account based on shared representation which we propose differs from those of other authors in that it does not assume that a particular modality is critical. Damasio (2003) emphasize the importance of somatosensory representation, stating that it is only once our brain reaches a somatosensory representation of the body state of the person we observe that we understand the emotion he/she are undergoing. We, on the other hand, believe that somatosensory representations are important for understanding the somatosensory sensations of others, but are not more important than motor or emotional structures. The current proposal represents an extension from our own previous proposals (e.g., Gallese et al., 2004), where we emphasized the motor aspect of understanding other people. We believe that motor representations are essential for understanding the actions of others, but somatosensory and emotional structures are equally important for empathy. Each modality (actions, sensations, and emotions) is understood and shared in our brain using its own specific circuitry. The joint neural representation of actions, emotions, and sensations that result from the recruitment of these various shared representations are the intuitive key to understanding other people, without requiring that they have to pass necessarily though a somatosensory or motor common code to be interpreted.

Of course many social situations are complex, and involve multiple modalities: witnessing someone hitting his finger with a hammer contains an action, an emotion, and a sensation. In most social situations, the different above-mentioned shared circuits thus work in concert.

Next to being potentially helpful in understanding what occurs in the mind of other people, the combination of shared circuits we propose have important implications for learning and culture. Culture is unthinkable without the capacity to learn from other people through observation. Currently, we have a much better understanding of the neural basis of trial and error learning than we have of social learning. During social learning, we typically witness another individual perform a certain action and receive a certain reward or punishment following this action. While in principle, it would be possible for the brain to develop a specific mechanism for learning by observation, in the light of the model we propose, there is no need for such a separate mechanism: shared circuits will activate both a representation of the action and a representation of its consequences in the brain of the observer 'as if' the observer had experienced the episode himself. The brain mechanisms of trial and error learning that modify the propensity for the action based on the consequences can then perform its business 'as usual,' but on the simulated actions and consequences.

Once shared circuits have transformed the actions, sensations and emotions of others into our own representations of actions, sensations and emotions, understanding other people boils down to understanding ourselves – our own actions, sensations, and emotions. This aspect will return in relation to theory of mind.

5 Demystifying Shared Circuits Through a Hebbian Perspective

For many readers, the existence of single neurons responding to the sight, sound, and execution of an action – to take a single example – remains a very odd observation. How can single neurons with such marvelous capacities emerge? The plausibility of a shared circuit account of social perception stands and falls with our capacity to give a plausible explanation of how such neurons can emerge.

Shared circuits and mirror neurons could be inborn or they might be acquired through learning. The problem with the idea of genetically prewired mirror neurons is that the genome would need to specify the individual connections between neurons in STS and F5 with similar properties (e.g., a vision-of-grasping neuron in STS has to be wired with an execution-of-grasping neuron in F5 and so on). It is difficult to imagine how the genome would suffice to specify such a vast amount of connections. In addition, such genetic predisposition could not explain how mirror neurons can be acquired for novel skills, for example, piano playing or opening a can of pop (Bangert et al., 2006; Gazzola et al., 2006; Lahav et al., 2007). Recently, we have developed a hypothesis termed the 'Hebbian perspective' that attempts to provide a plausible account of how shared circuits could emerge from canalized Hebbian learning. This account is described in details elsewhere (Keysers & Perrett, 2004; Del Giudice, Manera, & Keysers, submitted), but we will describe the gist of it here. Given the restriction on the total amount of citations in this chapter, the reader should consult (Keysers & Perrett, 2004; Del Giudice et al., submitted) for references supporting the claims made in this chapter.

Monkey and human infants spend most of their waking time watching themselves perform actions (White, Castle, & Held, 1964). Each time, the infant's hand wraps around an object, and brings it toward it, a particular set of neural activities overlaps in time: neurons in the premotor cortex responsible for the execution of this action are active at approximately the same time as the audiovisual neurons in the STS responding to the sight and sound of the same action. Given that STS and F5 are connected through PF, this means that ideal Hebbian learning conditions (Bi & Poo, 2001) are met: what fires together wires together. As a result, the synapses going from STS vision-of-grasping neurons to PF and F5 execution-of-grasping neurons will be strengthened as the grasping neurons at all three levels will be repeatedly coactive. After repeated self-observation, neurons in F5 receiving the enhanced input from STS will fire at the mere sight of grasping. Given that many neurons in the STS show reasonably viewpoint-invariant responses, responding in similar ways to views of a hand taken from different perspective, the sight of someone else grasping in similar ways then suffices to activate F5 mirror neurons. All that is required for the emergence of such mirror responses is the availability of connections between STS–PF–F5 that can show Hebbian learning, and there is evidence that Hebbian learning can occur in many places in the neocortex (Bi & Poo, 2001). In this perspective, what is genetically predisposed are general connections between temporal, parietal, and premotor regions and the tendency of infants to observe their own actions. These predispositions then 'canalize' learning, that is, provide the right structure and the right experiences, in a way that is ideal for Hebbian learning (Del Giudice et al., submitted). The precise wiring of a grasping observation neuron in the STS and a grasping execution neuron in F5 is not genetically predisposed but a result of the canalized Hebbian learning.

Arguments in favor of the idea that the MNS for actions originates from self observation stems from the fact that children's understanding of the actions of others depend on their own expertise in executing these motor skills, with a training improving infant's grasping expertise improving the infant's capacity to understand similar actions in others (Sommerville, Woodward, & Needham, 2005). In addition, the existence of auditory shared circuits for the sound of actions that are unlikely to be genetically predisposed (e.g., opening a coca-cola can or playing the piano) suggests that shared circuits can be acquired (Bangert et al., 2006; Gazzola et al., 2006; Lahav et al., 2007). Also during action observation, acquired motor expertise modifies the response in shared circuits (Calvo-Merino et al., 2005, 2006).

This proposal does not preclude the possibility that although part of the MNS may originate from Hebbian learning, part of it may still be inborn (Meltzoff & Moore, 1977).

The same Hebbian argument can be applied to the case of sensations and emotions. While seeing ourselves being touched, somatosensory activations overlap in time with visual descriptions of an object moving towards and touching our body. After Hebbian association the sight of someone else being touched could trigger somatosensory activations (Keysers et al., 2004;

Blakemore et al., 2005). While seeing our own actions, the visual description of our actions overlap in time not only with motor, but also somatosensory activity relating to what it feels like to perform the action (Nishitani & Hari, 2000; Grezes et al., 2003; Raos et al., 2004; Gazzola et al., 2006, 2007a; Gazzola & Keysers, submitted).

The presence of multimodal responses in the STS and shared circuits is essential for understanding how shared circuits could learn to represent cases where we cannot see our own actions. Associating the sight of someone's lip movements with our own lip movements is an important step in language acquisition. How can we link the sight of another individual's mouth producing a particular sound with our own motor programs given that we cannot usually see our own mouth movements? While seeing other individuals producing certain sounds with their mouth, the sound and sight of the action are correlated in time, and can lead to STS multimodal neurons. During our own babbling attempts to produce sounds, the sound and the motor program are correlated in time. As the sound will recruit multimodal neurons in the STS, the established auditory-motor connections will also link the sight of other people producing similar sounds to our motor program. The visual information thereby rides on the wave of the auditory associations (Keysers & Perrett, 2004).

The case of emotional facial expressions presents similar difficulties. How can the sight of a disgusted facial expression trigger our own emotion of disgust, despite the fact that we do not normally see our own disgusted facial expression? First, disgust can often have a cause that will trigger simultaneous disgust in many individuals (e.g., a disgusting smell): ones own disgust then correlates directly with the disgusted facial expression of others. Second, in parent-child relationships, facial imitation is a prominent observation (e.g., Stern, 2000). In our Hebbian perspective, this imitation means: the parent acts as a mirror for the facial expression of the child, leading again to the required correlation between the child's own emotion and the sight of similar facial expression. What is genetically predisposed might again be the parental tendency to imitate the facial expression of the child, thereby generating the right environmental stimuli for the infant's brain to engage in Hebbian learning.

To summarize, Hebbian association (a simple and molecularly relatively well-understood process), could predict the emergence of shared circuits. The value of this hypothesis lays in showing that shared circuits *could* emerge from learning without requiring for shared circuits to be fully determined by genetics. The extent to which genetics do preprogram this wiring still remains to be investigated.

6 Shared Circuits and Communication

The brain appears to spontaneously transform the visual and auditory descriptions of the actions, sensations, and emotions of others into neural representations normally associated with our execution of similar actions, and our experience of similar sensations and emotions. This transformation represents an intuitive and powerful form of communication: it transmits the experience of doing and feeling from one brain to another. This simple form of communication has obvious adaptive value: being able to peek into someone else's mind, and share his/her experiences renders constructive social interactions faster and more effective. For instance, sharing the disgust of a conspecific probing a potential source of food will prevent the observer from tasting potentially damaging items.

Most forms of communication have a fundamental problem: the sender transforms a content into a certain transmittable form according to a certain encoding procedure. The receiver then receives the encoded message, and has to transform it back into the original content. How can the receiver learn to decode the message? When we learn a spoken language we spend years of our life guessing this encoding/decoding procedure. For the case of actions, the shared circuits we propose use the correlation in time in the STS–PF–F5 circuit during self observation to determine the reciprocal relationship between the motor and the somatosensory representation of actions and their audio-visual consequences. Similar procedures may apply to sensations and emotions. The acquired reciprocal relationships can then serve as a rough Rosetta stone that lays out fundamental relationships between our own inner world and that of others and can provide important keys for deciphering the encoded messages hidden in other individual's behaviors.

The link between shared circuits and language remain a matter of debate. Rizzolatti and Arbib (1998) first claimed that mirror neurons for hand actions could be essential for the development of gestural language. More recently, the idea has been put forward that the combination of auditory and visual mirror neurons could provide an important step in the evolution of spoken language (Kohler et al., 2002; Keysers et al., 2003; Gazzola et al., 2006; Keysers & Gazzola, 2006). This latter idea has found support from the observation that the putative auditory mirror system in humans is more left lateralized than the visual mirror system (Aziz-Zadeh et al., 2004; Gazzola et al., 2006), and that only the left premotor cortex thus has extensive auditory and visual mirror neurons in humans, in accord with the left hemisphere dominance of language in humans.

As we will see below, mirror systems have to integrate with more cognitive knowledge to explain the complexities of communication and social interactions. Clearly, mirror neurons per se do not explain language, yet they could be an important pre-adaptation, evolved under evolutionary pressures to control ones own behavior and predict that of others, but later exapted for the purpose of language (Keysers & Perrett, 2004; Keysers & Gazzola, 2006).

7 Simulation and Theory of Mind – A Hypothesis

Social cognitions are not restricted to the simulations that shared circuits provide. Explicit thoughts exist in humans and clearly supplement these automatic simulations. If we see someone smile, we activate our premotor, somatosensory, and insular cortex as if we would be smiling, especially if we are empathic. If the person smiling is a politician promising us a tax cut, we become skeptical: experience has taught us that politician's promises are often more instrumental than sincere. Similarly, if we see a person cry, shared circuits would lead us to share her sorrow. The added knowledge of knowing that she has just received a love letter stating that the man she loves is leaving his wife to move in with her, can significantly modify our interpretation of her tears, realizing that they are probably tears of joy. There are many examples of our daily life in which we clearly use explicit thoughts to guide our interpretation of the social stimuli we observe, utilizing our knowledge of the rules that govern social interactions in addition to the intuitive use of shared circuits.

The words 'Theory of Mind' (ToM) and 'Mentalizing' have often been used to describe the set of cognitive skills involved in thinking about the mind of others, in particular about their beliefs (Frith & Frith, 2003). People are considered to have a ToM if they are able to deal with the fact that other people can have beliefs that differ from reality, a capacity that is tested with so-called false belief tasks such as the famous 'Sally and Anne' test (Baron-Cohen, Leslie, & Frith, 1985). In that test, an observer sees Sally hide an object in a basket. Sally then goes away for a while, and unbeknown to her, Anne moves the object from the basket into a box. Sally then returns, and the observer is asked: "Where will Sally [first] look for her object?" If the observer answers, "in the basket, because she doesn't know that Anne moved it," the observer is thought to have a ToM. If the answer is "in the box," the observer failed. Children from the age of 4 years pass this test, while autistic individuals often fail the test even in their teens (Baron-Cohen et al., 1985). ToM tasks usually activate medial prefrontal (mPFC) regions of the brain and the temporo-parietal junction (TPJ).

In contrast to authors that see simulation and ToM as competing perspectives of social cognition (Saxe, 2005), we believe that one of the challenges of current social neuroscience is to examine the link between simulation and ToM and have recently proposed a hypothesis on the nature of this relationship (Fig. 3 adapted from (Keysers & Gazzola, 2007)). From a first person perspective, we have experiences (e.g., "we eat an oyster gone bad and feel bad") and thoughts about these experiences that are based on an analysis of our introspection (e.g., "I probably feel so bad because I ate a rotten oyster...."). The former cause activations in premotor, parietal, and insular regions, while the latter are linked to activity in the insula and mPFC (Critchley, Wiens, Rotshtein, Ohman, & Dolan, 2004). What is important is that we do not always engage in the interoceptive processes that turn the lower-level representations of what we feel into higher-level thoughts about these states that can be verbally communicated. Indeed, subjects vary in how good they are at introspecting their states, with alexithymia measuring interindividual differences in this capacity. At first sight, this distinction between (a) states and (b) conscious thoughts about states with introspection going from a to b, seems entirely detached from social cognitions: introspection is a process that only deals with ourselves. In contrast, we propose that a very similar concept maybe the key to examining the relationship between simulation and theory of mind.

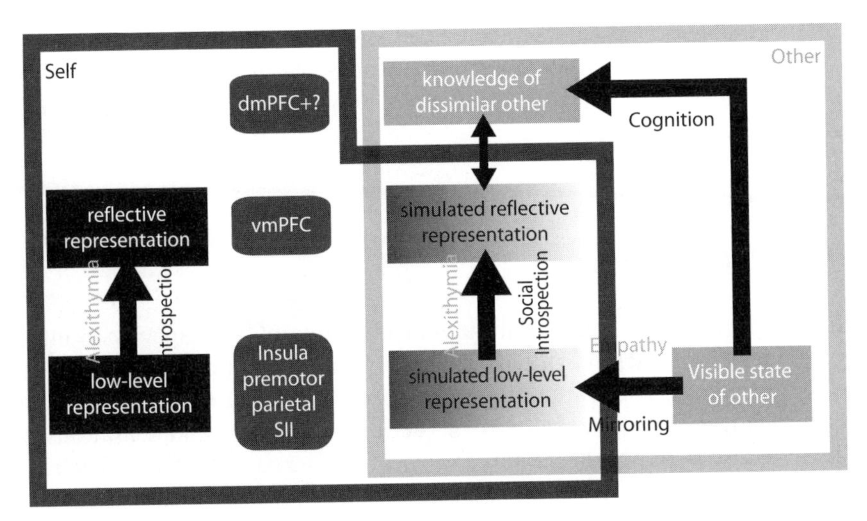

Fig. 3 Illustration of the model linking theory of mind and simulation. The self is shown in *dark grey*, the other in *light grey*, and candidate brain areas thought to implement representation shown in *rounded dark grey boxes*. During our own experiences, pre-reflective representations can lead, through introspection, to reflective representations (*left*). While witnessing the states of others, mirroring leads to activations that simulate pre-reflective representations of our own bodily states. A process of social introspection, utilizing the mechanisms of introspection, activates representations that simulate reflective representations of our own bodily states. A more cognitive route leads to more abstract knowledge about the other that escapes from the constraints of our own experiences

While we witness what other people do or what happens to other people, shared circuits transform these events into a representation of our own actions, emotions, and sensations. As a result, we have a simulated inner state that mirrors the state of the people we observe. Based on the brain locations involved in this spontaneous mirroring, these simulated states are akin to the low-level representations of our own actions, emotions, and sensations. If asked to mentalize about what goes on in the mind of other people, as is the case in ToM tasks, as an observer we have two options: either we rely entirely on abstract, un-embodied knowledge about other people (e.g., "politicians lie") or we take advantage of the simulation process, and introspect the state of our own lower level representations, which during social cognition not only represents our state but also the simulated states of the people we observes. This introspection then allows us to interpret the inner state of other individuals just as we would interpret our own states. In support of this idea, Mitchell, Macrae, and Banaji (2006) found that reflecting about yourself and reflecting about people you believe to be similar to yourself indeed activate overlapping regions of the mPFC (introspection = social introspection) whilst thinking about people you believe to be too different from yourself activates a separate region of the mPFC in analogy to our disembodied cognitive route.

Along these lines, ToM and simulation are complementary processes, and empathy could measure how well we transform the states of others into our own states, while alexithymia how well we can reflect about our own states or the simulated states of others.

8 Overall Conclusions

While 15 years ago, a neuroscientific understanding of our capacity to intuitively understand other individuals was unthinkable, scientific discoveries over the last decade start to outline the skeleton of a unifying theory of intuitive social cognition. While we witness the actions, emotions, and sensations of other individuals, our brain spontaneously recreates a pattern of neural activity that resembles that while we perform similar actions or have similar sensations and emotions. At the neural level, this appears to occur through networks of brain areas shared between the experience of actions, sensations, and emotions and the observation of these states in others. These shared activations are stronger in more empathic and weaker in more autistic individuals, suggesting that these mechanisms maybe the neural basis of our intuitive empathy for other individuals. Lesions in regions involved in our own emotions or actions reduce our capacity to understand similar aspects of the mind of other individuals, supporting the view that shared circuits are necessary for an intuitive understanding of other individuals. Such shared circuits need not be mysterious: they could arise from Hebbian associations during self observation. While the interaction of these intuitive and spontaneous processes and more deliberate mentalizing processes remain to be understood, shared circuits start to give a plausible neural basis for our remarkable connection to the people that surround us.

Acknowledgments CK was financed by a VIDI grant by NWO and a Marie Curie Excellence Grant. We thank Marco del Giudice for help in developing the idea of canalized Hebbian learning, and Marc Thioux for helpful comments on the manuscript.

References

Adolphs, R., Damasio, H., Tranel, D., Cooper, G., & Damasio, A. R. (2000). A role for somatosensory cortices in the visual recognition of emotion as revealed by three-dimensional lesion mapping. *Journal of Neuroscience*, 20, 2683–2690.

Adolphs, R., Tranel, D., & Damasio, A. R. (2003). Dissociable neural systems for recognizing emotions. *Brain and Cognition*, 52, 61–69.

Amiez, C., Kostopoulos, P., Champod, A. S., & Petrides, M. (2006). Local morphology predicts functional organization of the dorsal premotor region in the human brain. *Journal of Neuroscience*, 26, 2724–2731.

Amunts, K., Schleicher, A., Burgel, U., Mohlberg, H., Uylings, H. B., & Zilles, K. (1999). Broca's region revisited: Cytoarchitecture and intersubject variability. *Journal of Comparative Neurology*, 412, 319–341.

Avenanti, A., Bueti, D., Galati, G., & Aglioti, S. M. (2005). Transcranial magnetic stimulation highlights the sensorimotor side of empathy for pain. *Nature Neuroscience*, 8, 955–960.

Avenanti, A., Paluello, I. M., Bufalari, I., & Aglioti, S. M. (2006). Stimulus-driven modulation of motor-evoked potentials during observation of others' pain. *Neuroimage*, 32, 316–324.

Aziz-Zadeh, L., Iacoboni, M., Zaidel, E., Wilson, S., & Mazziotta, J. (2004). Left hemisphere motor facilitation in response to manual action sounds. *European Journal of Neuroscience*, 19, 2609–2612.

Baker, C. I., Keysers, C., Jellema, T., Wicker, B., & Perrett, D. I. (2001). Neuronal representation of disappearing and hidden objects in temporal cortex of the macaque. *Experimental Brain Research*, 140, 375–381.

Bangert, M., Peschel, T., Schlaug, G., Rotte, M., Drescher, D., Hinrichs, H., et al. (2006). Shared networks for auditory and motor processing in professional pianists: Evidence from fMRI conjunction. *Neuroimage*, 30, 917–926.

Baron-Cohen, S., Leslie, A. M., & Frith, U. (1985). Does the autistic child have a "theory of mind"? *Cognition*, 21, 37–46.

Bekkering, H., Wohlschlager, A., & Gattis, M. (2000). Imitation of gestures in children is goal-directed. *Quarterly Journal of Experimental Psychology A*, 53, 153–164.

Bi, G., & Poo, M. (2001). Synaptic modification by correlated activity: Hebb's postulate revisited. *Annual Review of Neuroscience*, 24, 139–166.

Blakemore, S. J., Bristow, D., Bird, G., Frith, C., & Ward, J. (2005). Somatosensory activations during the observation of touch and a case of vision-touch synaesthesia. *Brain*, 128, 1571–1583.

Botvinick, M., Jha, A. P., Bylsma, L. M., Fabian, S. A., Solomon, P. E., & Prkachin, K. M. (2005). Viewing facial expressions of pain engages cortical areas involved in the direct experience of pain. *Neuroimage*, 25, 312–319.

Buccino, G., Binkofski, F., Fink, G. R., Fadiga, L., Fogassi, L., Gallese, V., et al. (2001). Action observation activates premotor and parietal areas in a somatotopic manner: An fMRI study. *European Journal of Neuroscience*, 13, 400–404.

Buccino, G., Lui, F., Canessa, N., Patteri, I., Lagravinese, G., Benuzzi, F., et al. (2004a) Neural circuits involved in the recognition of actions performed by nonconspecifics: An FMRI study. *Journal of Cognation Neuroscience*, 16, 114–126.

Buccino, G., Vogt, S., Ritzl, A., Fink, G. R., Zilles, K., Freund, H. J., et al. (2004b) Neural circuits underlying imitation learning of hand actions: An event-related fMRI study. *Neuron*, 42, 323–334.

Bufalari, I., Aprile, T., Avenanti, A., Di Russo, F., & Aglioti, S. M. (2007). Empathy for pain and touch in the human somatosensory cortex. *Cereb Cortex* 17 (11), 2553–2561.

Calder, A. J., Keane, J., Manes, F., Antoun, N., & Young, A. W. (2000). Impaired recognition and experience of disgust following brain injury. *Nature Neuroscience*, 3, 1077–1078.

Calvo-Merino, B., Glaser, D. E., Grezes, J., Passingham, R. E., & Haggard, P. (2005). Action observation and acquired motor skills: An FMRI study with expert dancers. *Cerebral Cortex*, 15, 1243–1249.

Calvo-Merino, B., Grezes, J., Glaser, D. E., Passingham, R. E., & Haggard, P. (2006). Seeing or doing? Influence of visual and motor familiarity in action observation. *Current Biology*, 16, 1905–1910.

Carr, L., Iacoboni, M., Dubeau, M. C., Mazziotta, J. C., & Lenzi, G. L. (2003). Neural mechanisms of empathy in humans: A relay from neural systems for imitation to limbic areas. *Proceedings of the National Academy of Science U S A*, 100, 5497–5502.

Chaminade, T., Franklin, D. W., Oztop, E., & Cheng, G. (2005). Motor interference between humans and humanoid robots: Effect of biological and artificial motion. Proceedings of 2005 4th IEEE International Conference on Development and Learning: 96–101.
Critchley, H. D., Wiens, S., Rotshtein, P., Ohman, A., & Dolan, R. J. (2004). Neural systems supporting interoceptive awareness. *Nature Neuroscience*, 7, 189–195.
Damasio, A.R. (2003). *Looking for Spinoza: Joy, Sorrow and the Feeling Brain*. New York, NY: Hartcourt.
Dapretto, M., Davies, M. S., Pfeifer, J. H., Scott, A. A., Sigman, M., Bookheimer, et al. (2006). Understanding emotions in others: Mirror neuron dysfunction in children with autism spectrum disorders. *Nature Neuroscience*, 9, 28–30.
Davis, M. (1983). Measuring individual differences in empathy: Evidence for a multidimensional approach. *Journal of Personality and Social Psychology*,, 44 113.
Decety, J., Grezes, J., Costes, N., Perani, D., Jeannerod, M., Procyk, E., et al. (1997). Brain activity during observation of actions. Influence of action content and subject's strategy. *Brain*, 120 (Pt 10), 1763.
Decety J., & Sommerville, J.A. (2003). Shared representations between self and other: A social cognitive neuroscience view. *Trends in Cognitive Sciences 7*, 527–533.
Del Giudice, M., Manera, V., & Keysers, C. (in press). Programmed to learn? The ontogeny of Mirror Neurons. *Developmental Science*.
Fadiga, L., Craighero, L., Buccino, G., & Rizzolatti, G. (2002). Speech listening specifically modulates the excitability of tongue muscles: A TMS study. *European Journal of Neuroscience*, 15, 399–402.
Fogassi, L., Ferrari, P. F., Gesierich, B., Rozzi, S., Chersi, F., & Rizzolatti, G. (2005). Parietal lobe: From action organization to intention understanding. *Science*, 308, 662–667.
Foldiak, P., Xiao, D., Keysers, C., Edwards, R., & Perrett, D. I. (2004). Rapid serial visual presentation for the determination of neural selectivity in area STSa. *Progress in Brain Research*, 144, 107–116.
Frith, U., & Frith, C. D. (2003). Development and neurophysiology of mentalizing. *Philosophical Transactions of the Royal Society of London Series B, Biology Science*, 358, 459–473.
Fujii N., Hihara, S., & Iriki, A. (in press) Social cognition in premotor and parietal cortex. *Social Neuroscience*
Gallese, V., Fadiga, L., Fogassi, L., & Rizzolatti, G. (1996). Action recognition in the premotor cortex. *Brain*, 119 (Pt 2), 593–609.
Gallese V., & Goldman, A. (1998). Mirror neurons and the simulation theory of mind-reading. *Trends in Cognitive Sciences 2*, 493–501.
Gallese, V., Keysers, C., & Rizzolatti, G. (2004). A unifying view of the basis of social cognition. *Trends in Cognitive Science*, 8, 396–403.
Gangitano, M., Mottaghy, F. M., & Pascual-Leone, A. (2001). Phase-specific modulation of cortical motor output during movement observation. *Neuroreport*, 12, 1489–1492.
Gazzola, V., Aziz-Zadeh, L., & Keysers, C. (2006). Empathy and the Somatotopic Auditory Mirror System in Human. *Current Biology*, 16, 1824–1829.
Gazzola V., Keysers, C. (submitted). The observation and execution of actions share motor and somatosensory voxels in all tested subjects: Single subject analysis of unsmoothed fMRI data. *Cerebral Cortex*.
Gazzola, V., Rizzolatti, G., Wicker, B., & Keysers, C. (2007a). The anthropomorphic brain: The mirror neuron system responds to human and robotic actions. *Neuroimage*, 35, 1674–1684.
Gazzola, V., van der Worp, H., Mulder, T., Wicker, B., Rizzolatti, G., & Keysers, C. (2007b). Aplasics born without hands mirror the goal of hand actions with their feet. *Current Biology*, 17, 1235–40.
Gergely, G., Bekkering, H., & Kiraly, I. (2002). Rational imitation in preverbal infants. *Nature*,, 415 755.

Grafton, S. T., Arbib, M. A., Fadiga, L., & Rizzolatti, G. (1996). Localization of grasp representations in humans by positron emission tomography. 2. Observation compared with imagination. *Experimental Brain Research*, 112, 103–111.

Grefkes, C., Geyer, S., Schormann, T., Roland, P., & Zilles, K. (2001). Human somatosensory area 2: Observer-independent cytoarchitectonic mapping, interindividual variability, and population map. *Neuroimage*, 14, 617–631.

Grezes, J., Armony, J. L., Rowe, J., & Passingham, R. E. (2003). Activations related to "mirror" and "canonical" neurones in the human brain: An fMRI study. *Neuroimage*, 18, 928–937.

Hamilton, A. F., & Grafton, S. T. (2006). Goal representation in human anterior intraparietal sulcus. *Journal of Neuroscience*, 26, 1133–1137.

Hari, R., Forss, N., Avikainen, S., Kirveskari, E., Salenius, S., & Rizzolatti, G. (1998). Activation of human primary motor cortex during action observation: A neuromagnetic study. *Proceedings of the National Academy of Science U S A*, 95, 15061–15065.

Heiser, M., Iacoboni, M., Maeda, F., Marcus, J., & Mazziotta, J. C. (2003). The essential role of Broca's area in imitation. *European Journal of Neuroscience*, 17, 1123–1128.

Hietanen, J. K., & Perrett, D. I. (1993). Motion sensitive cells in the macaque superior temporal polysensory area. I. Lack of response to the sight of the animal's own limb movement. *Experimental Brain Research*, 93, 117–128.

Hietanen, J. K., & Perrett, D. I. (1996). Motion sensitive cells in the macaque superior temporal polysensory area: Response discrimination between self-generated and externally generated pattern motion. *Behavior Brain Research*, 76, 155–167.

Hutchison, W. D., Davis, K. D., Lozano, A. M., Tasker, R. R., & Dostrovsky, J. O. (1999). Pain-related neurons in the human cingulate cortex. *Nature Neuroscience*, 2, 403–405.

Iacoboni, M., Koski, L. M., Brass, M., Bekkering, H., Woods, R. P., Dubeau, et al. (2001). Reafferent copies of imitated actions in the right superior temporal cortex. *Proceedings of the National Academy of Science U S A*, 98, 13995–13999.

Iacoboni, M., Molnar-Szakacs, I., Gallese, V., Buccino, G., Mazziotta, J. C., & Rizzolatti, G. (2005). Grasping the intentions of others with one's own mirror neuron system. *PLoS Biology*,, 3 e79.

Iacoboni, M., Woods, R. P., Brass, M., Bekkering, H., Mazziotta, J. C., & Rizzolatti, G. (1999). Cortical mechanisms of human imitation. *Science*, 286, 2526–2528.

Jabbi, M., Swart, M., & Keysers, C. (2007). Empathy for positive and negative emotions in the gustatory cortex. *Neuroimage*, 34, 1744–1753.

Jackson, P. L., Meltzoff, A. N., & Decety, J. (2005). How do we perceive the pain of others? A window into the neural processes involved in empathy. *Neuroimage*, 24, 771–779.

Keysers, C., & Gazzola, V. (2006). Towards a unifying neural theory of social cognition. *Progress in Brain Research*, 156, 383–406.

Keysers, C., & Gazzola, V. (2007). Integrating simulation and theory of mind: From self to social cognition. *Trends in Cognitive Science*, 5, 194–196.

Keysers, C., Kohler, E., Umilta, M. A., Nanetti, L., Fogassi, L., & Gallese, V. (2003). Audiovisual mirror neurons and action recognition. *Experimental Brain Research*, 153, 628–636.

Keysers, C., & Perrett, D. I. (2004). Demystifying social cognition: A Hebbian perspective. *Trends in Cognitive Science*, 8, 501–507.

Keysers, C., Wicker, B., Gazzola, V., Anton, J. L., Fogassi, L., & Gallese, V. (2004). A touching sight: SII/PV activation during the observation and experience of touch. *Neuron*, 42, 335–346.

Keysers, C., Xiao, D. K., Foldiak, P., & Perrett, D. I. (2001). The speed of sight. *Journal of Cognitive Neuroscience*, 13, 90–101.

Kilner, J. M., Paulignan, Y., & Blakemore, S. J. (2003). An interference effect of observed biological movement on action. *Current Biology*, 13, 522–525.

Kohler, E., Keysers, C., Umilta, M. A., Fogassi, L., Gallese, V., & Rizzolatti, G. (2002). Hearing sounds, understanding actions: Action representation in mirror neurons. *Science*, 297, 846–848.

Lahav, A., Saltzman, E., & Schlaug, G. (2007). Action representation of sound: Audiomotor recognition network while listening to newly acquired actions. *Journal of Neuroscience*, 27, 308–314.

Logothetis, N. K. (2003). The underpinnings of the BOLD functional magnetic resonance imaging signal. *Journal of Neuroscience*, 23, 3963–3971.

Matelli, M., Camarda, R., Glickstein, M., & Rizzolatti, G. (1986). Afferent and efferent projections of the inferior area 6 in the macaque monkey. *Journal of Comparative Neurology*, 251, 281–298.

Meltzoff, A. N., & Moore, M. K. (1977). Imitation of facial and manual gestures by human neonates. *Science*, 198, 74–78.

Mitchell, J. P., Macrae, C. N., & Banaji, M. R. (2006). Dissociable medial prefrontal contributions to judgments of similar and dissimilar others. *Neuron*, 50, 655–663.

Molnar-Szakacs, I., Kaplan, J., Greenfield, P. M., & Iacoboni, M. (2006). Observing complex action sequences: The role of the fronto-parietal mirror neuron system. *Neuroimage*, 33, 923–935.

Montagna, M., Cerri, G., Borroni, P., & Baldissera, F. (2005). Excitability changes in human corticospinal projections to muscles moving hand and fingers while viewing a reaching and grasping action. *European Journal of Neuroscience*, 22, 1513–1520.

Morrison I., Downing P. E. (2007) Organization of felt and seen pain responses in anterior cingulate cortex. *Neuroimage*, 37, 642–651.

Morrison, I., Lloyd, D., di Pellegrino, G., & Roberts, N. (2004). Vicarious responses to pain in anterior cingulate cortex: Is empathy a multisensory issue? *Cognitive, Affective, & Behavioral Neuroscience*, 4, 270–278.

Nishitani N., & Hari, R. (2000). Temporal dynamics of cortical representation for action Proceedings of the National Academy of Sciences, USA., 97 913.

Pelphrey, K. A., Morris, J. P., Michelich, C. R., Allison, T., & McCarthy, G. (2005). Functional anatomy of biological motion perception in posterior temporal cortex: An FMRI study of eye, mouth and hand movements. *Cerebral Cortex*, 15, 1866–1876.

Penfield, W., & Faulk, M. E., Jr. (1955). The insula; further observations on its function. *Brain*, 78, 445–470.

Phillips, M. L., Young, A. W., Scott, S. K., Calder, A. J., Andrew, C., Giampietro, V., et al. (1998). Neural responses to facial and vocal expressions of fear and disgust. *Proceedings of Biological Science*, 265, 1809–1817.

Phillips, M. L., Young, A. W., Senior, C., Brammer, M., Andrew, C., Calder, A. J., et al. (1997). A specific neural substrate for perceiving facial expressions of disgust. *Nature*, 389, 495–498.

Pizzamiglio, L., Aprile, T., Spitoni, G., Pitzalis, S., Bates, E., D'Amico, S., et al. (2005). Separate neural systems for processing action- or non-action-related sounds. *Neuroimage*, 24, 852–861.

Press, C., Bird, G., Flach, R., & Heyes, C. (2005). Robotic movement elicits automatic imitation. *Cognitive Brain Research*, 25, 632–640.

Puce, A., & Perrett, D. (2003). Electrophysiology and brain imaging of biological motion. *Philosophical Transactions of the Royal Society of London Series B, Biology Science*, 358, 435–445.

Raos, V., Evangeliou, M. N., & Savaki, H. E. (2004). Observation of action: Grasping with the mind's hand. *Neuroimage*, 23, 193–201.

Rijntjes, M., Dettmers, C., Buchel, C., Kiebel, S., Frackowiak, R. S., & Weiller, C. (1999). A blueprint for movement: Functional and anatomical representations in the human motor system. *Journal of Neuroscience*, 19, 8043–8048.

Rizzolatti, G., & Arbib, M. A. (1998). Language within our grasp. *Trends in Neuroscience*, 21, 188–194.

Rizzolatti G., & Craighero, L. (2004). The mirror-neuron system. *Annual Review of Neuroscience, 27*, 169–192.

Saxe, R. (2005). Against simulation: The argument from error. *Trends in Cognitive Science*, 9, 174–179.

Saygin, A. P., Wilson, S. M., Dronkers, N. F., & Bates, E. (2004). Action comprehension in aphasia: Linguistic and non-linguistic deficits and their lesion correlates. *Neuropsychologia*, 42, 1788–1804.

Seltzer, B., & Pandya, D. N. (1978). Afferent cortical connections and architectonics of the superior temporal sulcus and surrounding cortex in the rhesus monkey. *Brain Research*, 149, 1–24.

Seltzer, B., & Pandya, D. N. (1994). Parietal, temporal, and occipital projections to cortex of the superior temporal sulcus in the rhesus monkey: A retrograde tracer study. *Journal of Computational Neurology*, 343, 445–463.

Singer, T., Seymour, B., O'Doherty, J., Kaube, H., Dolan, R. J., & Frith, C. D. (2004). Empathy for pain involves the affective but not sensory components of pain. *Science*, 303, 1157–1162.

Singer, T., Seymour, B., O'Doherty, J. P., Stephan, K. E., Dolan R. J., & Frith, C. D. (2006). Empathic neural responses are modulated by the perceived fairness of others. *Nature*, 439, 466–469.

Small, D. M., Gregory, M. D., Mak, Y. E., Gitelman, D., Mesulam, M. M., & Parrish, T. (2003). Dissociation of neural representation of intensity and affective valuation in human gustation. *Neuron*, 39, 701–711.

Sommerville, J. A., Woodward, A. L., & Needham, A. (2005). Action experience alters 3-month-old infants' perception of others' actions. *Cognition*,, 96 B1–11.

Stern, D.N. (2000). *The Interpersonal world of the infant: A View from Psychoanalysis and Development Psychology* (2nd Ed.). New York: Basic Books.

Subiaul, F., Cantlon, J. F., Holloway, R. L., & Terrace, H. S. (2004). Cognitive imitation in rhesus macaques. *Science*, 305, 407–410.

Umilta, M. A., Kohler, E., Gallese, V., Fogassi, L., Fadiga, L., Keysers, C., et al. (2001). I know what you are doing. A neurophysiological study. *Neuron*, 31, 155–165.

van der Gaag, C., Minderaa, R., & Keysers, C. (2007). The BOLD signal in the amygdala does not differentiate between dynamic facial expressions. *Social Cognitive and Affective Neuroscience*, 2, 93–103.

van der Gaag, C, Minderaa, R., & Keysers, C. (2007) Facial expressions: What the mirror neuron system can and cannot tell us. Social Neuroscience 2:179–222.

Verhagen, J. V., Kadohisa, M., & Rolls, E. T. (2004). Primate insular/opercular taste cortex: Neuronal representations of the viscosity, fat texture, grittiness, temperature, and taste of foods. *Journal of Neurophysiology*, 92, 1685–1699.

Vogt, B.A, & Laureys, S. (2005). Posterior cingulate, precuneal and retrosplenial cortices: Cytology and components of the neural network correlates of consciousness. *Progress in Brain Research, 150,* 205–217.

Wheaton, K. J., Thompson, J. C., Syngeniotis, A., Abbott, D. F., & Puce, A. (2004). Viewing the motion of human body parts activates different regions of premotor, temporal, and parietal cortex. *Neuroimage*, 22, 277–288.

White, B. L., Castle, P., & Held, R. (1964). Observations on the development of visually-directed reaching. *Child Development*, 35, 349–364.

Wicker, B., Keysers, C., Plailly, J., Royet, J. P., Gallese, V., & Rizzolatti, G. (2003). Both of us disgusted in My insula: The common neural basis of seeing and feeling disgust. *Neuron*, 40, 655–664.

Williamson, R. A., & Markman, E. M. (2006). Precision of imitation as a function of preschoolers' understanding of the goal of the demonstration. *Developmental Psychology*, 42, 723–731.

Wohlschlager, A., Gattis, M., & Bekkering, H. (2003). Action generation and action perception in imitation: An instance of the ideomotor principle. *Philosophical Transactions of the Royal Society of London Series B, Biology Science*, 358, 501–515.

Reflections on the Mirror Neuron System: Their Evolutionary Functions Beyond Motor Representation

Lindsay M. Oberman and V.S. Ramachandran

Abstract The discovery of the mirror neuron system by Rizzolatti and his colleagues began with a chance observation, but has since inspired 15 years of research into the properties, functions, and dysfunction of this system. Though much has been learned about this system of neurons, many questions still remain. What defines a mirror neuron? What is the extent of its involvement in social behaviors? How do these neurons develop? What is the potential for therapeutic interventions targeting these neurons? We aim to explore these questions in the following chapter and suggest that the current characterization of this system maybe more restrictive than necessary. We suggest that mirror neurons are endowed with the precise properties allowing for complex remapping from one domain into another, which may lead to behaviors which arguably distinguish humans from all the other animals, namely our abilities to interact socially, understand others thoughts and emotions, communicate using complex language, and the ability to reflect on ourselves.

Keywords Mirror Neuron · Function · Development · Therapy

1 The Discovery

"This structure has novel features which are of considerable biological interest." This quote, published in the journal Nature in 1953, refers to the discovery of the double-helical structure of DNA. To say that it has been of "considerable biological interest" is certainly an understatement. James Watson and Francis Crick's discovery of the structure of DNA led to a revolution in medical research that still continues today. Why is this story important? What does the discovery of the structure of DNA have to do with the mirror neuron system? We suggest that the discovery of the mirror neuron system will do for psychology what DNA has done for biology; provide a unifying theory at the

L.M. Oberman
Department of Neurology, Harvard University, Boston, MA, USA

J.A. Pineda (ed.), *Mirror Neuron Systems*, DOI: 10.1007/978-1-59745-479-7_2,

level of anatomical hardware as opposed to more general 'black-box' theories. In the case of DNA the 'structural logic' of the molecule, the complementary strands of the helix, dictates the functional logic of heredity. Analogously, the discovery of neurons in the premotor cortex that respond to both observed and executed actions similarly provided the mechanism that may underlie a host of seemingly unrelated cognitive and social abilities including imitation, theory of mind, empathy, language. However, we would caution that one must not fall into the trap of ascribing too much to a single set of circuits. Furthermore, if proper development of this neural system is necessary for normal development of these abilities, the discovery of mirror neurons may also provide a candidate neural basis for autism, a developmental disorder affecting the very abilities thought to be mediated by the mirror neuron system.

As is the case of many great discoveries, the mirror neuron system was first observed while Rizzolatti and his colleagues were studying something completely different. In the late 1980s, he and his colleagues had identified neurons in the macaque premotor cortex which responded when the monkey performed specific actions, such as grasping a peanut. While investigating the specificity of the responses of neurons, using single unit electrophysiology in region F5 of premotor cortex, the experimenter noted that certain neurons would respond not only when the monkey, for example, grasped the peanut, but also when the monkey observed the experimenter grasping the peanut (see Fig. 1). Likewise another neuron would respond both when the monkey put the peanut in his mouth and when the monkey observed the experimenter putting the peanut in his mouth. (Di Pellegrino, Fadiga, Fogassi, Gallese, & Rizzolatti, 1992).

As this was the first account of this new class of neurons, it was difficult to determine their exact function. The original paper, however, did provide some preliminary suggestions, for example their potential role in action recognition as well as gesture and language comprehension. Though the authors admit that the observations documented in the original paper could not prove such claims, the neurons seemed to possess precisely the properties that would be necessary to mediate such abilities.

Four years later, in a paper in the journal Brain, these neurons were dubbed 'mirror' neurons as their response to observed actions appeared to 'mirror' the response of performed actions. Also in this paper, the authors propose possible functional roles of these 'mirror' neurons. These roles included those proposed in the original paper, namely understanding the meaning of actions in the macaque and human language comprehension as well as a new potential function, imitation (Gallese, Fadiga, Fogassi, & Rizzolati, 1996). These are good suggestions and an argument can be made that the known properties of these neurons, specifically their response to both observed and performed actions, endow them with precisely the properties that would allow for such abilities, but how can we be sure that they *are* the underlying mechanism driving these abilities? Could other abilities also utilize such a system? Is the term 'mirror neuron' exclusive to the motor system, or can the use of the phrase 'mirror neuron' be generalized in a meaningful way to other neural systems? If other systems do, in fact, utilize a mirror-like algorithm,

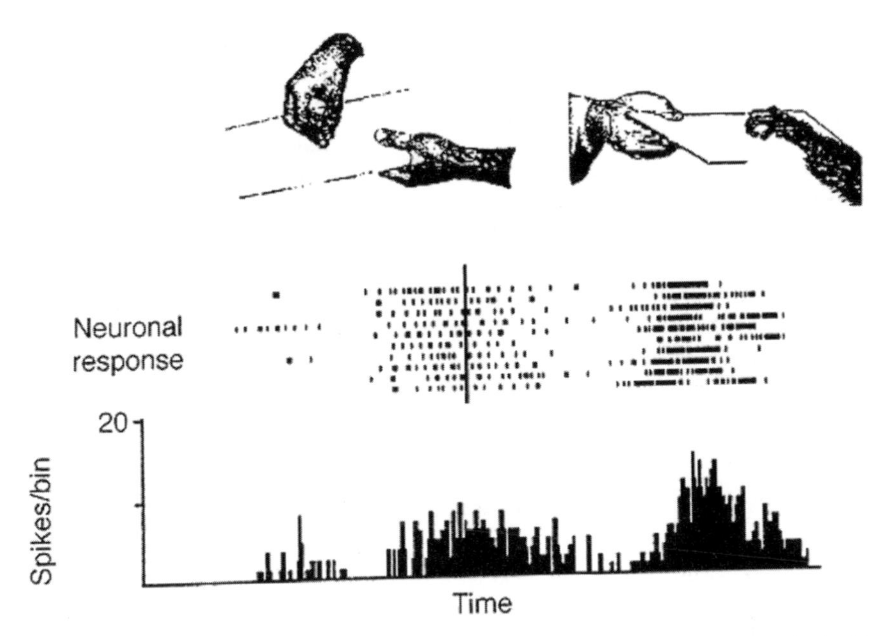

Fig. 1 An example of a mirror neuron while the experimenter grasps a piece of food with his hand then moves it toward the monkey, who, at the end of the trial, grasps it. The neuron discharges during observation of the grip, ceases to fire when the food is given to the monkey, and discharges again when the monkey grasps it (from Rizzolatti and Arbib, 1998)

the functions of such a system may extend far beyond action understanding. We submit that 'mirror neurons' may play a large role in our social evolution including development of culture and development of the self. Attributing so many functions to a single (or small set of) mechanisms may seem farfetched, but in science multiple effects do not necessarily imply multiple causes. For example diabetes has a single cause, decreased insulin production or sensitivity, but is protean in its manifestations. The resulting symptoms include retinal damage, gangrene, heart disease, and renal damage. At the same time one must apply this principle carefully, especially in biology, given the somewhat messy, happenstance nature of organic evolution.

The aim of this chapter is to present a specific argument that is part of the current zeitgeist regarding the mirror neuron system, but has not been explicitly stated before. We suggest that mirror neurons fundamentally mediate the mapping of one dimension or representation in a brain map onto a completely different dimension. Thus, the mirror neurons in the motor system remap a visual representation of observed actions onto the observer's motor representation. In a similar sense language processing may involve a remapping of visual and auditory representations onto motor representations while social processes such as empathy and theory of mind might involve a remapping of visual representations onto limbic and premotor brain regions in order to have an embodied representation of mental states. Whether or not all of these crossmodal neurons are called 'mirror neurons' is just a matter of semantics. We propose that the computational algorithm in each of these scenarios maybe analogous to that which is traditionally ascribed to mirror neurons. And maybe a fundamental feature of brain function that is "repeated" for multiple purposes once it was "discovered" in evolution. How this type of system develops has yet to be extensively studied. Some researchers have suggested that these types of neurons are simply a result of Hebbian processes (e.g., every time the monkey moves it sees its own hand move, thus over time motor neurons begin to respond to visual perception of actions). This is an open question that we propose maybe answered through a series of well-designed experiments. Finally as we have suggested in the past that this shared representation between two domains can be exploited to develop therapeutic interventions for a number of clinical disorders including phantom limb pain, stroke, and autism.

2 Possible Functions of the Mirror Neuron System

2.1 *Action Understanding*

Upon the discovery of the mirror neuron system, the first proposed function of the system was action 'understanding.' At first, one might believe that a motor neuron would not be necessary for action understanding, why not simply use highly sophisticated visual areas including structures like MT and STS would be sufficient for this task and indeed some of these regions respond to 'point-light walkers' that convey human movements with a minimum of form information (Allison, Puce, & McCarthy, 2000; Grezes et al., 2001; Grossman et al., 2000; Vaina, Solomon, Chowdhury, Sinha, & Belliveau, 2001) Visual processing is thus sufficient for recognizing inanimate actions or actions that are not part of the observers motor repertoire (Buccino et al., 2004; Di Pellegrino et al., 1992). However, animate action understanding goes beyond simple visual recognition. The observer has to not only recognize that the observed agent is performing an action, it also needs to differentiate this action from other similar actions, understand the intention or goal of the action, and use this information to formulate an appropriate response. This final step, motor planning, is a well-established function of premotor cortex, but mirror neurons may also be involved in previous stages of action recognition and understanding of the actor's goal/intention.

Behavioral studies as early as those performed by Darwin indicate that when individuals are in the presence of others, the observer tends to synchronize his or her movements to match those of the others (Condon & Ogston, 1967; Darwin, 1872/1965; Kendon, 1970). Thus, the observation of actions appears to facilitate the performance of the same action. Conversely, it also appears that the performance of action facilitates the recognition of a similar action (Reed & Farah, 1995). In their study, if the observer was moving his own arm, he was more likely to recognize that the confederate moved her arm than her leg and vice versa. Thus, it appears that recognition of actions may, at least in part, depend on a motor system. Furthermore, there is much evidence that some mirror neurons are sensitive to exactly how an action is performed, while others are sensitive to the overall goal or intention of the action.

A proportion of mirror neurons in the macaque are broadly congruent, meaning they respond to the performance of an action and the observation of an action with a similar goal even if the exact physical properties of the action differs, while a separate group of neurons are strictly congruent, meaning they only respond when the observed action corresponds both in goal and in physical properties of the action (Gallese, Fadiga, Fogassi, & Rizzolati, 1996). In fact, a later study finds that the final action does not even need to be observed if the macaque has sufficient information to create a mental representation of the action, the mirror neurons will fire. For example, as shown in Fig. 2, Umilta and colleagues (2001) recorded from single unit electrodes and found that a subset of mirror neurons becomes active when the final part of the action, depicting the hand grasping the object, is obscured from view just prior to the action. This implies that the monkey's premotor mirror neurons are able to internally generate the motor representation of the action, even when the visual image of the action is not currently present.

A recent human Functional Magnetic Resonance Imaging (fMRI) study also provides evidence for the sensitivity of the mirror neuron system to goals and intentions. Iacoboni et al. (2005) showed participants videos of four different types of actions. The first video showed a person grasping objects in the absence of any context. The second video depicted scenes containing objects in a context with no actions. The third video showed someone grasping a cup with the intention to drink, and the fourth video showed the same action but in a different context that implied the intention to clean. Results suggested that the posterior part of the inferior frontal gyrus and the adjacent portion of the ventral premotor hand area (both within regions thought to be part of the human premotor mirror neuron system) were more active in the two intention conditions as compared with the other two nonintention videos. Additionally, the drinking intention condition resulted in significantly more activation than the cleaning intention videos. These findings suggest that, like that of the monkey, the human premotor mirror neuron system is sensitive to the underlying intention that motivates perceived actions.

2.2 *Imitation*

Another proposed function of the mirror neuron system was for action imitation. Positron Emission Tomography (PET) studies have demonstrated that when human subjects are instructed to observe actions with the intent to imitate, as opposed to remember, the premotor mirror neuron system is selectively involved in the imitation condition (Decety et al., 1997). Additionally, recent reports suggest that the superior temporal sulcus, a region primarily thought to be

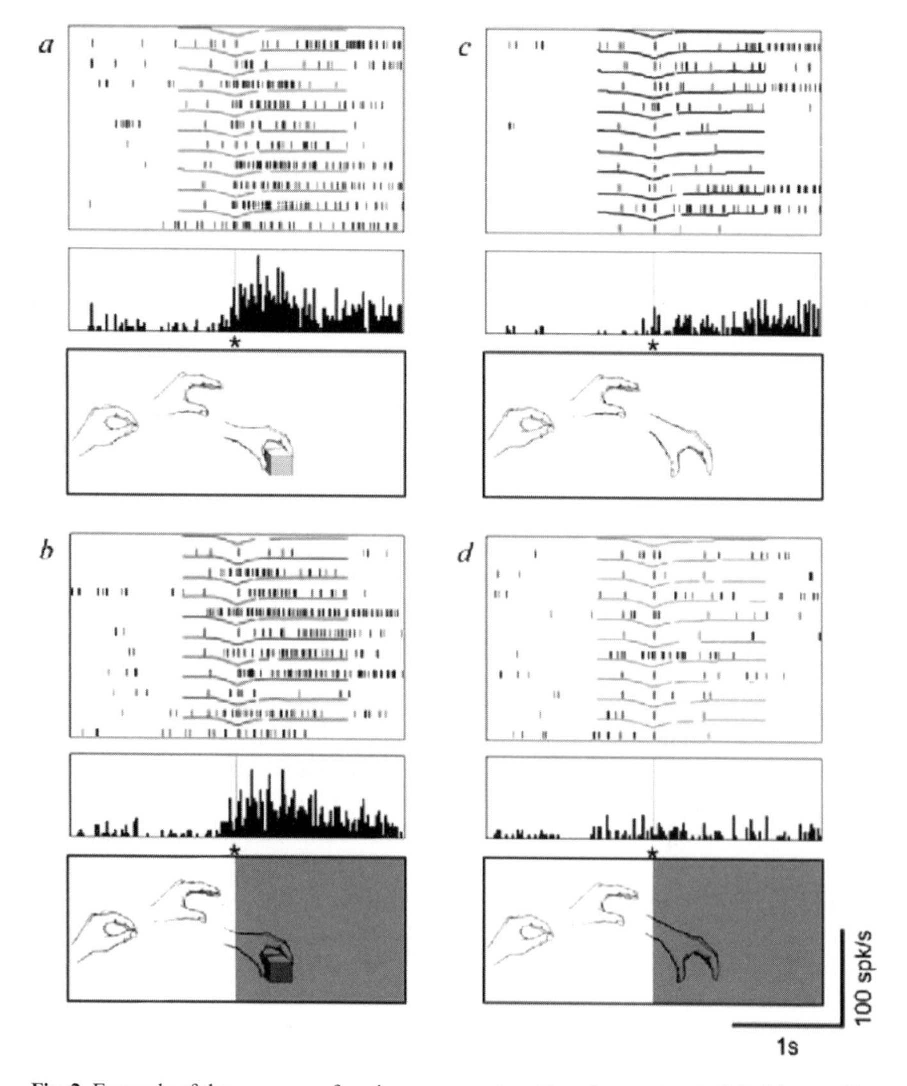

Fig. 2 Example of the response of a mirror neuron to action observation in full vision and in a hidden condition but not in mimed conditions (from Umilta et al., 2001)

involved in biological action perception, but recently added to the list of regions thought to contain mirror neurons, responds to the imitation of an action. This activation is greater during imitation than during control motor tasks and continues to respond even when the subject's view of his or her hand is obscured.

Critics of the mirror neuron hypothesis are quick to point out that though some primates such as orangutans naturally imitate in the wild, macaques and many other primates do not spontaneously show imitation behaviors (Visalberghi & Fragaszy, 2001; Whiten & Ham, 1992) but do have mirror neurons. To this criticism, there are two logical responses. First, simply because an animal does not do something (e.g., imitate) does not mean that they are incapable of this behavior. It may simply not be appropriate in the animal's culture. Also, there is some debate over what *is* imitation. Perhaps many studies utilize an anthropomorphic measure of imitation and miss the types of imitation which maybe more species specific. Secondly, even if it is agreed that macaques and other primates do not imitate, this does not negate the possibility that the mirror neuron system subserves imitation for humans. Perhaps the mirror neuron system became more sophisticated in the human allowing for integration with a wider range of brain structures or the cultural environment was such that imitation became evolutionarily useful to the human. Instead of the brain having to evolve a new system for this purpose it adapted an existing system which possessed the necessary properties to mediate this behavior. Either way, one can easily imagine how a neuron that remaps a visual perception onto a motor representation would be useful for action imitation. Having a mirror neuron system may have been an evolutionarily necessary step that led to the ability to imitate in humans (Arbib, 2005; Ramachandran, 2000) and indirectly led to the 'great leap forward' in human cultural evolution that occurred 100,000 years ago.

2.3 Language

Based on the location of the premotor mirror neuron system in a region thought to be involved in both hand and mouth actions, and its proximity to speech production regions in the human, early theories emerged suggesting a role for mirror neurons in language and specifically speech perception. A leading theory of speech perception may also explain why a speech production area might be involved in speech perception. This theory, termed the *motor theory of speech perception* (Liberman & Mattingly, 1985) suggests that the objects of speech perception are not the acoustic signals (as proposed by Ohala, 1996) but rather the phonetic gestures, motor commands that signal movements of the mouth, lips, and tongue in specific configurations. Thus speech perception and speech production are intimately linked.

Neuromagnetic and neuroimaging studies provide support for the involvement of the mirror neuron system in speech perception. A Transcranial Magnetic Stimulation (TMS) study performed by Fadiga, Craighero, Buccino, and Rizzolatti (2002) found that the application of a magnetic pulse over left motor cortex resulted in a greater motor evoked potential in the tongue when participants listened to speech sounds (which required tongue movements to produce) as compared to nonspeech sounds or speech sounds that did not require the tongue. Similarly, Watkins, Strafella, and Paus (2003) found that both hearing and watching speech sounds being produced resulted in a greater motor evoked potential in the lips when TMS was applied to primary motor cortex. FMRI studies have also shown specific activation of speech production areas during the listening of speech sounds (Wilson, Saygin, Sereno, & Iacoboni, 2004).

The second claim of the motor theory of speech perception is that speech perception and speech production share an underlying mechanism. Along these lines, recent evidence suggests that the superior temporal sulcus as well as the inferior frontal gyrus respond to both the sight and sound of human speech (Calvert & Campbell, 2003). Given these findings, researchers are beginning to draw connections from the action-related mirror neuron system to the utilization of this system for communication. If the purpose of simulating the action with mirror neurons is to understand the observed action, one would theorize that activating the speech production areas while listening to language would lead to better understanding of the verbalization.

At its most basic level, the mirror neuron system is a multisensory system converting observed (either visual or auditory) stimuli into a sensorimotor representation. One task which highlights the involvement of sensorimotor processes in language and may support the role of mirror neuron-like algorithms is the 'Bouba-Kiki task'. This task was originally which described by German-American psychologist Wolfgang Kohler (1929, 1947) requires participants to name nonsense shapes. When this task is performed on normal adults, results suggest that an overwhelming majority of participants match the sound of the name with the visual form of the nonsense shape. For example, if they are shown Fig. 3 and asked to identify which of the shapes is 'bouba' and which is 'kiki', 95% of people will pick the jagged shape as kiki and the rounded amoeboid shape as bouba (Ramachandran & Hubbard, 2001). Ramachandran and Hubbard suggest that this effect is the result of multisensory neurons (potentially mirror neurons) mapping certain sound inflections and phonemic representations with contours of visual stimuli (Ramachandran & Hubbard, 2001).

We have previously suggested that the reason why neurotypical individuals pick the jagged shape as kiki and the rounded amoeboid shape as bouba is that

Fig. 3 Demonstration of bouba and kiki. Because of the rounded contours of the visual shape, subjects tend to map the name bouba onto the figure on the left, while the sharp inflection of the figure on the right make it more like the sharp auditory inflection of kiki (from Ramachandran & Hubbard, 2001)

the sharp changes in visual direction of the lines in the right-hand figure mimics the sharp phonemic inflections of the sound kiki, as well as the sharp inflection of the tongue on the palate while the rounded shape of the figure on the left mimics the more smooth phonemic inflections of the sound bouba and the motor movements of the lips (Ramachandran & Hubbard, 2001). This type of multisensory integration maybe based on a three-way cross-activation between auditory sensory representations and motor representations in Broca's area creating a natural bias toward mapping certain sound contours onto certain vocalizations.

Anecdotal examples of this natural tendency can be observed in the motor gestures performed when referring to something small or large. Speaking the words such as 'little', 'petite', or 'teeny' results in an unconscious narrowing of the vocal tracts and lips while the words 'large' or 'enormous' results in the opposite effect. Also, when referring to 'you' speakers produce a partial outward pout with my lips (as in English 'you', French 'tu' or 'vous' and Tamil 'thoo'), whereas when referring to 'me', my lips and tongue move inwards (as in English 'me', French 'moi' and Tamil 'naan'). Such phenomena, we suggest blend imperceptibly into such universal phenomena as synesthesia (Ramachandran & Hubbard, 2001) and metaphor. Taken collectively, these findings support a shared mechanism for speech production and speech perception in both auditory and visual domains. Whether this shared mechanism is the mirror neuron system that mediates action understanding is still up for debate, but we suggest that the neural computations are operationally analogous. A recent theory first proposed by our laboratory (Ramachandran & Hubbard, 2001) and further elaborated by Arbib (2005) attempts to trace the evolutionary stages required for the motor based mirror neuron system to mediate human language. Arbib (2005) suggests that the mirror neuron system, which mediated action comprehension in monkeys and higher primates, went through six stages of evolution that led to its eventual role in language comprehension in the modern human.

1. The mirror neuron system may originally have been for the purpose of recognition of grasping actions through a simulation mechanism in early primates. We have previously suggested that that prehensility in arboreal monkeys required cross-modal abstraction between visual orientation of branches with 'motor' orientation involving proprioception and action, thereby paving the way for the emergence of mirror neurons (Ramachandran & Hubbard, 2001)
2. This primal system evolved in the chimpanzee to support simple imitation for object-directed grasping.
3. In the early homonid, this simple system may have become elaborated to include complex imitation that allowed for learning of novel actions that could be approximated by variants of actions that were already part of the observer's repertoire.
4. The early hominid then developed a system of protosigns, which was a manual-based communication system. This leap from imitation for the

sake of instrumental goals to imitation for the sake of communication was likely when the specificity of mirror neuron system for object-directed actions was lost.

5. Once protosign had evolved, this provided the scaffold for which protospeech could develop. Once an individual learned a conventional gesture, that gesture could be paired with a vocalization (protospeech). For example, the motor algorithms required for hand gestures may 'spillover' into lip and tongue movements which, if combined with guttural utterances, would lead to protospeech.
6. The final stage, language, it is argued, was mediated more by cultural rather than biological evolution in *Homo sapiens* (Ramachandran, 2001).

2.4 Empathy and Theory of Mind

The mechanism mediating the ability to understand other's mental states has long been of interest to psychologists. When the announcement was made that mirror neurons may have the ability to remap other's motor states onto the observer's motor representations many proponents of similar theories of mental attribution were eager to claim mirror neurons as the biological basis for whatever behavioral phenomenon they were investigating. The first study to actually test whether the mirror neuron system was involved in mental state attribution used fMRI while participants observed and performed emotional facial expressions and noted that multiple regions including inferior frontal cortex, superior temporal cortex, insula, and amygdala all showed increased activity in both the observe and perform conditions when compared to rest (Carr, Iacoboni, Dubeau, Mazziotta, & Lenzi, 2003). This study indicates that there is an overlapping region of premotor cortex that responds both when subjects are asked to interpret emotional facial expressions and when they observe hand actions. Thus, perhaps emotion recognition is also mediated by premotor mirror neurons. However, the premotor cortex is not the only region that possesses these types of neurons.

Another brain region that may contain neurons with 'mirror-like properties' for internal mental states is the medial prefrontal cortex, Brodmann's Area 9. Though not traditionally thought of as part of the mirror neuron system, this area responds both when subjects are asked to make judgments regarding their own abilities, personality traits, and attitudes (Johnson et al., 2002; Kelley et al., 2002) and when they are asked to attribute intentions to characters in a comic strip (Brunet, Sarfati, Hardy-Bayle, & Decety, 2000) or infer another person's knowledge about a familiar or unfamiliar object (Goel, Grafman, Sadato, & Hallett, 1995). In a recent study, subjects were asked to evaluate their own emotional responses to a picture and to infer the mental state of the individual in the picture (Ochsner et al., 2004). The medial prefrontal cortex responded during both conditions. Thus, it seems conceivable that the same region of the brain that is involved in representing the mental state of ourselves is also involved in inferring the mental states of others. Similar to the shared representation for the perception and performance of actions mediated by premotor mirror neurons, a system of neurons in the medial prefrontal cortex may serve to create a mirror-like shared representation for the experience and perception of mental states.

Thirdly, it appears that emotional centers in the brain may also possess mirror-like systems. In three fMRI studies (Morrison, Lloyd, Di Pellegrino, & Roberts, 2004; Singer et al., 2004; Wicker et al., 2003), empathy for specific emotions activated networks of cerebral cortex similar to the actual experience of that emotion. Both the experience of disgust (while inhaling foul smelling odorants) and the observation of others performing facial expressions of disgust activates the same regions of the insula and the anterior cingulate cortex (Wicker et al., 2003). Both the experience of a physically painful stimulus and the knowledge that a loved one is experiencing the same painful stimulus activates the anterior insula and rostral anterior cingulated cortex bilaterally (Singer et al., 2004). These areas were also correlated with individual empathy scores, indicating that the more an individual was able to use this shared network of cortices, the better his or her ability to empathize with others. Similarly, another study (Morrison et al., 2004) found that receiving a painful pin prick and watching a stranger receive the same pin-prick activated dorsal anterior cingulate cortex.

Lesion studies also speak to shared representations for emotions. For example, damage in the amygdala, for example, appears to impair both the expression and recognition of fear (Adolphs et al., 1999; Adolphs, Tranel, Damasio, & Damasio, 2002; Sprengelmeyer et al., 1999). Similarly, damage to the insula and basal ganglia results in a paired impairment in the experience and recognition of disgust (Calder, Keane, Cole, Campbell, & Young, 2000). Finally, both the experience and recognition of anger appear to depend on the dopamine system, with a dopamine antagonist impairing both processes (Lawrence et al., 2002).

2.5 *Self Representation*

It has been previously suggested (Gallese, 2004) that mirror neurons are capable of creating an internal simulation of others' actions in order to adopt the others point of view in interpreting the action. We suggest that the converse might also be true, that self awareness may result from mirror-like systems capable of creating an internal simulation of your own self from the point of view of others. Thus, the mirror neuron mechanism, the same algorithm that originally evolved to help the observer adopt *another's* point of view was turned inward to facilitate self awareness. It is possible that 'other awareness' may have evolved first and then counterintuitively, as often happens in evolution, the same ability was exploited to model ones own mind, although there is a 'chicken or egg' aspect to this that remains, as yet, unresolved. It may not be coincidental that we use phrases like 'self conscious' when we really mean that we are conscious of others being conscious of ourselves. Or say "I am reflecting" when we mean we are aware of ourselves thinking.

In other words the ability to turn inward to introspect or reflect maybe a sort of metaphorical extension of the mirror neurons ability to read others minds. It is often tacitly assumed that the uniquely human ability to construct a 'theory of other minds' or 'TOM' (seeing the world from the others point of view; "mind reading", figuring out what someone is up to, etc.) must come after an already pre-existing sense of self. We argue that the exact opposite maybe true; TOM may have evolved *first* in response to social needs and then later, as an unexpected bonus, came the ability to introspect on one's own thoughts and intentions. As a note of clarification, we are not arguing that mirror neurons are *sufficient* for the emergence of self, only that they likely play a pivotal role.

Does the mirror neuron theory of self make other predictions? Given our discovery that autistic children have deficient mirror neurons and correspondingly deficient TOM, we would predict that they would have a deficient sense of self and difficulty with introspection. The same might be true for other neurological disorders involving damage to the inferior parietal lobule/Temporo-Parieto-Occipital (TPO) junction (which are known to contain mirror neurons) and parts of the frontal lobes should also lead to a deficiency of certain aspects of self awareness. It has recently been shown that if a conscious awake human patient has his parietal lobe stimulated during neurosurgery, he will sometimes have an 'out of body' experience, as if he was a detached entity watching his own body from up near the ceiling. We suggest that this arises because of a dysfunction in the mirror neuron system in the parieto-occipital junction caused by the stimulating electrode. Additional neurological evidence for the role of the mirror neuron system in self representations is a condition called anosognosia. Some years ago we examined a patient with anosognosia who had a lesion in his right parietal lobe and vehemently denied the paralysis. Remarkably the patient also denied the paralysis of another patient sitting in an adjacent wheelchair (Ramachandran & Rogers-Ramachandran, 1996a, 1996b)!

3 What Is a Mirror Neuron?: Beyond Semantics

When mirror neurons were first described, they were characterized based both on their location and on their response properties. Mirror neurons were neurons in area F5 of the monkey premotor cortex that responded to both observed and performed actions. Then overtime neurons with similar response patterns to action observation and execution were found in regions of inferior parietal cortex (Fogassi, Gallese, Fadiga, & Rizzolatti, 1998, Gallese, Fogassi, Fadiga, & Rizzolatti, 2002). Iacoboni and colleagues (2001) specifically lay out a definition of what is a 'mirror neuron.' First, mirror neurons must primarily be motor neurons. Next, they must respond greater to imitation as compared to a control motor task. Lastly, they must respond to the observation of actions. Thus, despite the finding that the superior temporal sulcus contains regions that respond when the individual performs an action outside of sight because this region is primarily a visual region, which also responds to action execution and not vice versa, it can only be said that STS has neurons with 'mirror properties' and not 'mirror neurons.'

Though mirror neurons are not limited to premotor cortex, it does appear, at least for some individuals that a 'mirror neuron' must primarily be a motor neuron. We would respectfully disagree. Simply because the motor system was where mirror neurons were first identified does not give it exclusive right to this type of neuron. By suggesting that mirror neurons must first be primarily motor neurons needlessly restricts this type of neuron whose properties are clearly useful to a number of domains. In fact, we have already suggested in the previous sections that other regions of the brain may possess neurons that perform analogous computations for other domains (e.g., emotion and theory of mind).

Hence the question, what is a mirror neuron? Perhaps a more useful definition would focus on its response properties rather than the exact domain in which it is responding. Similar to the immune system which is not spatially localized but has one function and response pattern in various parts of the body, perhaps the mirror neuron system serves to connect our own representations with those of others across multiple domains and more generally mapping one dimension onto another in order to abstract what is common to them. Thus, we would like to submit that the term 'mirror neuron' should be used for any neuron which is capable of remapping an observed state in another onto the observers own representation of that state. Whether the state is a change in body position (action) or an internal state is no longer relevant for this current definition.

Under this new definition, to solve the problem of understanding others motor states, the brain has evolved a 'mirror' system in motor regions of the brain. Likewise, to solve the problem of understanding others' mental and emotional states, the brain has evolved a 'mirror' system in medial prefrontal and limbic structures. Despite this underlying similarity, historically these abilities have been studied in isolation, and consequently, multiple theories have been developed to explain one or two of these phenomena. However, given that these abilities all seem to be impaired in a single clinical syndrome, namely autism, it is not only parsimonious but also clinically necessary to speak of the mirror neuron system on a functional rather than anatomical level.

Barsalou (1999) argues a similar point in his perceptual symbol systems (PSS) framework. The first tenet of PSS is that knowledge (or internal representations) regarding perceptions, actions, and introspective states are represented in the same systems in which they are experienced. According to PSS, neurons, called 'conjunctive neurons,' receive efferent copies of the input signal from all of the senses (including vision, audition, olfaction, gestation, haptics, proprioception, and introspection) and store it for future cognitive use. Once these so-called conjunctive neurons are established, they can be activated by cognitive simulation in the absence of bottom-up input. 'Simulators' are distributed networks of conjunctive neurons that activate in response to a specific category. They are responsible for integrating the modality-specific information and forming a supramodal representation of a concept. Thus, Barsalou's 'simulators' maybe the functional equivalent of 'mirror neurons' for any given domain.

4 How Do Mirror Neurons Develop?

There are a variety of ways on which unique mechanism could be set up. It is possible that this type of system is 'hard wired' by genes. Others believe it is acquired through learning. Evidence for both an innate mirror system and a learned mirror system exist and will be discussed in turn.

30 years ago, studies designed by Meltzoff and Moore (1977) provided evidence for neonatal imitation in infants as young as a few hours old. Specifically, these infants were shown to imitate mouth opening, tongue protrusion, and hand opening. The researchers also suggest that the pattern of imitation is not likely due to either conditioning or innate releasing mechanisms. They suggest that this early imitation implies that human neonates have an innate ability to equate their own unseen behaviors with gestures they see others perform. By our current definition, 'mirror neurons' maybe mediating this remapping. Consider the first time an infant mimics a mother's smile. Instead of being based on mirror neurons, this could be a reflex in response to a smile—like a sneeze in response to pepper. One way to find out would be to test whether infants can mimic an asymmetrical smile they have never seen before; this would eliminate the 'reflex' explanation and implicate a hard wired mirror neuron–like mechanism. Vocal mimicry might involve similar mechanisms. An internal brain template of the mother's vocalization might be set up during the first exposure. Through repeated attempts to match a trial vocalization (whether actually voiced or entirely internal) to the template, the baby's brain might 'learn' how to set up mirror neurons even without the benefit of continued external feedback. On the other hand, if the imitation of a sound is immediate upon the first exposure, it is more likely that these mirror neurons are hardwired rather than learned.

The second possibility is that mirror neurons represent a form of associative learning. For example, every time the monkeys motor command neuron fired to reach for a peanut the visual appearance of the monkeys own hand reaching activated visual neurons in its brain so that the firing of the two neurons (motor and visual) become linked through Hebbian association. The net result is that the motor neuron itself is then activated by the visual image of peanut grabbing, even if the visual image is of another monkey's hand. Studies finding modulation of the mirror neuron system based on familiarity and experience with the action (Calvo-Merino, Glaser, Grezes, Passingham, & Haggard, 2005) suggest that this system does seem to 'learn' based on the person's own actions.

This Hebbian association hypothesis has been suggested by those who argue against those that claim that mirror neurons are performing a complex remapping of other's representations onto one's own motor system. One logical question, however, is if only a portion of F5 neurons have 'mirror' properties, why do these neurons 'learn' while others do not? If they were purely a result of Hebbian associations, one would predict that all neurons in that region would have 'mirror' properties, but this is not the case. This shows that there are specialized mechanisms and constraints that characterize the subset of the population we refer to as mirror neurons. Additionally, the Hebbian hypothesis cannot account for the facial mimicry literature as the brain receives no visual feedback when the infant smiles on which to build an association. It is still possible that the mirror neuron system may not be mediating this ability; however, the Hebbian hypothesis is no better at explaining this behavior.

The third possibility is that mirror neurons may take time to develop and may require certain preliminary mechanisms before being fully functional. This third option does not speak to whether the development results in the motor neuron being converted into a 'mirror neuron' which is capable of doing a complex self-other algorithm or a motor neuron simply responding to the visual stimulus as a result of Hebbian associative processes. Thus, we propose that the question of innateness and the question of development are empirically separable and the necessary studies to answer these questions have yet to be conducted.

In order to answer the question of innateness, one could record from area F5 in the macaque (a region already known to contain 'mirror neurons') in a new born macaque and expose the monkey to several actions including ones that the monkey will likely be exposed to early, (e.g., peanut breaking, grasping, etc.) as well as novel actions that the monkey is capable of performing, but is unlikely to have any evolutionary innate releasing mechanism set-up for. If neurons in F5 respond to both the familiar and novel actions the first time they are presented then that would argue for an innate system that does not depend on Hebbian association mechanisms. If F5 neurons only respond to the familiar actions then the same argument could be made for these findings as was made for the Meltzoff findings that the brain is 'hard wired' to respond to certain evolutionarily relevant actions. Though this finding would speak to the innateness of mirror neurons, one could argue that it is not doing the complex algorithm argued by proponents of the mirror neuron system. Finally, if no F5 neurons respond to the observation of any actions in the newborn monkey this would argue against the mirror neurons mediating the neonatal behavioral imitation found by Meltzoff and Moore (1977). Additionally, it would argue against mirror neurons being innate. This final finding would not be able to speak to whether the development of the mirror neuron system was based on Hebbian mechanisms or complex self-other algorithms.

To test whether 'mirror neurons' are actually a new type of neuron capable of creating a self-other metarepresentation or simply a motor neuron which has made an associative link to the visual representation of that action a different type of study would need to be conducted. One possible study would be to record from a F5 'mirror neuron' in an adult macaque while he watches either another monkey grasp a peanut or himself reaching for a peanut. In the self condition it would be important that the monkey not actually reach for the peanut (enlisting other motor and sensory systems), but instead present the monkey with an optically reversed image so that the inactive monkey has the visual perception of its 'own' hand moving. If it is true that these neurons are set-up through Hebbian associative processes, the 'self' condition should elicit a greater response than the other condition, as this 'egocentric view' is what the association was built on. If the metarepresentation hypothesis is correct, however, the 'other' condition should elicit a greater response. This study, to our knowledge, has yet to be conducted.

5 The Mirror Neuron System as a Target for Therapeutic Interventions

As future studies further characterize the functions of the mirror neuron system, therapeutic interventions can be developed that take advantage of its specific properties and functions. Though not explicitly designed to tap into the mirror neuron system, therapeutic interventions involving the patient looking at his own actions in a mirror (see Fig. 4) have been applied for the treatment of phantom limb pain (Ramachandran & Hirstein, 1998; Ramachandran & Rogers-Ramachandran, 1996a, 1996b), stroke (Altschuler

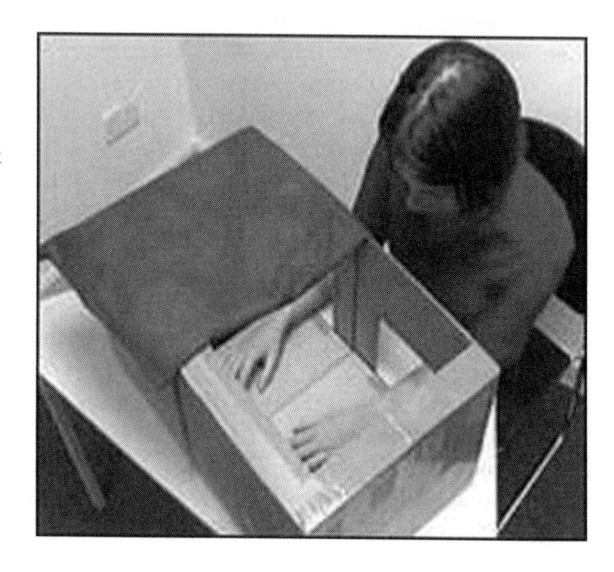

Fig. 4 Example of the therapeutic intervention that involves patients looking at their own actions in a mirror. Though not explicitly designed to tap into the mirror neuron system, such interventions have been applied to the treatment of phantom limb pain, stroke, and even for rehab of finger movements after restorative hand surgery (photo courtesy of Dublin Psychoprosthetics Group)

et al., 1999; Dohle et al., 2006; Sütbeyaz, Yavuzer, Sezer, & Koseoglu, 2007), RSD (McCabe et al., 2003) and even for rehab of finger movements after restorative hand surgery. All of these conditions have been shown to improve in a number of patients after they were asked to move their unaffected hand while looking in a mirror. The efficacy of these rehabilitation procedures, first introduced by our laboratory in the early 1990s, have now been confirmed in a number of double blind controlled trials and have ushered in a new era of rehabilitation as well as a new view of brain function that places emphasis on cross-sensory interactions and abstraction.

A recent paper suggests that this subjective improvement maybe partially mediated by premotor mirror neurons. A study by Giraux and Sirigu (2003) finds activation in premotor mirror neuron regions during the observation of virtually produced actions of the patient's phantom limb that with training expands to other regions of premotor and motor cortex. Thus, perhaps the improvements reported in these cases may have been a result of mirror neurons using their visual properties to facilitate activity in motor cortex. A similar explanation maybe used for the positive results gleaned from mirror therapy with stroke patients.

Additionally, recent studies suggest that the mirror neuron system may also play a role in the behavioral symptoms of autism spectrum disorders (for a review see Oberman & Ramachandran, 2007). If the behavioral symptoms of autism are related to a dysfunction of the mirror neuron system, a neurofeedback training paradigm could be designed to 'rev up' the functioning of this system and in turn improve some of the behavioral impairments which maybe caused by this dysfunction. Preliminary studies along these lines are currently underway (Pineda, personal communication). Additionally, it has been suggested that the mirror neuron system dysfunction in autism maybe a result of faulty connectivity patterns such that long-range connections are reduced and unnecessary short-range connections are facilitated, resulting in less effective transmission of the signal within the mirror neuron system. Transcranial Magnetic Stimulation (TMS) is a candidate therapeutic technique capable of changing neural connectivity patterns. Studies using TMS to investigate neural plasticity in autism and potentially change the connection patterns in these children are currently underway (Pascual-Leone, personal communication). Finally, as it pertains to autism, we have suggested that another novel therapeutic approach might rely on correcting chemical imbalances that disable the mirror neurons in individuals with autism. Specialized neuromodulators may enhance the activity of mirror neurons involved in emotional responses. According to this hypothesis, the partial depletion of such chemicals could explain the lack of emotional empathy seen in autism, and therefore researchers should look for compounds that stimulate the release of the neuromodulators or mimic their effects on mirror neurons. One candidate for investigation is Methylenedioxymethamphetamine (MDMA), better known as ecstasy, which has been shown to foster emotional closeness and communication. It is possible that researchers maybe able to modify the compound to develop a safe, effective treatment that could alleviate at least some of autism's symptoms. Another candidate is prolactin, a hormone known to promote social affiliation in animal studies.

6 Conclusions

The primary goal of this chapter was to put forth some new suggestions regarding the functions and implications of the mirror neuron system in the hopes of advancing our understanding. We accept the possibility that some of our suggestions may need to be revised as new information becomes available, but if one never questions current perspectives then science can never advance. We suggest that the existence of mirror neurons endowed with the properties allowing for complex remapping from one domain into another has endless potential not only to neuroscientists, but to clinicians, anthropologists, sociologists, and psychologists. It provides a foundation from which we as humans are distinct from all the other animals in our abilities to interact socially, understand others thoughts and emotions, communicate using complex language, and the ability to reflect on our self. While one must be careful not to be carried away ascribing too many functions to a unitary mechanism it is not a bad place to start in science, modifying or even rejecting the idea as further evidence comes in.

References

Adolphs, R., Tranel, D., Hamann, S., Young, A. W., Calder, A. J., Phelps, E. A., et al. (1999). Recognition of facial emotion in nine individuals with bilateral amygdala damage. *Neuropsychologia, 37*, 1111–1117.

Adolphs, R., Tranel, D., Damasio, H., & Damasio, A. (2002, December 15). Impaired recognition of emotion in facial expressions following bilateral damage to the human amygdala. *Nature, 372*, 669–672.

Allison, T., Puce, A., & McCarthy, G. (2000). Social perception from visual cues: Role of the STS region. *Trends in Cognitive Science, 4*, 267–278.

Altschuler, E. L., Wisdom, S. B., Stone L., Foster, C., Galasko, D., Llewellyn, D. M. E., et al. (1999), Rehabilitation of hemiparesis after stroke with a mirror. *Lancet, 353*, 2035–2036.

Arbib, M. A. (2005). From monkey-like action recognition to human language: An evolutionary framework for neurolinguistics. *Behavioral Brain Sciences, 28*, 105–167.

Barsalou, L. W. (1999). Perceptual symbol systems. *Behavioral and Brain Sciences, 22*, 577–660.

Brunet, E., Sarfati, Y., Hardy-Bayle, M. C., & Decety, J. (2000). A PET investigation of the attribution of intentions with a nonverbal task. *Neuroimage, 11*, 157–166.

Buccino, G., Lui, F., Canessa, N., Patteri, I., Lagravinese, G., Benuzzi, F., et al. (2004). Neural circuits involved in the recognition of actions performed by nonconspecifics: An fMRI study. *Journal of Cognitive Neuroscience*, 16, 114–126.

Calder, A. J., Keane, J., Cole, J., Campbell, R., & Young, A. W. (2000). Facial expression recognition by people with Mobius syndrome. *Cognitive Neuropsychology*, 17, 73–87.

Calvert, G. A., & Campbell, R. (2003). Reading speech from still and moving faces: The neural substrates of visible speech. *Journal of Cognitive Neuroscience*, 15, 50–70.

Calvo-Merino, B., Glaser, D. E., Grezes, J., Passingham, R. E., & Haggard, P. (2005). Action observation and acquired motor skills: An fMRI study with expert dancers. *Cerebral Cortex*, 15, 1243–1249.

Carr, L., Iacoboni, M., Dubeau, M. C., Mazziotta, J. C., & Lenzi, G. L. (2003). Neural mechanisms of empathy in humans: A relay from neural systems for imitation to limbic areas. *Proceedings of the National Academy of Sciences, USA*, 100, 5497–5502.

Condon, W. S., & Ogston, W. D. (1967). A segmentation of behavior. *Journal of Psychiatric Research*, 5, 221–235.

Darwin, C. (1965). *The expression of emotions in man and animals*. Chicago: University of Chicago Press. (Original work published 1872)

Decety, J., Grezes, J., Costes, N., Perani, D., Jeannerod, M., Procyk, E., et al. (1997). Brain activity during observation of actions. *Brain*, 120, 1763–1777.

Di Pellegrino, G., Fadiga, L., Fogassi, L., Gallese, V., & Rizzolatti, G. (1992). Understanding motor events: A neurophysiological study. *Experimental Brain Research*, 91, 176–180.

Dohle, C., Puellen, J., Nakaten, A., Kuest, J., Rietz, C., Wullen, T., et al. (October, 2006). The effect of mirror therapy in the acute phase of stroke recovery. Poster Session Presented at the 36th Annual Meeting of the Society for Neuroscience, Atlanta, GA.

Fadiga, L., Craighero, L., Buccino, G., & Rizzolatti, G. (2002). Short communication: Speech listening specifically modulates the excitability of tongue muscles: A TMS study. *European Journal of Neuroscience*, 15, 399–402.

Fogassi, L., Gallese, V., Fadiga, L., & Rizzolatti, G. (November, 1998). Neurons responding to the sight of goal directed hand/arm actions in the parietal area PF (7b) of the macaque monkey. Poster Session Presented at the 28th Annual Meeting of the Society for Neuroscience, Los Angeles, CA.

Gallese, V. (2004). *Intentional attunement: The mirror neuron system and its role in interpersonal relations*. Retrieved January 22, 2007, from http://www.interdisciplines.org/mirror/papers/1

Gallese, V., Fadiga, L., Fogassi, L., & Rizzolati, G. (1996). Action recognition in the premotor cortex. *Brain*, 119, 593–609.

Gallese, V., Fogassi, L., Fadiga, L., & Rizzolatti, G. (2002). Action representation and the inferior parietal lobule. In W. Prinz & B. Hommel (Eds.), *Attention & performance XIX: Common mechanisms in perception and action* (pp. 247–66). Oxford, England: Oxford University Press.

Giraux, P., & Sirigu, A. (2003) Illusory movements of the paralysed limb restore motor cortex activity. *Neuroimage*, 20, S107–111.

Goel, V., Grafman, J., Sadato, N., & Hallett, M. (1995). Modeling other minds. *NeuroReport*, 6, 1741–1746.

Grezes, J., Fonlupt, P., Bertenthal, B., Delon-Martin, C., Segebarth, C., & Decety, J. (2001). Does perception of biological motion rely on specific brain regions? *Neuroimage*, 13, 775–785.

Grossman, E., Donnelly, M., Price, R., Pickens, D., Morgan, V., Neighbor, G., et al.. (2000). Brain areas involved in perception of biological motion. *Journal of Cognitive Neuroscience*, 12, 711–720.

Iacoboni, M., Koski, L., Brass, M., Bekkering, H., Woods, R. P., Dubeau, M.-C., et al. (2001). Re-afferent copies of imitated actions in the right superior temporal cortex. *Proceedings of the National Academy of Sciences, USA*, 98, 13995–13999.

Iacoboni, M., Molnar-Szakacs, I., Gallese, V., Buccino, G., Mazziotta, J. C., & Rizzolatti, G. (2005). Grasping the intentions of others with one's own mirror neuron system. *PLoS Biology*, 3, e79.

Johnson, S. C., Baxter, L. C., Wilder, L. S., Pipe, J. G., Heiserman, J. E., & Prigatano, G. P. (2002). Neural correlates of self-reflection. *Brain*, 125, 1808–1814.

Kelley, W. M., Macrae, C. N., Wyland, C. L., Caglar, S., Inati, S., & Heatherton, T. F. (2002). Finding the self? An event-related fMRI study. *Journal of Cognitive Neuroscience*, 14, 785–794.

Kendon, A. (1970). Movement coordination in social interaction: Some examples described. *Acta Psychologica*, 32, 101–125.
Kohler, W. (1929). *Gestalt psychology*. New York: Liveright.
Kohler, W. (1947). *Gestalt psychology* (2nd ed.). New York: Liveright.
Lawrence, A. D., Calder, A. J., McGowan, S. W., & Grasby, P. M. (2002). Selective disruption of the recognition of facial expressions of anger. *Neuroreport*, 13, 881–884.
Liberman, A. M., & Mattingly, I. G. (1985). The motor theory of speech perception revised. *Cognition*, 21, 1–36.
McCabe, C. S., Haigh, R. C., Ring, E. F., Halligan, P. W., Wall, P. D., & Blake, D. R. (2003). A controlled pilot study of the utility of mirror visual feedback in the treatment of complex regional pain syndrome (type 1). *Rheumatology*, 42, 97–101.
Meltzoff, A. N., & Moore, M. K. (1977). Imitation of facial and manual gestures by human neonates. *Science*, 198, 75–78.
Morrison, I., Lloyd, D., Di Pellegrino, G., & Roberts, N. (2004). Vicarious responses to pain in anterior cingulate cortex: Is empathy a multisensory issue? *Cognitive, Affective, and Behavioral Neuroscience*, 4, 270–278.
Oberman, L.M., & Ramachandran, V.S. (2007). The simulating social mind: The role of simulation in the social and communicative deficits of autism spectrum disorders. *Psychological Bulletin, 133*, 310–327.
Ochsner, K. N., Knierim, K., Ludlow, D., Hanelin, J., Ramachandran, T., & Mackey, S. (2004). Reflecting upon feelings: An fMRI study of neural systems supporting the attribution of emotion to self and other. *Journalof Cognitive Neuroscience*, 16, 1746–1772.
Ohala, J. J. (1996). Speech perception is hearing sounds, not tongues. *Journal of the Acoustical Society of America*, 99, 1718–1725.
Ramachandran, V. S. (2000). Mirror neurons and imitation learning as the driving force behind "the great leap forward" in human evolution. Retrieved August 25, 2006, from http://www.edge.org/documents/archive/edge69.html
Ramachandran, V. S., & W. Hirstein (1998), The perception of phantom limbs: The D.O. Hebb lecture, *Brain,* 9(121), 1603–1630.
Ramachandran, V.S., & Hubbard, E.M. (2001). Synaesthesia – A window into perception, thought and language. *Journal of Consciousness Studies*, *8*(12), 3–34.
Ramachandran, V. S., & Rogers-Ramachandran, D. (1996a, August 8). Denial of disabilities in anosognosia. *Nature,* 382, 501.
Ramachandran, V. S., & Rogers-Ramachandran, D. (1996b). Synaesthesia in phantom limbs induced with mirrors. *Proceedings of the Royal Society of London*, 263, 377–386.
Reed, C. L., & Farah, M. J. (1995). The psychological reality of the body schema: A test with normal participants. *Journal of Experimental Psychology: Human Perception and Performance*, 21, 334–343.
Singer, T., Seymour, B., O'Doherty, J., Kaube, H., Dolan, R. J., & Frith, C. D. (2004, February 20). Empathy for pain involves the affective but not sensory components of pain. *Science*, *303*, 1157–1162.
Sprengelmeyer, R., Young, A. W., Schroeder, U., Grossenbacher, P. G., Federlein, J., Buttner, T., et al. (1999). Knowing no fear. *Proceedings: Biological Sciences*, *266*, 2451–2456.
Sütbeyaz, S., Yavuzer, G., Sezer, N., & Koseoglu, F. (2007) Mirror therapy enhances lower-extremity motor recovery and motor functioning after stroke: A randomized controlled trial. *Archives of Physical Medicine and Rehabilitation*, *88*, 555–559.
Umilta, M. A., Kohler, E., Gallese, V., Fogassi, L., Fadiga, L., Keysers, C., et al. (2001). I know what you are doing: A neurophysiological study. *Neuron,* 31(1), 155–165.
Vaina, L. M., Solomon, J., Chowdhury, S., Sinha, P., & Belliveau, W. J. (2001). Functional neuroanatomy of biological motion perception in humans. *Proceedings of the National Academy of Science, USA*, *98*, 11656–11661.

Visalberghi, E., & Fragaszy, D. (2001). Do monkeys ape? Ten years after. In K. Dautenhahn & C. Nehaniv (Eds.), *Imitation in animals and artifacts* (pp. 471–499). Boston: M.I.T. Press.

Watkins, K. E., Strafella, A. P., & Paus, T. (2003). Seeing and hearing speech excites the motor system involved in speech production. *Neuropsychologia*, *41*, 989–994.

Whiten, A., & Ham, R. (1992). On the nature and evolution of imitation in the animal kingdom: Reappraisal of a century of research. In P. B. J. Slater, J. S. Rosenblatt, C. Beer, & M. Milinski (Eds.), *Advances in the study of behavior* (pp. 239–83). San Diego, CA: Academic Press.

Wicker, B., Keysers, C., Plailly, J., Royet, J. P., Gallese, V., & Rizzolatti, G. (2003). Both of us disgusted in my insula: The common neural basis of seeing and feeling disgust. *Neuron*, *40*, 655–664.

Wilson, S. M., Saygin, A. P., Sereno, M. I., & Iacoboni, M. (2004). Listening to speech activates motor areas involved in speech production. *Nature Neuroscience*, *7*, 701–702.

Part II
Developmental Aspects

The Neurophysiology of Early Motor Resonance

François Champoux, Jean-François Lepage, Marie-Christine Désy,
Mélissa Lortie, and Hugo Théoret

Abstract The discovery of mirror neurons in the monkey brain and the subsequent description of a mirror neuron system (MNS) in humans have triggered unprecedented interest in the neural basis of social cognition. Although the mechanism matching action observation and execution is now well documented in human adults, much remains to be learned about its normal development. In light of accumulating evidence suggesting a link between impaired motor resonance and developmental disorders such as autism spectrum disorder, it is of paramount importance to better understand how the MNS develops and under what specific conditions it appears. Until recently, speculation about the existence of a functioning MNS in the infant brain relied primarily on evidence of imitative behavior shortly after birth. Here, we describe studies that have tackled the issue of detailing MNS properties in the immature brain. Through a variety of neuroimaging techniques, a preliminary description of neurophysiological mechanisms underlying early imitation and action understanding is starting to emerge. We will see, however, that although important first steps have been taken, much remains to be learned about early mirroring processes and their relationship to social behavior.

Keywords Development · Neurophysiology · Motor resonance · Infant · Imitation · Action observation

1 Introduction

The discovery of mirror neurons in the premotor cortex of macaque monkeys (Gallese, Fadiga, Fogassi, & Rizzolatti, 1996; Rizzolatti, Fadiga, Gallese, & Fogassi, 1996) provided the initial impetus for the unprecedented interest in the

H. Théoret
Psychology Department and CHU Sainte-Justine, University of Montreal, Montreal, Quebec, Canada
e-mail: hugo.theoret@umontreal.ca

J.A. Pineda (ed.), *Mirror Neuron Systems*, DOI: 10.1007/978-1-59745-479-7_3,

neural correlates of social cognition by suggesting an intuitive account of how we understand other's actions, intentions, and emotions. Although mirror neurons per se have not yet been discovered in humans, the existence of a mirror neuron system (MNS) has been revealed through a combination of behavioral, imaging, and lesion data (see Rizzolatti and Craighero (2004) for a review of the existing literature). Despite striking advances in our understanding of the adult mirroring circuitry in recent years, a comprehensive description of *immature* motor matching mechanisms has been lacking. Here, we wish to review data suggesting the existence of direct-matching mechanisms in the infant brain as well as describing recent experiments that tackle the difficult issue of detailing MNS properties early in development. A better understanding of motor resonance in infants and children, and its relationship to social behavior, is not only critical to our understanding of typical human development but may also provide critical insight into developmental disorders characterized by social impairments.

2 The Emergence of an Observation-Execution Matching System: Grasping from Birth?

Motor imitation can provide insight into MNS function because it requires the contribution of a neuronal circuitry similar to that solicited by a perceived motor act. In human adults, imitation involves regions traditionally associated with the MNS, particularly the pars opercularis of the inferior frontal gyrus (BA44; Iacoboni et al., 1999). Additionally, three supplemental lines of evidence suggest a role for a direct matching mechanism in imitation: (i) a dysfunctional MNS has been described in a clinical population with well-documented imitative deficits (Dapretto et al., 2006; Oberman et al., 2005; Théoret et al., 2005); (ii) patients with a lesion in Broca's area (BA44) display imitation impairments (Saygin, Wilson, Dronkers, & Bates, 2004); and (iii) repetitive TMS-induced virtual lesions to Broca's area induce imitation deficits (Heiser, Iacoboni, Maeda, Marcus, & Mazziotta, 2003). Given this body of knowledge, it is expectedly toward motor imitative competencies that neuroscientists first looked for the presence of a mirror system early in development (Meltzoff & Decety, 2003). Because the issue of neonatal imitation is discussed elsewhere in this book, we will only briefly review relevant data in the following sections.

2.1 *Finger/Hand Movement Imitation in Newborns*

Finger/hand movements have been shown to be more frequent following observation of hand closings and openings in 14 day-old infants (Meltzoff & Moore, 1977). Recent evidence also supports the existence of imitation-like behaviors shortly after birth. In a group of 39 neonates aged between 2 and 96 hours, Nagy and collaborators (2005) reported an increase in the number of finger movements, complete and incomplete, following observation of index finger protrusions.

2.2 *Facial and Orofacial Gestures Imitation in the Newborn*

Neonatal imitation has been primarily investigated with facial movement. Indeed, imitative abilities in the newborn have been reported using mouth opening (Fontaine, 1984; 2002; Heimann, Nelson, & Schaller, 1989; Heimann & Schaller, 1985; Kugiumutzakis, 1999; Legerstee, 1991; Maratos, 1982; Meltzoff & Moore, 1977, 1983, 1989, 1992), emotional expressions (Field, Woodson, Greenberg, & Cohen, 1982, Field et al., 1983; Field, Goldstein, Vaga-Lahr, & Porter, 1986), lip and cheek movement (Fontaine, 1984; Kugiumutzakis, 1999; Melzoff and Moore, 1997; Reissland, 1988), eye blinking (Fontaine, 1984; Kugiumutzakis, 1999), and tongue protrusion (Meltzoff & Moore, 1983, 1989, 1997). Of great importance to the issue at hand is the fact that imitative behavior has been reported as early as 42 min after delivery (Meltzoff & Moore, 1977, 1983), where infants execute orofacial gestures that match those performed by an adult model.

2.3 *Imitation of Vocal Gestures in Newborns*

Vocal utterances have also been used in the examination of neonatal imitation. Research on vocal perception demonstrates that very young infants are able to imitate vocal stimuli (Kuhl & Meltzoff, 1982, 1996; Kuhl, Williams, & Meltzoff, 1991). Infants as early as 18–20 weeks of age have been shown to preferentially look at a face executing the utterance /a/ or /i/ when it is accompanied by a congruent sound (Kuhl & Meltzoff, 1982). Moreover, infants as young as 12 weeks of age have been shown to match the vowels /a/, /i/, or /u/ to those articulated by an adult (Kuhl & Meltzoff, 1996).

Chen, Striano, and Rakoczy (2004) have demonstrated that infants can correctly imitate vocal gestures at an even earlier age, even in the absence of visual information. Orofacial gestures of twenty-five newborns aged between 24 hours and 7 days (mean = 3 days) were monitored while they listened to vocalizations (/a/ and /m/). Infants produced significantly more mouth clutching than mouth opening during the /m/ condition, and significantly more mouth opening than mouth clutching during the /a/ condition. These finding suggest the presence of a modality-independent observation-execution matching system shortly after birth.

2.4 *Neonatal Imitation: A Reflex-Like Phenomenon?*

Some authors have suggested that the propensity of neonates to imitate is a reflex-like phenomenon resulting from an innate releasing mechanism, whereby an individual's responsiveness to certain stimuli is genetically programmed (Anisfeld, 1996; Anisfeld et al., 2001; Hayes & Watson, 1981). The idea that newborns are incapable of true imitation was at first partly motivated by the belief that newborns were unable to initiate voluntary action, and that arm movements were purposeless and reflexive (Piaget, 1952). It is also commonly known that the occurrence of imitative behaviours greatly decreases around the second month of life (Abravanel & Sigafoos, 1984; Field et al., 1986; Fontaine, 1984), to reappear in more complex fashion at around one year of age (Meltzoff & Moore, 1992). With respect to these arguments, it has been shown that neonates are in fact able to produce directed movements at will (van der Meer, 1997; van der Meer, van der Weel, & Lee, 1995) and it has been proposed that the diminishing propensity of infants to imitate may reflect the fact that older infants initiate social interactions more dynamically than newborns that are simply responding to social cues (Meltzoff & Moore, 1992).

2.5 *Methodological Limitations*

The possible reflex-like nature of early imitation can also be debated on methodological grounds. Neonatal imitation has been primarily investigated through the use of limited orofacial gestures, creating two important concerns. First, facial actions have a high spontaneous production rate that makes it hard to relate the occurrence of the behavior to the experimental condition (Heimann et al., 1989; Jones, 1996). Second, it has been argued that tongue protrusions, which are the most reliably reproduced behaviors, may not be consequent to a specific observed action (Jones, 1996, 2006). Indeed, tongue protrusions have been shown to be elicited by non-biological stimuli such as pen or ball movement (Jacobson, 1979). Non-vocal auditory stimuli (i.e., music) have also been shown to elicit tongue protrusions in 4-week-old infants (Jones, 2006). Furthermore, many other behaviors such as lip and jaw movements have been characterized by some authors as simple forms of oral exploration (Rochat, 1983; Pechaux, Lepecq, & Salzarulo, 1988; Jones, 2006) .

3 Neurophysiological Evidence of Early Emergence

Whereas a more thorough discussion of the debate surrounding the nature of early imitative ability is beyond the scope of this chapter, it is clear that inferring the existence of a functional infant MNS based on evidence of imitation in neonates maybe overreaching. It is clear that neurophysiological evidence is needed to determine the maturational course of resonance mechanisms and determine the extent to which MNS properties maybe present at birth. Because of methodological limitations inherent to the study of the infant brain, few data are available regarding the existence of a direct matching mechanism in the newborn, infant and child brain. Nevertheless, recent research efforts have provided insight into the immature MNS and can serve as a stepping stone for future endeavors.

3.1 *Subdural Recordings*

We recently reported the pattern of electronecephalography (EEG) activity recorded from a subdural grid electrode during execution and observation of hand and arm movements in a 36-month old child undergoing surgery for intractable epilepsy (Fecteau et al., 2004). Single photon emission computed tomography (SPECT) imaging revealed a left peri-central focal lesion. A 64-contact subdural grid electrode was surgically implanted over the left fronto-parietal cortex.

Recordings for the experiment were performed during a one-hour period, during which the patient was sitting in her hospital bed. She was instructed to (1) draw with her right hand (*execution*); (2) watch passively as an experimenter standing next to the bed drew with his right hand (*observation*); and (3) rest with her eyes open (*rest*). Following the experiment, video of the session was reviewed and sequences that met experimental criteria (uninterrupted sequence of 5 seconds; no overt speech or vocalizations; attention to the drawing hand) were analyzed. In total, 5 *rest*, 5 *execution,* and 4 *observation* sequences were analyzed. Extraoperative cortical mapping was performed during the implantation of the grid electrode. For this purpose, stimulation was done by generating trains of electrical pulses at selected pairs of adjacent contacts with careful observation of responses.

The extraoperative study revealed two contact sites at which electrical stimulation elicited quick redirection of head and gaze toward the right hand. This area was thus putatively defined as the somatosensory representation of the right hand. Stimulation of all other contact sites elicited no overt reponse. Spectral analysis showed six contact sites at which the alpha band absolute power was significantly lower in the *execution* condition compared to *rest*. These sites overlapped, or were just anterior to, the contacts that when stimulated gave rise to orientation toward the hand. Further analysis revealed that activity related to the *observation* and *rest* conditions also differed significantly at two of these contact sites. These two contacts were also in the putative sensorimotor area of the participant and were the only activated sites in both the *observation* and *execution* conditions.

This single-case study suggested that a direct-matching motor mechanism was present in a specific region of a child's sensorimotor cortex as early as 36 months after birth. Indeed (i) electrical stimulation of a specific area of cortex resulted in sensorimotor reactions in the contralateral hand; (ii) execution of contralateral hand movement was associated with activations in this functionally defined area; and (iii) parts of that sensorimotor area was also activated by mere observation of hand movements executed by the experimenter.

3.2 Electrophysiological Data

The presence of an observation-execution matching mechanism in typically developing children was also reported using scalp-EEG. We investigated mu rhythm modulation in a group of children that were either executing or passively observing a series of hand gestures (Lepage & Théoret, 2006). Found in the alpha range of the EEG signal (8–12 Hz), the mu rhythm is strongly suppressed during the performance of contralateral acts, reflecting desynchronization of neuronal populations. Importantly, mu rhythm suppression has been reported during the *observation* of motor actions (Muthukumaraswamy, Johnson, & McNair, 2004), possibly reflecting a basic resonance mechanism at the level of the motor cortex. In an extensive review of the topic, Pineda (2005) suggested that the mu rhythm is reflective of an integrative process which, just like the MNS, "translates seeing and hearing into doing."

To verify the existence of a neurophysiological mechanism matching observation and execution of actions in the brain of children, we measured mu rhythm amplitude during the execution and observation of hand grasping movements in a group of 15 children aged between 52 and 133 months (mean: 99,3) (Lepage and Théoret, 2006). Each participant was presented with three possible videos: (i) an adult's extended hand moving horizontally; (ii) an adult's hand preforming a precision grip; and (iii) an adult's static hand. Participants also executed self-paced precision grips similar to those seen in the grip-observation condition.

We found that the typical decrease in mu rhythm amplitude that occurs at central sites (electrodes C3–C4) during action execution was also present during observation of the same movement. Specifically, mu rhythm suppression was maximal during the execution of a hand movement, followed by observation of a grasping movement, observation of flat hand movement, observation of a static hand, and then rest. This pattern of response is strikingly similar to that observed in adults (Muthukumaraswamy et al., 2004). Indeed, the preferential response to object-oriented movements observed in our group of children is also present in adult and is similar to that of monkeys, where mirror neurons are responsive only to transitive, goal-directed gestures (Rizzolatti & Craighero, 2004). This is particularly relevant in light of the fact that in children, imitation is guided by goals. In a study of pre-school children (Bekkering, Wohlschlager, & Gattis, 2000), participants were asked to copy movements performed by an experimenter. It was found that reducing the number of possible goals decreased the numbers of errors made by the participants and that total absence of goals (targets) also increased imitative performance. It was later reported that a similar mechanism operates in adults where the presence of a goal favorably modulated imitation in both error patterns and response time (Wohlschlager & Bekkering, 2002). As it is now well established that the MNS in humans is intimately linked to imitative behavior (e.g., Iacoboni et al., 1999), the importance of goals for imitation might underlie MNS preference for these types of movement.

3.3 Functional Magnetic Resonance Imaging (fMRI)

The existence of a mechanism matching observation and execution of motor actions in children has also been evidenced with hemodynamic responses during imitative behavior. In a recent fMRI study (Dapretto et al., 2006), 10 typically developing children (mean: 12.4 years of age) were presented with images of emotional faces, which the participants had to imitate or passively observe. Both imitation and observation of emotional expressions produced activations in a network of brain areas similar to that of adults, including the inferior frontal gyrus, which has been traditionally associated with the MNS.

3.4 Near Infrared Spectroscopy

Any developmental theory of MNS function must aim at determining the earliest manifestations of direct-matching mechanisms in the human brain. In addition, direct comparisons between an infant and an adult's brain response to action observation need to be performed to address relevant issues related to the effects of experience on MNS function. Shimada and Hiraki (2006) used near infrared spectroscopy (NIRS) to evaluate motor cortex activity in adults and 6- to 7-month old infants during the observation of goal-directed hand actions. NIRS is a neuroimaging technique particularly well-suited for investigations of the infant brain that uses light in the near infrared range to measure blood flow and infer brain activity.

Shimada and Hiraki (2006) presented adult and infant participants with videos or live actions of a human model manipulating an object. Functional identification of the participants' motor areas brains was achieved by having them perform repetitive hand openings and closings (adults) or during free play sessions (infants). In both adults and infants, the same channels that recorded activation during action execution also picked up activity increases during the passive observation condition. These data clearly showed that observation and execution of hand actions produced robust activations in motor cortical areas of an infant's brain. This led the authors to suggest that "6-month olds are capable of internally simulating novel movements performed by others by assembling primitive (component) movements in their motor repertoire" (Shimada & Hiraki, 2006).

Following on that one may then wonder how MNS activity in the immature brain is be related to the *understanding* of others' behavior. In adults, eye-movements that occur during performance of an action match those occurring during the observation of a similar action performed by others (Flanagan & Johansson, 2003). Indeed, the goal of an observed action appears to guide an observer's eye movements. Falck-Ytter, Gredeback, and von Hofsten (2006) addressed the issue of whether the development of such predicting abilities was dependent upon the development of action, such that prediction of a model's action should follow the acquisition of the specific behavior to be predicted. Eye movements were recorded in a group of adults, 12-month-old infants, and 6-month-old infants while they observed a model performing a motor sequence that is acquired at around 7–9 months of life. The authors found that whereas adults and 12 month-old infants use a direct-matching system to predict the goal of an observed action, 6 month-olds do not. Because infants develop specific social cognition abilities believed to be underlied by the MNS at around 8–12 months of life, Falck-Ytter et al. (2006) suggest that the existence of proactive goal-directed eye movements at 12 months but not at 6 months is a strong claim for the presence of a functioning MNS around that time frame.

Evidently, the next important issue to tackle is the determination of the specifics of MNS development at even earlier stages of life. Internal representations of actions are present at birth since fetuses are known to perform a variety of mouth gestures in utero (Prechtl, 1986). Significantly, these are the same kinds of movements that neonates can imitate very shortly after birth (Meltzoff & Moore, 1977). Whether mere *experience* with a specific gesture is a sufficient condition for the development of a direct-matching system at the motor level remains unknown. Obviously, even if such a system was present at birth, much room would be available for refinements and changes through learning and experience. One can speculate, for example, that goal-predicting features of the MNS would be dependent upon the learning of associations between a baby's own actions and goals.

This is evidenced by the fact that infants' experience as agents underlies their ability to view actions as goal-directed. For example, Sommerville, Woodward, and Needham (2005) tested the hypothesis that action *experience* contributes significantly to action *interpretation*. In a group of 3-month-old infants, the authors found that following a few minutes of experience in specific goal-directed behavior, the participants could transfer this newly acquired knowledge to their perception of actions in others. Importantly, infants who only had prior visual experience with the action did not subsequently apply the newfound knowledge to infer goals. Here again, data suggest that infants possess a shared motor representation of self and other that can readily be built upon to promote complex behavioral and cognitive schemes.

4 The Flexible Nature of the MNS

The impact of experience and learning on the MNS cannot be overstated, even if the existence of a direct-matching mechanism at birth was to be demonstrated. In a classic demonstration of mirror neuron function, Kohler and collaborators showed that some macaque monkey neurons fired when an animal broke a peanut, observed an experimenter breaking a peanut or when it was exposed to the sound of a peanut being broken (Kohler et al., 2002). No one would argue that the association between the action of cracking a peanut and the specific sound that it creates is genetically pre-programmed. Indeed, it is obvious that mirror neurons cannot achieve such specific pairings at birth. More realistically, an adult macaque could have learned to associate the action of his own hand breaking the peanut and the sound it creates by observing and practising this action many times, coupled with remembrance of contextual information (attendance to food).

Another striking example of experience and learning effects on mirror neurons is that of Ferrari, Rozzi, & Fogassi (2005). They showed that monkey mirror neurons can extend their firing response to the observation of actions made with tools after long visual exposure to tool actions. These data suggest that through experience and learning, single mirror cells properties were modified because tool-responding mirror neurons began to fire only after months of training. This suggests that an association between hand and tool use was slowly built over time, resulting in some neurons being actually more sensitive to tool actions than hand actions. This drastic modification of single cell function supports the idea that whatever the nature of early direct-matching mechanisms during infancy, interactions with the environment constantly modify the initial MNS configuration to adjust to the requirements of social life.

The inherent malleability of the MNS has also been thoroughly evidenced in the human brain. For example, Calvo-Merino, Glaser, Grezes, Passingham, and Haggard (2005) used a clever design to investigate the effect of motor experience on MNS activity by having expert dancers in different styles (ballet or capoeira) watch videos depicting kinematically similar dance sequences in each of the two styles. In line with the idea that the presence of a specific motor repertoire acquired through practice facilitates MNS activity when observing actions related to that skill, activation in classical mirror areas (premotor areas) was stronger when dancers observed dance sequences in which they were expert. A subsequent study showed that such learning effects could occur rapidly. Cross, Hamilton, & Grafton (2006) measured brain activity with fMRI at different time points while expert dancers learned new dance sequences (5 hours a week for 5 weeks). In close correspondence with the Calvo-Merino et al. (2005) study, it was found that typical MNS areas were activated when dancers observed the newly acquired skill. Significantly, a dancer's own perceived ability to perform the newly acquired sequence was significantly correlated with activity in MNS areas.

Practice effects have also been reported using auditory material. As was previously mentioned, the monkey brain comprises audiovisual mirror neurons that respond to both sound and visual representations of an action (Kohler et al., 2002). As such, one may wonder if the auditory components of the MNS are also prone to learning effects, which would have important ramifications related to theories of language acquisition (Arbib, 2005). In a design reminiscent of that of Cross et al. (2006), naive subjects were trained to play a simple piano sequence. Following 5 days of practice, participants reached a plateau-level of performance, indicating that the musical piece had been learned. Three different musical sequences were then presented to participants while their brain was scanned with fMRI: (i) the sequence of notes that was learned; (ii) a completely different sequence of notes; and (iii) the sequences of notes that were learned in a different order. Results revealed that although all sequences produced similar activations in auditory cortices, passive listening to the trained piece was associated with activations in classical MNS areas (predominantly the Inferior frontal gyrus (IFG)). Interestingly, listening to a musical sequence that comprises the same notes as the learned piece did not activate the IFG. This reinforces the link between sound-actions associations that result from practice.

This brief description of learning effects in the adult MNS underscores an important point: although neurophysiological and behavioral evidence suggest the presence of a functioning MNS early in development, it does not preclude numerous factors from modifying the way the brain responds to observed actions. For example, one can only look at the important modifications that occur in the parietal cortex (a core MNS area) of the adolescent brain (Blakemore & Choudhury, 2006) to be convinced of the dynamic nature of MNS development. These data are supported at the behavioral level by the fact that from adolescence to adulthood, the correspondence between the execution and imagination of actions significantly an increase, suggesting that action representation itself goes through significant changes in adolescent years (Choudhury, Charman, Bird, & Blakemore, 2007).

5 Conclusions

The data reviewed in this chapter constitute a first step toward a clear understanding of the neuronal events that underlie the normal development of the MNS. The few studies that have addressed the existence of mirror matching mechanisms in the infant brain tend to support the notion that a rudimentary MNS appears very early in development. Specifically, NIRS data show that a resonance mechanism in motor cortical areas maybe in place as early as 6–7 months after birth. When one considers behavioral evidence suggesting the presence of imitative abilities earlier than that, as well as data showing an infant's ability at predicting action goals, it is tempting to conclude that motor behaviors in utero may serve as a building block for the precocious appearance of a functioning MNS. However, in the absence of neurophysiological evidence describing motor resonance in newborns, and in light of the long-standing debate regarding the exact nature of imitation-like behaviors in neonate, one must remain cautious.

Rapid advances in the field of social neuroscience have led to unprecedented interest in the neuronal mechanism underlying social mirroring. A better understanding of mirror system function in the newborn and infant and a thorough description of its developmental course is of paramount importance in light of the possible relationship between MNS impairments and some neurodevelopmental disorders. Indeed, the establishment of critical periods for proper MNS development would be invaluable for the development of diagnostic and intervention tools that acknowledge the strong link between imitation and mirror matching mechanisms. Moreover, because the MNS is believed to mediate the understanding of actions and emotions when they occur in others, it has been repeatedly proposed that many aspects of social cognition depend on embodied simulation mechanisms (Gallese, Keysers, & Rizzolatti, 2004). According to this view, the MNS is the basic component enabling the creation of a 'bridge' between others and ourselves. Following on this, it could be hypothesized that social cognitive abilities such as theory of mind and empathy would follow similar developmental courses as those observed in the MNS. This is particularly relevant in light of preliminary data showing that imitation of emotional facial expressions is associated with greater MNS activity in children (aged 9.5–10.5 years) scoring higher on behavioral measures of empathy and interpersonal ability (Pfeifer, Iacoboni, Mazziotta, & Dapretto, 2005).

Future studies will certainly disentangle the complex relationship between the development of MNS function and social cognition. There is no doubt that a thorough understanding of resonance mechanisms in newborns, infants, adolescents, and adults are crucial for an understanding of social cognition that reflects the richness of interindividual interactions.

Acknowledgments We gratefully acknowledge financial support from the Canadian Institutes of Health Research, the National Sciences and Engineering Research Council of Canada, and the Fonds de Recherche en Santé du Québec.

References

Abravanel, E., & Sigafoos, A. D. (1984). Exploring the presence of imitation during early infancy. *Child Development*, *55*, 381–392.

Anisfeld, M. (1996). Only tongue protrusion modelling is matched by neonates. *Developmental Review*, *16*, 149–161.

Anisfeld, M., Turkewitz, G., Rose, S., Rosenberg, F., Sheiber, F., & Couturier-Fagan, D. (2001). No compelling evidence that newborns imitate oral gestures. *Infancy*, *2*, 111–122.

Arbib, M. (2005). From monkey-like action recognition to human language: An evolutionary framework for neurolinguistics. *Behavioral & Brain Science*, *28*, 105–124.

Bekkering, H., Wohlschlager, A., & Gattis, M. (2000). Imitation of gestures in children is goal-directed. *Quarterly Journal of Experimental Psychology A*. 2000 Feb, *53*(1), 153–64.

Blakemore, S. J.,& Choudhury, S. (2006). Development of the adolescent brain: Implications for executive function and social cognition. *Journal of Child Psychology and Psychiatry*, *47*, 296–312.

Calvo-Merino, B., Glaser, D. E., Grezes, J., Passingham, R. E., & Haggard, P. (2005). Action observation and acquired motor skills: An fmri study with expert dancers. *Cerebral Cortex*, *15*, 1243–1249.

Chen, X., Striano, T., & Rakoczy, H. (2004). Auditory-oral matching behavior in newborns. *Developmental Science*, *7*, 42–47.

Choudhury, S., Charman, T., Bird, V., & Blakemore, S. J. (2007). Development of action representation during adolescence. *Neuropsychologia*, *45*, 255–262.

Cross, E. S., Hamilton, A. F., & Grafton, S. T. (2006). Building a motor simulation de novo: Observation of dance by dancers. *Neuroimage, 31*, 1257–1267.
Dapretto, M., Davies, M. S., Pfeifer, J. H., Scott, A. A., Sigman, M, Bookheimer, S. Y., et al. (2006). Understanding emotions in others: Mirror neuron dysfunction in children with autism spectrum disorders. *Nature Neuroscience, 9*, 28–30.
Falck-Ytter, T., Gredeback, G., & von Hofsten, C. (2006). Infants predict other people's action goals. *Nature Neuroscience, 9*, 878–879.
Fecteau, S., Carmant, L., Tremblay, C., Robert, M., Bouthillier, A., & Theoret, H. (2004). A motor resonance mechanism in children? Evidence from subdural electrodes in a 36-month-old child. *Neuroreport, 15*, 2625–2627.
Ferrari, P. F., Rozzi, S., & Fogassi, L. (2005). Mirror neurons responding to observation of actions made with tools in monkey ventral premotor cortex. *Journal of Cognitive Neuroscience, 17*, 212–226.
Field, T. M., Woodson, R., Greenberg, R., & Cohen, D. (1982). Discrimination and imitation of facial expression by neonates. *Science, 218*, 179–181.
Field, T. M., Woodson, R., Cohen, D., Greenberg, R., Garcia, R., & Collins, E. (1983). Discrimination and imitation of facial expressions by term and preterm neonates. *Infant Behavior and Development, 6*, 485–489.
Field, T. M., Goldstein, S., Vaga-Lahr, N., & Porter, K. (1986). Changes in imitative behavior during early infancy. *Infant Behavior and Development, 9*, 415–421.
Flanagan J. R., & Johansson R. S. (2003). Action plans used in action observation. *Nature, 424*, 769–771.
Fontaine, R. (1984). Imitative skills between birth and six months. *Infant Behavior and Development, 7*, 323–333.
Gallese, V., Fadiga, L., Fogassi, L., & Rizzolatti, G. (1996). Action recognition in the premotor cortex. *Brain, 119*, 593–609.
Gallese, V., Keysers, C., & Rizzolatti, G. (2004). A unifying view of the basis of social cognition. *Trends in Cognitive Science, 8*, 396–403.
Hayes, L., & Watson, J. (1981). Neonatal imitation: Fact or artifact. *Developmental Psychology, 177*, 660–665.
Heimann, M., & Schaller, J. (1985). Imitative reactions among 14-sup-21 day old infants. *Infant Mental Health Journal, 6*, 31–39.
Heimann, M., Nelson, K. E., & Schaller, J. (1989). Neonatal imitation of tongue protrusion and mouth opening: Methodological aspects and evidence of early individual differences. *Scandinavian Journal of Psychology, 30*, 90–101.
Heimann, M. (2002). Notes on individual differences and the assumed elusiveness of neonatal imitation. In: A. N. Meltzoff & W. Prinz, (Eds.), *The imitative mind: Development evolution, and brain bases* (pp. 74–84). Cambridge: Cambridge University Press.
Heiser, M., Iacoboni, M., Maeda, F., Marcus, J., & Mazziotta, J. C. (2003). The essential role of Broca's area in imitation. *European Journal of Neuroscience, 17*, 1123–1128.
Iacoboni, M., Woods, R. P., Brass, M., Bekkering, H., Mazziotta, J. C., & Rizzolatti G. (1999). Cortical mechanisms of human imitation. *Science, 286*, 2526–2528.
Jacobson, S. W. (1979). Matching behavior in the young infant. *Child Development, 50*, 425–430.
Jones, S. S. (1996). Imitation or exploration? Young infants' matching of adults' oral gestures. *Child Development, 67*, 1952–1969.
Jones, S. S. (2006). Exploration or imitation? The effect of music on 4-week-old infants' tongue protrusions. *Infant Behavior and Development, 29*, 126–130.
Kohler, E., Keysers, C., Umilta, M. A., Fogassi, L., Gallese, V., & Rizzolatti, G. (2002). Hearing sounds, understanding actions: Action representation in mirror neurons. *Science, 297*, 846–848.
Kugiumutzakis, G. (1999). Genesis and development of early infant mimesis to facial and vocal models. In: J. Nadel & G. Butterworth (Eds.), *Imitation in infancy* (pp. 36–59). Cambridge: Cambridge University Press.

Kuhl, P. K., & Meltzoff, A. N. (1982). The bimodal perception of speech in infancy. *Science, 218*, 1138–1141.
Kuhl, P. K., & Meltzoff, A. N. (1996). Infant vocalizations in response to speech: Vocal imitation and developmental change. *Journal of Acoustical Society of America, 100*, 2425–2438.
Kuhl, P. K., Williams, K. A., & Meltzoff, A. N. (1991). Cross-modal speech perception in adults and infants using nonspeech auditory stimuli. *Journal of Experimental Psychology Human Perception Perform, 17*, 829–840.
Legerstee, M. (1991). The role of person and object in eliciting early imitation. *Journal of Experimental Child Psychology, 51*, 423–433.
Lepage, J., & Théoret, H. (2006). EEG evidence for the presence of an observation-execution matching system in children. *European Journal of Neurology, 23*, 2505–2510.
Maratos, O. (1982). Trends in the development of imitation in early infancy. In: T. G. Bever (Ed.), *Regressions in mental development: Basic phenomena and theories* (pp. 81–101). Hillsdale, NJ: Erlbaum.
Meltzoff, A. N., & Decety, J. (2003). What imitation tells us about social cognition: A rapprochement between developmental psychology and cognitive neuroscience. *Philosophical Transactions of the Royal Society B: Biological Sciences, 358*, 491–500.
Meltzoff, A. N., & Moore, M. K. (1977). Imitation of facial and manual gestures by human neonates. *Science, 198*, 74–78.
Meltzoff, A. N., & Moore, M. K. (1983). Newborn infants imitate adult facial gestures. *Child Development, 54*, 702–709.
Meltzoff A. N., & Moore, M. K. (1989). Imitation in newborn infants: Exploring the range of gestures imitated and the underlying mechanisms. *Developmental Psychology, 25*, 954–962.
Meltzoff, A. N., & Moore, M. K. (1992). Early imitation within a functional framework: The importance of person identity, movement, and development. *Infant Behavior and Development, 15*, 479–505.
Meltzoff, A. N., & Moore, M. K. (1997). Explaining facial imitation: A theoretical model. *Early Developmental and Parenting, 6*, 179–192.
Muthukumaraswamy, S. D., Johnson, B. W., & McNair, N. A. (2004). Mu rhythm modulation during observation of an object-directed grasp. *Brain Research Cognitive Brain Research, 19*, 195–201.
Nagy, E., Compagne, H., Orvos, H., Pal, A., Molnar, P., Janszky, I., et al. (2005). Index finger movement imitation by human neonates: Motivation, learning, and left-hand preference. *Pediatric Research, 58*(4), 749–753.
Oberman, L. M. H., Edward, M., McCleery, J. P., Altchuler, E. L., Ramachadran, V. S., & Pineda, J. A. (2005). Eeg evidence for mirror neuron dysfunction in autism spectrum disorders. *Cognitive Brain Research, 24*, 190–198.
Pechaux, M. G., Lepecq, J. C., & Salzarulo, P. (1988). Oral activity and exploration in 1–2 month-old infants. *British Journal of Developmental Psychology, 6*, 245–256.
Pfeifer, H., Iacoboni, M., Mazziotta, C., & Dapretto, M. (2005). Mirror neuron system activity in children and its relation to empathy and interpersonal competence. In: Abstract Viewer/Itineraray Planner. Soc Neurosci Abstr, 660.24.
Piaget, J. (1952). *The origins of intelligence in children.* New York: International Universities Press.
Pineda, J. A. (2005). The functional significance of mu rhythms: Translating "seeing" and "hearing" into "doing". *Brain Research Review, 50*, 57–68.
Prechtl, H. F. (1986). Prenatal motor development. In M. G. Wade & H. T. Whiting (Eds.), *Motor development in children: Aspects of coordination and control* (pp. 53–64). Dordecht: Martinus Nijhoff.
Reissland, N. (1988). Neonatal imitation in the first hour of life: Observations in rural Nepal. *Developmental Psychology, 24*, 464–469.

Rizzolatti, G., & Craighero, L. (2004). The mirror-neuron system. *Annual Review Neuroscience, 27*, 169–192.

Rizzolatti, G., Fadiga, L., Gallese, V., & Fogassi, L. (1996). Premotor cortex and the recognition of motor actions. *Brain Research Cognitive Brain Research, 3*, 131–141.

Rochat, P. (1983). Oral touch in young infants: Response to variations of nipple characteristics in the first months of life. *International Journal of Behavioral Development, 6*, 123–133.

Saygin, A. P., Wilson, S. M., Dronkers, N. F., & Bates, E. (2004). Action comprehension in aphasia: Linguistic and non-linguistic deficits and their lesion correlates. *Neuropsychologia, 42*, 1788–1804.

Shimada, S., & Hiraki, K. (2006). Infant's brain responses to live and televised action. *Neuroimage, 32*, 930–939.

Sommerville, J. A., Woodward, A. L., & Needham, A. (2005). Action experience alters 3-month-old infants' perception of others' actions. *Cognition, 96*, B1–11.

Théoret, H., Halligan, E., Kobayashi, M., Fregni, F., Tager-Flusberg, H., & Pascual-Leone, A. (2005). Impaired motor facilitation during action observation in individuals with autism spectrum disorder. *Current Biology, 15*, R84–85.

van der Meer, A. L., van der Weel, F. R., & Lee, D. N. (1995). The functional significance of arm movements in neonates. *Science, 267*, 693–695.

van der Meer, A. L. (1997). Keeping the arm in the limelight: Advanced visual control of arm movements in neonates. *European Jounral of Pediatric Neurology, 1*, 103–108.

Wohlschlager, A., & Bekkering, H. (2002). Is human imitation based on a mirror-neurone system? Some behavioural evidence. *Experimental Brain Research, 143*, 335–341.

The Rational Continuum of Human Imitation

Derek E. Lyons

Abstract Mirror neuron's common coding of action perception and action production raises exciting possibilities for understanding imitation's neural basis. Thus far, however, theorizing on this score has been hampered by over-simplification of imitation's complex cognitive reality. This chapter provides a starting point for addressing this problem, synthesizing a broad range of behavioral research into a theoretically unified conception of human imitation. Particular attention is paid to the problem of reconciling selective imitation, in which children rationally omit unnecessary aspects of an adult's behavior, with over-imitation, in which they seem to do exactly the opposite. The apparent contradiction is resolved by refining the notion of selective imitation to its underlying computational basis, and showing that this computational substrate in fact predicts observed patterns of overimitation. Selective imitation and overimitation are thus re-conceptualized as different sides of the same coin—two rational social learning strategies whose superficial opposition conceals a deeper mechanistic and conceptual commonality.

Keywords Resonance · Selective imitation · Overimitation · Tools · Intentionality

1 The Rational Continuum of Human Imitation

Mirror neurons arguably rank among the most intriguing scientific discoveries of the past decade, and have been the source of much debate in both rarified neuroscience journals and the popular press alike. This unusual degree of interest can be credited to mirror neuron's intriguing functional properties—easy to understand conceptually but rich in their potential implications.

D.E. Lyons
Department of Psychology, Reed College, 3203 S.E. Woodstock Blvd., Portland, USA
e-mail: derek.lyons@reed.edu

J.A. Pineda (ed.), *Mirror Neuron Systems*, DOI: 10.1007/978-1-59745-479-7_4,

Specifically, mirror neurons implement a suggestive visuomotor symmetry: the same populations of mirror neurons that are activated by the performance of a particular goal-directed action (for example, grasping an object) are also activated by the observation of another agent performing that action (Gallese, Fadiga, Fogassi, & Rizzolatti, 1996; Rizzolatti, Fadiga, Fogassi, & Gallese, 1996). This equivalent treatment of action perception and action production has generated much speculation about the role that mirror neurons might play in imitation. It is clear that mirror neurons are not simply a dedicated 'imitation circuit,' insofar as the only species in which they have been directly observed—the rhesus macaque—does not engage in anything approximating true imitation (Tomasello & Call, 1997). Yet even if imitation is not mirror neurons' phylogenetically primary function, the unique manner in which they straddle the visuomotor divide continues to make some sort of connection to imitation plausible—at least in humans, where these neurons have also been inferred to exist (Rizzolatti & Craighero, 2004).[1] One widely-held theory is that mirror neurons originally arose as a means of inferring other agents' intentions (Lyons, Santos, & Keil, 2006; Rizzolatti & Craighero, 2004)—something that many nonhuman primates do quite well (Lyons & Santos, 2006)—and that they were recruited to support imitation later in primate evolution, that is, after our divergence from Old World monkeys. On this view, even though mirror neurons are not alone sufficient to support imitation in species such as macaques, they may have an important role to play in human imitation.

Rizzolatti and colleagues (Rizzolatti, Fogassi, & Gallese, 2001) have proposed one possible mechanism by which a mirror system might support our species' unique imitative capacity. The authors speculate that when humans observe another individual executing a novel motor action, the mirror system may perform the function of 'recognizing and segmenting' that action 'into strings of discrete elements, each of which is a motor act in the observer's repertoire' (p. 668). In other words, the mirror system may drive imitative learning by automatically segmenting observed behavior into a finite vocabulary of motor 'atoms' that we are already capable of producing. In this manner the mirror system could absorb much of the computational overhead of imitative learning, reducing imitation to a matter of replaying a sequence of familiar motoric subcomponents after they have been automatically extracted from visual input. This account, though obviously meant to be speculative and conceptual, is consistent with fMRI data showing that the learning of new motor patterns occurs in the same regions of the brain that have been implicated in the human mirror system (Rizzolatti & Craighero, 2004).

---

[1] Mirror neurons have not been directly observed in humans as they have in macaques. Evidence for the existence of similar neurons in humans, though increasingly compelling, is all derived from much more indirect measures such as fMRI. We will thus use the deliberately loose term 'mirror system' to refer to the mechanisms in the human brain that are inferred to correspond to macaque mirror neurons.

For all of its appeal, however, the above theory also illustrates a conceptual problem that is pervasive in the mirror neuron literature: that of oversimplifying imitation. In particular, this theory and others like it reduce the encoding of observed actions to an automatic 'resonance' process. Observed actions are conceptualized as triggering resonance in the mirror system of the observer, and this resonance is in turn thought to provide the basic representation from which imitation can occur. While this way of speaking maps naturally onto the physiological properties of mirror neurons (and may indeed be accurate at a suitably fine level of granularity), it leaves one with the impression that imitation is a low-level process that occurs without much cognitive mediation; it characterizes imitation as a matter of plugging an input perception stream directly into an output motor stream without the need for any interpretation or transformation. This characterization is quite inaccurate. Even though our direct experience of imitating often feels effortless and automatic, imitation is in fact a highly interpretive process that involves a surprising degree of inference making and cognitively-mediated representation.

The interpretive complexity of imitation is aptly illustrated by the phenomenon of *selective imitation*—particularly as it occurs in children. From as early as 12 months of age, children use subtle social and contextual clues to infer which aspects of an observed behavior merit careful reproduction and which aspects can be safely ignored (Schwier, Van Maanen, Carpenter, & Tomasello, 2006). Children, in other words, often imitate in a surprisingly discriminating way, copying only the portions of observed actions that are sensible given their immediate social and practical context. From the very first stages of human life then, imitation is far more sophisticated than simply piping uninterpreted visual input directly into motor output. Rather, as we will soon see in more detail, our imitative capacity appears to involve encoding observed actions in a sophisticated hierarchical manner, making inferences about how multiple goals and situational constraints combine to determine an actor's behavior, and then using those inferences to map from the observed behavior hierarchy to our own motor reproduction.

A full and complete theory of mirror neurons' involvement in imitation needs to take this complexity into account, replacing placeholder constructs such as 'resonance' with concepts and mechanisms that do better justice to the level of imitative sophistication that we see even in very young children. The goal of this chapter is to support this process of adjusting neurophysiological theory to better coincide with cognitive reality. In particular, this chapter focuses on the challenge of articulating a unified computational framework for conceptualizing imitation, one that captures as much of the empirical ground truth as possible. If such a framework can be achieved, it will provide a much better basis for neurophysiological theorizing than the oversimplified sketches of imitation that are often presently employed.

There is a difficulty, however, in pulling the full range of empirical findings together into a unified conception of imitation. In particular, a full theory of imitation needs to account not only for rational imitation as we have previously described it, but also for the seemingly opposing phenomenon of *overimitation* (Lyons, Young, & Keil, 2007). That is, though children imitate with great rational selectivity in some contexts, in other situations they imitate in ways that seem deeply *irrational*, persistently copying actions that are obviously irrelevant or counterproductive. This seeming contradiction poses a significant problem for a cohesive account of human imitation, and by extension for the prospect of understanding the mirror system's role in this competence. This chapter will attempt to resolve this tension by demonstrating that both selective imitation and overimitation can be conceptualized as points on a single continuum of rational imitative strategy. By better understanding this continuum and the key computations that define it, we will arrive at a much better starting point for crafting plausible and empirically-grounded accounts of the mirror system's involvement in imitation.

The chapter begins by reviewing the literature on both selective imitation and overimitation as they occur in young children, highlighting the apparent contradiction that these opposing phenomena create. Existing theories of overimitation are then discussed. The substance of these theories, essentially that overimitation is a byproduct of implicit social demands or imitative habit, is argued to be unsatisfying, and a more compelling theoretical union between these opposing trends is sought. The search for such a unifying viewpoint begins with a more careful computational analysis of selective imitation, framed in terms of the hierarchical organization of behavior. It is argued that the key computation underlying selective imitation is that of determining the appropriate level in an observed behavior hierarchy to switch from imitation to emulation. The ways in which humans appear to make this determination are then surveyed, focusing particularly on the role that domain-specific inferences may play. One domain in particular—that of tools and novel artifacts—is argued to present special challenges for imitative learning. It is shown that the causal opacity of our species' advanced tool use can make emulation within this domain risky precise imitation, even in situations where some of a model's actions appear to be unnecessary, may in fact be more rational than it first appears. Overimitation is then shown to arise as a logically appropriate response to these considerations, thus closing the loop and unifying two seeming poles of imitative behavior within a common conceptual framework. The chapter closes with a brief reflection on the important role that this kind of cognitive theorizing can play in the construction of empirically realistic neurophysiological models of imitation.

2 Two Poles of an Imitation Dichotomy? Selective Imitation and Overimitation

In order to develop a unified conceptualization of human imitation, we will begin by considering two poles of a seeming imitative dichotomy: children's tendency to imitate with impressive rational selectivity in some contexts, and their equally salient tendency to overimitate unnecessary actions in others.

2.1 Selective Imitation

A vivid illustration of selective imitation comes from a clever study by Gergely and colleagues (Gergely, Bekkering, & Kiraly, 2002), which was based on a prior investigation of deferred imitation in 14-month-old infants (Meltzoff, 1988, 1995a). In the original experiment, infants watched an adult cause a novel 'lightbox' to illuminate by pushing on its translucent plastic top panel. Rather than pushing on the panel using his hands as might have been expected, the adult activated the lightbox by leaning forward from the waist and pushing the panel with his forehead. One week after seeing this curious demonstration, participants returned to the lab and were allowed to manipulate the lightbox themselves for the first time. Fully two-thirds of participants attempted to activate the lightbox using their forehead just as the adult had done, despite the fact that age-matched infants in a control condition (who had not seen the adult's unusual action) used only their hands to manipulate the box.

On the surface this result seems far removed from any sort of selective imitation; if anything, the 14-month-olds tested seem to have been copying the adult quite blindly. However, Gergely and colleagues noted that when the adult in these prior studies used his forehead to activate the lightbox, his hands were visibly free. One rational interpretation of this scene, therefore, would be that the adult was deliberately *choosing* to use his forehead rather than his hands, perhaps because doing so conferred some non-obvious advantage. If infants were sensitive to this possibility, then perhaps their copying of the forehead-push action was not blind and reflexive but rather quite rational? In order to test this hypothesis, Gergely et al. replicated the original experiment with a clever twist. Fourteen-month-olds again watched an adult (who now had a blanket draped across her shoulders) activating a novel lightbox by pushing on it with her forehead. For one group of participants, the *hands-free* group, the adult used her forehead in this manner while her hands rested flat on the table; as in the original study then, she appeared to *choose* to use her forehead for the task despite the fact that her hands were free. Contrastingly, for a second group of participants, the *hands-occupied* group, the adult pushed the lightbox with her forehead while simultaneously using her hands to keep the aforementioned blanket wrapped tightly around herself (as though she was cold); thus, in this case, the adult appeared to use her forehead for the task only because her hands were already in use.

How did participants respond to this subtle change? Gergely and colleagues found that in the hands-free condition, 69% of participants used their forehead to activate the lightbox, precisely replicating the findings of the original study. In the hands-occupied condition, however, only 21% of children used their forehead; the vast majority instead chose the more efficient route of pushing the button with their hands. In the formal parlance of social learning, children in the hands-occupied condition tended to *emulate* rather than imitate; that is, they tended to accomplish the previously demonstrated goal of activating the lightbox through their *own means*, rather than through the same awkward means that the adult had employed.[2] Thus, through this elegant manipulation, Gergely et al. succeeded in showing that even 14-month-olds approach imitation in a rational and selective way. Young children are not slavish imitators, hopelessly in thrall of whatever actions they happen to see adults perform. Rather, children are able to make sophisticated attributions about the competing intentions and physical constraints present in a scene, and can then use those attributions to separate superfluous aspects of observed action from those aspects that are likely to be functionally important.

This result is not an isolated one, but rather typifies a pattern that has been replicated repeatedly in the literature. In many different kinds of situations, infants and children show a remarkable degree of intelligent selectivity with regard to the actions that they imitate (e.g., Bekkering, Wohlschläger, & Gattis, 2000; Carpenter, Nagell, & Tomasello, 1998; Carpenter, Call, & Tomasello, 2005; Gleissner, Meltzoff, & Bekkering, 2000; Meltzoff, 1995b). Indeed, rationally selective imitation has now been documented in infants even younger than those tested by Gergely and colleagues. In a conceptual replication of the Gergely et al. study, for example, Schwier, Van Maanen, Carpenter, and Tomasello (2006) showed that 12-month-olds will also account for external constraints on an adult's behavior when imitating. Infants in this study watched as an adult 'helped' a stuffed toy dog enter a doghouse. In one condition (analogous to Gergely et al.'s *hands-free* condition), the dog entered the house through the chimney, despite the fact that the door was standing open. In a second condition (analogous to the *hands-occupied* condition) the dog tried to enter through the door, but finding it locked, went in through the chimney instead. In both conditions, the door to the house was then opened and infants were allowed to help the dog into the house for themselves. Just as in the Gergely et al. study, Schwier and colleagues found that infants were more likely to imitate putting the dog through the chimney when they had seen the adult freely choosing this route rather than being forced to pick it. Even at 12 months, infants imitate selectively rather than copying blindly.

As impressive as these results are, it is worth noting that the cognitive 'seeds' of this selective imitation ability appear to be present still earlier in life—even before the age at which infants are able to actually do much in the way of imitation. Using looking time measures, infants as young as 6.5 months of age have been shown to react with surprise when agents choose to accomplish simple goals in a rationally suboptimal way (for example, approaching a desired object via a circuitous trajectory rather than along a straight line) (Csibra, Gergely, Bíró, Koós, & Brockbank, 1999; Kamewari, Kato, Kanda, Ishiguro, &

[2] Different authors define the distinction between imitation and emulation in different ways. Following Tomasello (1996) and Csibra (2007) among others, we define imitation as reproducing a model's goal using the *same means* that the model employed; contrastingly, emulation is defined as reproducing a model's goal *using means of one's own devising*. See Call and Carpenter (2002) for a more detailed taxonomy of social learning mechanisms.

Hiraki, 2005; Phillips & Wellman, 2005; Sodian, Schoeppner, & Metz, 2004). It seems that our earliest appraisals of goal-directed action are evaluative rather than passive, a phenomenon that likely supports the rationally selective imitation that emerges around the end of the first year of life. From the very beginning then, imitation is no 'monkey see, monkey do' matter. Rather, the degree to which infants and young children will reproduce others' actions is generally modulated by a sophisticated appraisal of the physical and intentional context.

2.2 *Overimitation*

Puzzlingly though, as robust as children's selective imitation appears to be in the prior examples, there are equally vivid demonstrations of a very different phenomenon, one that we term *overimitation* (Lyons, Young, & Keil, 2007). A good illustration of overimitation comes from a recent study by Horner and Whiten (2005), in which the imitative tendencies of 3- and 4-year-old children were compared to those of similarly aged chimpanzees. Participants in both groups began the experiment by observing an adult as she extracted a reward from inside a novel object. The object itself was of trivial complexity—essentially a simple Plexiglas cube divided into upper and lower compartments by an internal floor. The reward, located in the lower compartment of the box, could be obtained quite directly by opening a door on the front of the box and reaching inside. This is not, however, what participants saw the adult do. Instead, the adult retrieved the reward in a much less efficient manner. She first uncovered an opening on the top of the box that led into its empty upper compartment, and then tapped a small wand inside this opening. Only after performing these two superfluous steps did she do the one thing that was actually causally necessary, namely opening the door on the front of the box and using the wand to extract the reward from the lower compartment.

Both the children and the chimps saw the adult performing these actions in two different conditions. In one condition the adult performed the above actions on a version of the box with opaque walls. In this condition the inefficiency of the experimenter's action sequence was hidden; there was no way for participants to tell that some of the experimenter's steps had been superfluous. In the second condition the adult performed the exact same actions on a box whose walls were *transparent*. Participants in this condition, therefore, could clearly see that the first two steps in the experimenter's retrieval sequence were not necessary for attaining the goal.

The question that Horner and Whiten asked was straightforward: what effect, if any, would the availability of causal information have on participants' imitation strategy? The selective imitation results that we have discussed thus far lead to a clear prediction for the children. In the opaque box condition, in which no causal information is available, we would expect children to copy all of the experimenter's actions. However, since it has been repeatedly shown that children's imitation is modulated by an assessment of an observed action's rationality, we would expect the children in the *transparent* box condition to omit the experimenter's unnecessary steps. To phrase our prediction more formally, we would expect children to *imitate* when no causal information is available, but to *emulate*—to reproduce the adult's goal via more efficient means—when some of the observed actions are visibly unnecessary.

Surprisingly, the chimpanzees conformed exactly to this prediction. Whereas chimps in the opaque condition copied all of the experimenter's actions approximately 60% of the time, in the clear condition they virtually never reproduced the visibly unnecessary steps.[3] In other words, the chimpanzees switched between imitation and emulation in a rationally optimal manner. Children's pattern of responding was a puzzlingly different story. In the opaque condition children showed a strong tendency to imitate all of the adult's actions, just as the chimps had done. In the clear condition, however, children *failed* to omit the now visibly unnecessary steps; they continued to copy all of the adult's actions, despite the fact that 2/3 of the action sequence was obviously superfluous. There was in fact no statistically significant difference between children's responses in the opaque and clear conditions.[4] Rather than imitating in a rational and selective way, children appeared to blindly copy all of the adult's actions without any regard for their causal necessity.

This curious failure of imitative selectivity is not an isolated incident. Children have been observed to overimitate in a variety of different contexts, carefully reproducing observed actions that are transparently unnecessary for the task at hand (Lyons et al., 2007; McGuigan, Whiten, Flynn, & Horner, 2007; Nielsen, 2006; Horner & Whiten, 2005; Call, Carpenter, & Tomasello, 2005; Carpenter, Call, & Tomasello, 2002; Want & Harris, 2002; Whiten, Custance, Gomez, Teixidor, & Bard, 1996; Nagell, Olguin, & Tomasello, 1993). As we have seen, children will sometimes overimitate in this way even in situations where chimpanzees correctly ignore the unnecessary steps (McGuigan et al., 2007; Nagell et al., 1993; Want & Harris, 2002; Whiten, 2005; Whiten et al., 1996). Interestingly, all of these reports of imitation involve children imitating actions on or with novel tools and artifacts, a fact that will become very important later in this chapter. At present though, we are left with the difficult question of how to interpret these findings. How can we reconcile the seemingly contradictory findings of selective imitation and overimitation within a single theoretical framework?

---

[3] This pattern of results was for chimps who participated in the opaque condition first, followed by the clear condition. When chimps were presented with the clear box first, they not only ignored the adult's irrelevant actions on that box, but also *generalized* to the opaque box later on. That is, once chimps had seen that the first two of the experimenter's actions were unnecessary on the clear box, they continued to ignore those actions even when they were performed on the opaque box.

[4] Nor did the order in which these conditions were presented alter children's responses.

2.3 Theories of Overimitation

One way of aligning these conflicting empirical trends is to focus on imitation's dual nature, as both (*a*) a means for learning about the world and (*b*) a form of social interaction. Uzgiris (1981) frames this dual nature nicely, emphasizing that whereas the *learning* function of imitation is to extract new knowledge from the behavior of others, the *social* function centers on 'the similarity relation that is established between the imitator and the model' (p. 3). Socially motivated imitation, in other words 'is a way of realizing the congruence that may exist between two individuals' (p. 3; see also Tomasello, Carpenter, Call, Behne, & Moll, 2004 for related arguments). Uzgiris goes on to make the important observation that, in socially driven imitation, 'the nature of the acts involved becomes less important' (p. 3). That is, when children approach imitation as a kind of social game, they are likely 'to perform imitations of most any act modeled as a way of participating' (p. 7).

Contemporary theorists such as Nielsen have now returned to this dual-function perspective in order to explain how children can oscillate so dramatically from cleverly selective imitation to seemingly blind overimitation. Nielsen (2006) argues that whereas younger infants and children 'copy primarily to satisfy cognitive motivations, to promote learning about events in the world,' by the time they reach their second birthday 'children are more motivated to copy to satisfy social motivations, to fulfill an interpersonal function of promoting shared experience with others' (p. 563). Overimitation is seen as a relatively insignificant byproduct of this latter social motivation. Children overimitate irrelevant actions not because they are confused in any theoretically significant way, but rather because they wish to engage in an imitative social interaction with the model. This view, which we will refer to as the *social game theory* of overimitation, essentially argues that selective imitation and overimitation are the social learning equivalent of apples and oranges. Though both phenomena are members of a loose superordinate category (i.e., varieties of imitation), they arise from fundamentally distinct and non-comparable kinds of mental processes—one driven by a desire to understand, and the other by a desire to relate.

Of course, it is also possible for socially motivated imitation to occur not as a voluntary way of maintaining a social interaction with an adult, but rather as a response to implicit social and task demands. Thus, the flip side of the social game theory of overimitation is what we might term the *task demand theory*. Horner and Whiten (2005) have argued for this view, positing that children in their study may have overimitated the adult's irrelevant actions 'because they saw the behaviour of the demonstrator as intentional, even if they did appreciate that some parts of the demonstration were causally irrelevant' (p. 179). Children may have overimitated, in other words, because they assumed that they were *supposed* to do so; the intentionality with which the adult performed the irrelevant actions may have constituted a powerful social task demand, leading children to conclude that they were intended to reproduce them.

A final potential explanation for overimitation, though difficult to formally test, offers considerable common-sense appeal. The *habit theory* of overimitation sees the effect as arising not from the distinction between learning-driven and socially-driven copying, but rather from the pervasive value of imitation in general. Specifically, Whiten and colleagues (1996) have argued that imitation is 'such a highly adaptive human strategy that it may frequently be employed by young children in situations where it is locally inefficient to do so' (p. 11). On this view, overimitation is a simple byproduct of the immense utility that we normally derive from imitating those around us. Imitation is such a valuable element in our cognitive toolkit that it 'remains habitual even in a specific situation in which less fidelity would actually afford more efficiency' (McGuigan et al., 2007).

Each of these theories is plausible in its own right, but they all share a common underlying premise. Specifically, each theory assumes that whereas selective imitation reflects the leading edge of children's critical reasoning abilities, overimitation reflects something a good deal more shallow—either a strategy for engaging and identifying with adults socially, or simply an occasional willingness to fall back on imitative habit. Selective imitation and overimitation are reconciled by essentially dismissing the latter as theoretically inconsequential. Yet there is another possibility. Perhaps the dichotomy between selective imitation and overimitation—so seemingly obvious on the surface—is actually false? The remainder of this chapter will argue that selective imitation and overimitation are not actually as different as they seem, and that the two phenomena are in fact endpoints on a common continuum of rational imitative strategy. Understanding this continuum, it will be argued,[5] allows us to synthesize a more theoretically unified view of human imitative behavior.

3 Selective Imitation and the Hierarchical Organization of Action

In order to argue that selective imitation and overimitation are more alike than different, we need to begin with a more careful specification of the former concept. In particular, we need to determine what sort of computation forms the underlying basis for selective imitation. Once this key computation has been specified, we can then see what it would predict about the situations in which children are known to overimitate. Let us begin this process by considering the input that all forms of imitation share: observed behavior.

Many theorists have argued that behavior has a fundamentally hierarchical organization (Byrne & Russon, 1998; Csibra, 2007). By allowing complex superordinate behaviors to be constructed from a finite vocabulary of simpler ones, hierarchies provide a natural and efficient means of organizing action.

[5] For empirical studies bearing on the theoretical arguments outlined here, see Lyons et al., 2007.

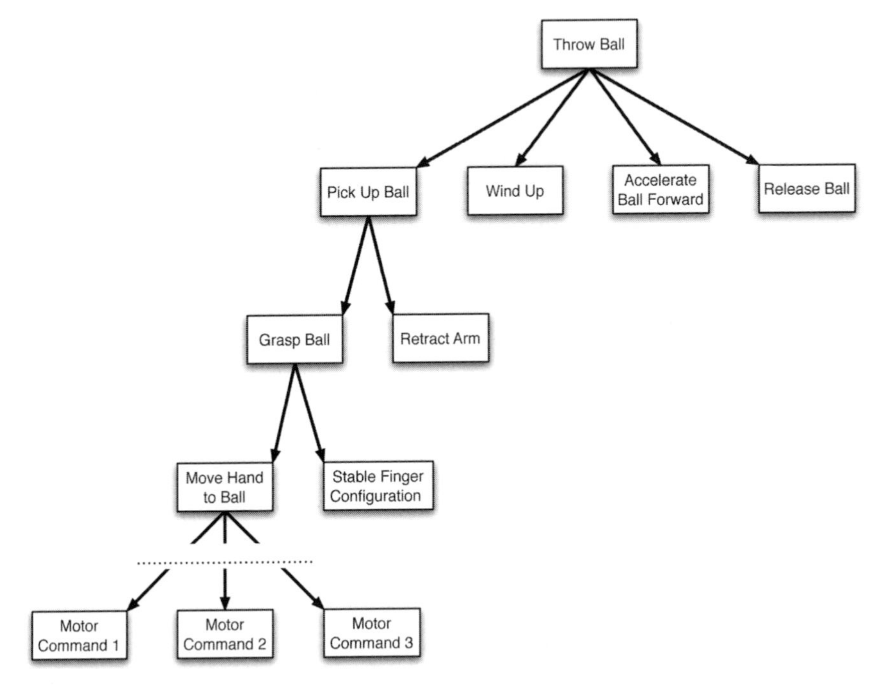

Fig. 1 Illustration of a behavior hierarchy corresponding to the act of throwing a ball. The top-level goal of throwing is decomposed into a sequence of sub-goals (pick up ball, wind up, accelerate ball forward, and release), each of which is in turn decomposed into sub-sub-goals. This process of decomposition can be continued until indivisible 'atomic' goals (such as contracting or relaxing particular muscles) are reached at the bottom-level of the hierarchy

Figure 1 provides a conceptual illustration of this point, showing a hypothetical behavioral hierarchy for the act of throwing a ball. The hierarchy decomposes the top-level goal of throwing into a series of sequential sub-goals (pick up the ball, wind up, accelerate the ball forward and release), each of which can in turn be decomposed into sub-sub-goals (e.g., to pick up the ball, grasp it and then retract the arm), and so on. This hierarchical decomposition could be continued down many more levels, specifying subordinate goals at ever finer grain sizes until eventually reaching 'atomic' goals minute enough to be directly implemented by the motor system (e.g., contracting or relaxing individual muscles).

The concept of hierarchical organization is not simply a matter of intellectual curiosity; it provides support for a much more precise understanding of imitation in general, and selective imitation in particular. As an example, imagine a child observing an adult who is attempting to throw a ball with a cast on her hand. The cast would particularly influence the portion of the adult's behavior hierarchy corresponding to the 'pick up ball' sub-goal. Being unable to employ the usual whole-hand grip, the adult might instead pick up the ball using a pincer grip between her thumb and index finger. Now imagine that the child is going to imitate the adult. In order to engage in 'true' imitation the child would have to reproduce the adult's top-level goal via the exact same means, including copying the unusual means of picking up the ball as well as any other idiosyncrasies of the display. The child would, in other words, have to reproduce the observed behavior hierarchy down to its bottom-most level.[6] Contrast this with what might occur if the child copied the adult in a selective manner. For example, a selectively imitating child might attribute the adult's unusual choice of grip to the constraints imposed by her cast, and thus choose to pick up the ball in the more efficient whole-handed manner. In terms of the behavior hierarchy, such a child would imitate down to the level of the 'grasp ball' sub-goal, but would then switch to reproducing that sub-goal in an *emulative manner*.

This simple example illustrates a more computationally formal definition of selective imitation. Whereas in 'true' imitation the imitator copies the model's actions as far down the observed behavior hierarchy as possible, in selective imitation the imitator *switches from imitation to emulation* midway down the hierarchy. Thus, the key computation that underlies selective imitation is that of selecting this switch point, or determining where in an observed behavior hierarchy to flip from imitation to emulation. As we have seen (and will discuss in more detail shortly), the switch point is generally selected by making a rational appraisal of the modeled action and its circumstances. In other words, the selective imitator uses cues derived from the larger social and physical context of the modeled behavior in order to judge how faithfully that behavior should be copied.

Selective imitation can have significant advantages, most notably in terms of computational efficiency. If the child knows how to accomplish a particular sub-goal himself, it is much easier to simply encode the identity of that sub-goal than it is to encode the particular means that the model used to accomplish it (Lyons et al., 2006). The former approach allows the child to leverage his existing knowledge, essentially using pre-existing behavioral 'subroutines' to represent the fine details of an observed action rather than attempting to extract all of them from viewing the model. This efficiency is not just important in the abstract, but may also enhance an observer's ability to learn from a demonstration. By 'outsourcing' familiar portions of a

[6] Of course, there is an important philosophical point to be raised at this juncture, namely that imitating down to the bottom level of any behavior hierarchy—down to the level of atomic motor primitives—is not practically possible. That is, as we progress down the hierarchy and subsidiary goals become progressively more fine-grained, discrepancies between the actor's behavior and that of the model are eventually inevitable (Csibra, 2007). For the purposes of this chapter then, when we speak of 'full' imitation we will mean imitation down to the lowest level of the hierarchy at which veridical reconstruction of the model's behavior is reasonably possible.

demonstration to existing behavioral subroutines, selective imitation conserves attentional resources for the aspects of a display that are truly novel. On a closely related note, selective imitation can also help to smooth over irrelevant details of a display that might otherwise introduce unnecessary noise into the learning process. Returning to the subject of the earlier example, imagine how difficult it would be to learn to throw a ball if you were attempting to reproduce the exact position of each of the thrower's fingers on the ball's surface each time she picked it up, as well as the exact trajectory traced by her arm during the wind up and release. All of this unnecessary data, varying randomly from trial to trail, would obscure the functionally important outlines of the task beneath a sea of trivial detail. Thus, picking the right point in the observed behavior hierarchy to switch from imitation to emulation can significantly improve the signal-to-noise ratio in the dataset that the learner is attempting to interpret.

Yet for all of its advantages, selective imitation is also a double-edged sword. While picking the *right* point in the observed behavior hierarchy to switch from imitation to emulation can make observational learning much easier, picking the *wrong* point can make it impossible. The reason for this is that emulation, much like other computational data compression algorithms, is a destructive process—information is always lost. All of the benefits associated with emulation, therefore, are contingent on the assumption that the details of the observed behavior that are being discarded are not actually important for achieving the desired outcome. Unfortunately, if the person who is imitating/emulating the model is doing so for the purpose of learning something new—as children so often are—then they are unlikely to know with certainty which details of the modeled behavior are functionally important and which are not. This is a central paradox of observational learning: learning a new skill requires attending to the right aspects of a model's behavior, yet it is often impossible to know for sure which aspects are the right ones without knowledge of the skill in question.

Clearly then, the key computation underlying selective imitation—that of picking the right point in an observed behavior hierarchy to switch from imitation to emulation—is far from trivial. How do children manage it? One part of the answer is that they often receive assistance from the adult model. As Gergely and Csibra have argued, there are many ways in which human observational learning incorporates specialized support for pedagogical knowledge transmission; adults offer subtle cues that help to direct children's attention to important aspects of a demonstration, and children show a strong predisposition to attend to these cues (Csibra, 2007; Csibra & Gergely, 2006; Gergely, Egyed, & Ildiko, 2007). Arguably though, such pedagogical specializations are only half of the story. In the next section we will argue that children go beyond the supports of pedagogy, making rational, independent inferences about when and to what degree selective imitation is appropriate. As previously foreshadowed, we will then show that the very same rational inference strategies that support children's selective imitation in some circumstances may also explain why they overimitate in others.

4 Picking the Right Level of Imitative Selectivity

Recent studies by Bekkering and colleagues (2000) provide interesting insight into the manner in which children make rational inferences about selective imitation. Utilizing a much-studied imitation paradigm (see Head, 1920), Bekkering et al. presented preschool-aged children with a simple copying task. Children were seated in front of an adult model, who instructed them to "Try to imitate me as if you were my mirror. You do what I do" (p. 155). The model then performed a series of simple actions for the child to copy. The critical items in this series were actions in which the model touched one of her ears with either her ipsilateral hand or her contralateral hand (all four combinations of ear and hand side were modeled during the experiment). When copying these actions children had no difficult picking their corresponding ear, touching it correctly on 96% of trials. However, they showed a systematic pattern of errors with regard to the hand that they touched their ear with. On 40% of the trials in which the adult touched her ear with the *contralateral* hand (thus reaching across the midline of her body), children touched their ear with their *ipsilateral* hand—a so-called contra-ipsi error.

These results replicate a pattern that has been observed in numerous prior experiments (Gordon, 1923; Schofield, 1976; Swanson & Benton, 1955; Wapner & Cirillo, 1968). In general, the contra-ipsi error had been interpreted not as a matter of cognitive interest, but rather as an artifact of incomplete neurophysiological development; it has been hypothesized that children simply have an initial deficit in their ability to produce cross-midline gestures that is naturally resolved in the course of neural development (Kephart, 1971). However, as Bekkering and colleagues point out, this view cannot explain why much older children make the same systematic mistakes as preschoolers. When tested with an analogous ear-touching procedure, children as old as 8-years-of-age have been shown to make contra-ipsi errors in the same manner as their younger counterparts (Schofield, 1976), a finding that strains the neurophysiological interpretation.

In order to get to the bottom of the contra-ipsi error, Bekkering et al. thus conducted a clever follow-up study. This second study used the same procedure as before with one small exception: whereas in the original experiment the ear that the adult reached for had varied across trials, in the follow-up she always reached for the same ear. According to the neurophysiological interpretation of the contra-ipsi error, this small procedural variation should not change the overall pattern of results. However, this is not what Bekkering et al. found. When the target of the adult's reach was consistent, children made contra-ipsi errors on only 1.9% of the trials—a performance that is more than 21 times more accurate than that observed in the original experiment.

This result was complemented with a third and final study, in which children watched an adult make both contralateral and ipsilateral pointing gestures to two different locations on a tabletop. In one condition these locations were marked by visible dots, while in a second condition they were unmarked. Bekkering and colleagues found that when children mirrored the adult in the 'dots present' condition, they committed the same contra-ipsi errors that were observed in the original experiment; they preferred to point using the ipsilateral hand even when the experimenter's gesture was made contralaterally. However, when the targets of the experimenter's pointing were *not* marked by visible dots, children mirrored much more accurately; in this 'dots absent' condition they were significantly more likely to reproduce the adult's contralateral gestures accurately (see also Gleissner et al., 2000 for analogous results).

What do these results suggest? Based on the second and third studies, Bekkering and colleagues hypothesized that the contra-ipsi error arises not from neurophysiological immaturity, but rather from children's difficulty in simultaneously encoding a model's goals at multiple levels of granularity. That is, they argued that because the mirroring task in both the first experiment and the 'dot' condition of the third experiment was relatively demanding, children omitted reproduction of lower-level sub-goals (i.e., the hand that was used to reach/point) in an effort to preserve the fidelity of the more salient top-level goal (the target of the reach/point). Contrastingly, when the model's actions were simplified—either by consistently reaching for the same ear in Experiment 2 or by seeming to point without a specific target in the 'dots absent' condition of Experiment 3—children were able to reproduce both the higher- and lower-level goals.

It seems clear that Bekkering et al.'s results are best interpreted in terms of the hierarchical organization of the model's behavior. However, rather than being a by product of insufficient processing resources, might the contra-ipsi error be the outcome of a rational computation? Is it possible that the contra-ipsi 'error' might not really an error at all, but instead an example of selective imitation? If this is true, then perhaps Bekkering and colleagues' results can provide important insight into the way that children meet the central challenge of selective imitation: knowing where to draw the line between imitating and emulating.

When viewed through the lens of selective imitation, the results of the Bekkering et al. studies suggest that children may determine how far down an observed behavior hierarchy to imitate by evaluating a measure of *information loss*. Conceptually, it is as though at each level of the observed behavior hierarchy children ask: 'if I were to switch over to emulation at this level in the hierarchy, accomplishing this level's sub-goal via my own means rather than those employed by the model, how much information in the original display would be lost? Would this degree of information loss be significant?' Consider Bekkering et al.'s first experiment as an example. Here the top-level goal is clear, namely touching a particular ear; the primary sub-goal is to touch that ear with a particular hand (an analogous decomposition applies to the 'dots present' condition of Experiment 3). When imitating, children begin from the top of this behavior hierarchy and evaluate how much information would be lost if the top-level goal was accomplished emulatively rather than imitatively. In this case, the information that emulation would discard is information at the level of the first sub-goal: the particular hand that the experimenter used. Is this information loss significant? From their behavior we can conclude that children answer this question in the negative, an answer that makes intuitive sense. That is, the particular hand that we use to reach for or grasp things is generally much less important than the target of the action. Interestingly, even 9-month-olds appear sensitive to a version of this common-sense observation, reacting with surprise when an adult favors consistency in the direction of reach over consistency in the reached-for object (Woodward, 1998). Since there is nothing in the adult's actions to indicate that this usual prioritization should be revised, children emulate the top-level goal through the maximally efficient means, thus reaching with their ipsilateral hand.

Contrast this case with children's behavior in the second Bekkering et al. experiment, in which the adult always reached for the *same* ear. In this case, the particular hand that the adult reached with constituted a much greater proportion of the information in the display. That is, though the handedness of a reach is generally less important than the target, if the target is *fixed* then handedness becomes the only piece of information that differentiates one trial from another. Since children were presented with many trials in succession, they appear to have made the reasonable inference that the specific hand used in each trial must be important. Thus, rather than discarding this evidently significant information by emulating the top-level goal ('touch left ear'), children imitated the specific manner in which the goal was achieved ('touch left ear with right hand'). A similar analysis applies to the 'dots absent' condition of Experiment 3. In that case too the adult's top-level goal (i.e., pointing) appeared to be fixed. That is, because the adult's gestures did not appear to have a specific visible target, the same abstract description—'point'—could be applied to all of them. Thus, the specific hand that was used to accomplish this task again cons a much greater proportion of the information available in any given iteration of the display. Because the identity of the hand used had greater informational prominence, children responded by imitating both the top-level goal and the sub-goal specifying how it was achieved.

A recent study by Carpenter et al. (2005) helps to further illustrate this theory of selective imitation. In this study, 12- and 18-month-old infants watched as an adult made a toy mouse move across a mat in one of two stylized ways: either by hopping or by sliding. The adult accompanied each style of movement with matching sound effects. The critical experimental variable pertained to the endpoints of the mouse's journey; on half of the trials the mouse was moving from one toy house to another, whereas on the other half of trials the mouse followed the same trajectory with no houses present. After each trial infants were given the verbal instruction 'Now you' and handed the mouse.

Based on the theory of selective imitation outlined above, how much of the adult's modeled behavior hierarchy would we expect infants to reproduce? Beginning with the top-level goal of moving the mouse from point A to point B, the theory predicts that infants will attempt to compute the significance of the information that would be lost if that goal was accomplished emulatively rather than imitatively. In other words, in relation to the *total* amount of information conveyed by the adult's actions, is the informational content of the hopping/sliding and of the accompanying sound effects—information that emulation would discard—significant? In the condition in which no houses are present (such that the mouse is hopping between two undifferentiated points on a blank mat), it is reasonable to argue that this style information *is* significant. That is, since the start and end of the mouse's journey appear to be arbitrary, the particular manner in which the mouse moves constitutes the majority of the information in the display. This calculation changes, however, in the case where the mouse moves between two houses. Here the start and endpoints of the mouse's journey are not arbitrary, but in fact largely define what the mouse is 'doing'; the mouse is not just hopping in a particular way but rather hopping for a particular purpose, that is, to reach the destination house. Since the majority of the information in the display is arguably conveyed by the journey's endpoints, the style of the mouse's movement has much less informational prominence. The prediction is therefore that children would be more likely to imitate selectively, omitting the style component of the adult's modeled action, when the houses are present compared to when they are absent. This is in fact just what Carpenter and colleagues observed. At both ages, infants were significantly less likely to reproduce the hopping/sliding style of movement and the accompanying sound effects when the houses were present.

5 Computing Informational Significance

The theory of selective imitation that we have proposed here seems to capture something important about the way in which children encode and recreate observed behavior hierarchies. There is, however, one point on which much more specificity is needed. It has been argued that children determine where in an observed behavior hierarchy to switch from imitation to emulation by considering the significance of the information that emulation would discard. Emulation only begins when the information that remains further down the hierarchy is deemed to be expendable. But how do children make these significance judgments? How do they decide how much information is too much information to discard? There is no simple, universal answer to this question. Rather, children seem to calculate informational significance in an impressively nuanced way, one that is responsive to many facets of an observed action's context. One such facet, which we will now consider, is the intentionality of the actor.

5.1 Intentional Cues

Consider a classic study by Meltzoff (1995b), in which 18-month-olds watched as an adult attempted to perform a simple task such as pulling the two ends of a dumbbell-shaped object apart. The unique thing about the adult's actions in this case was that they were unsuccessful. On the dumbbell object, for example, the adult tried three times to pull the ends of the object apart, each time failing when their hands slipped off one of the ends. This failure was not accompanied by any linguistic markers (i.e., 'oops!') or by any other facial or affective cues that might indicate failure. Nonetheless, children seemed to infer that the adult's intention had gone unfulfilled. When allowed to manipulate the objects for themselves, children showed a remarkably strong tendency to reproduce the *intended* action rather than the failed action that they had actually witnessed. Indeed, children in the experimental group (who saw only the beginning of the adult's failed intention) imitated the intended act just as frequently as children in a control group who actually saw the adult perform the complete action.

To put the study in the context of this chapter, what Meltzoff found was that children were *emulating* the top-level goal—despite not having seen it successfully completed—and discarding all other aspects of the model's actions. That is, children in this case determined that virtually *all* of the informational content of the display was insignificant, a judgment mediated by their understanding that most of the observed behavior was unrelated to the adult's true goal. This is a very clear example of the fact that children's determination of informational significance is not simply a matter of referring to a fixed decision threshold. Rather, children's definition of 'significant information' is highly context specific and strongly parameterized by the intentionality of the model.

Another excellent example of this intentional parameterization comes from work that has already been discussed, namely the lightbox study of Gergely et al. (2002). Recall that in that experiment, children observed an adult turning on a novel lightbox object by pushing on it with her forehead. Replicating a prior result from Meltzoff (1988, 1995a), children imitated the forehead push action when the adult appeared to choose it deliberately (i.e., when the adult's hands were free and could thus have been used to activate the lightbox). However, when the adult's hands were occupied by a secondary goal (holding a blanket around her shoulders) children imitated much more selectively, using their hands to push on the lightbox rather than their forehead. This result indicates that when assessing the informational significance associated with different components of a display, children are also sensitive to the constraints imposed by the actor's other simultaneous intentions. That is, in the hands-occupied condition, children discount the information conveyed by the model's use of her head because she lacked sufficient degrees of freedom to choose any other means. In the hands-free condition, on the other hand, the model's deliberate but highly counterintuitive use of her head is weighted as highly salient. Thus, children's ability to evaluate the significance of information in light of a model's intentional state is quite nuanced, taking account not only primary intentions but also the ramifications of secondary intentions as well.

However, though it is clearly very important, intentionality cannot be the only ingredient in children's calculation of informational significance. For example, compare how children responded to the hands-free condition just discussed to the pattern of results that was obtained in the houses-present condition of Carpenter et al.'s mouse study. In both of these cases the top-level goal was very clear, either activating the lightbox or moving the mouse to a new house. In both cases the adult accomplished this goal in an unusual and highly salient manner, either using her forehead rather than her hands to turn on the box or embellishing the mouse's motion with hops and sound effects. Furthermore, in both cases the adult appeared to select the unusual means by which she accomplished her goal quite deliberately; the choice of means was unconstrained by any external factor. Yet despite all of these similarities, children's responses to these two situations were quite different. Whereas children systematically omitted reproduction of the adult's means in the mouse case, instead emulating the top-level goal in their own way, they showed an equally robust pattern of precise imitation in the lightbox case. How can we account for this disparity?

The answer to this question begins with observing that there is a very important difference between the two displays, that being the *domain* of the task that the adult was performing. Whereas the adult's actions in the mouse case are best described as imaginary play, in the lightbox scenario the adult's actions are occurring in the domain of tools and artifacts. The target of the adult's actions, the lightbox, is a novel manmade artifact of which the child has no direct prior experience. The proposal is that this domain difference accounts for the discrepancy in children's responses to these two scenarios. In other words, when children evaluate the informational significance of different aspects of a display, their knowledge of a task's problem domain appears to have very important role to play. As we will see in the next section, these domain specific calculations constitute the conceptual bridge that links selective imitation to the seemingly opposing phenomenon of overimitation.

5.2 Domain Cues: What's Special About Tools?

At an intuitive level, the domain difference that is argued to distinguish the Gergely lightbox experiment from the Carpenter mouse experiment is easy to understand. We are used to thinking of tools and artifacts, particularly novel tools and artifacts, as things that often necessitate very precise imitation. This is because many of the artifacts that populate our human cultural world are not just complicated—they are in fact completely causally opaque. There is no way to understand how a computer or a cell phone works, for example, simply by carefully inspecting the physical device. Rather, in order to comprehend these objects, we need to enlist the support of more knowledgeable individuals. By carefully copying the behavior of experienced users, we are able to carry out complex operations that we could not have accomplished on our own. Correspondingly, because these artifacts are so causally opaque, selective imitation is something that we do only sparingly and with caution. We recognize that even though a particular action directed toward one of these objects may *appear* to be causally unnecessary, it may actually have significance that is hidden to us. In other words, we weigh the fact that the action was performed by a seemingly knowledgeable individual more heavily than causal hypotheses gleaned from our own inspection.

In this sense then, when children in Gergely et al.'s 'hands free' condition reproduce the model's use of her forehead, they are actually behaving in a very adult way. In the absence of any other information about the unfamiliar lightbox object, they privilege the actor's implied knowledge over their own potentially incomplete appraisal. In the Carpenter et al. mouse experiment, on the other hand, such caution is not necessary. Since the only objects involved are a stuffed mouse and simple toy houses, objects of a sort that are quite familiar to children, there is less risk of underestimating the causal importance of the adult's various actions. The only problem with this hypothesized domain sensitivity is that the children in the Gergely experiment are still on the cusp of infancy—just 14-months old; they are far too young to be making the sort of sophisticated domain attribution that we have described on the basis of their own experience. Thus, the question that we need to consider is whether such domain sensitivity could plausibily be driven by an innate, experience-independent mechanism.

In order to answer this question, we need to understand what is unique about human tool use.[7] What differentiates our use and understanding of tools from that of other nonhuman primates such as chimpanzees? Chimps will spontaneously use simple tools for a variety of purposes, including reaching and probing (e.g., using a stick to retrieve termites from inside a termite mound; Tomasello & Call, 1997), aggression (e.g., throwing rocks and sticks at conspecifics; Goodall, 1986), amplifying force (e.g., for nut cracking; Boesch & Boesch, 1990), and sopping up liquids (e.g., using leaves to soak up water in narrow crevices; Goodall, 1986). Moreover, Chimpangee tool use often appears to be impressively insightful; chimps are able to make simple modifications to tools in order to make them more functional, for example stripping leaves and twigs off of branches in order to improve their effectiveness for termite fishing (Tomasello & Call, 1997). Yet despite this sophistication, it is not clear that chimpanzees truly *understand* tools as tools. That is, chimpanzees do not save their tools for reuse, even in cases where they have expended effort to optimize the tool in question. Rather, chimp tool use is very much linked to the demands of an immediate situation, with tools being discarded after use (Csibra & Gergely, 2006).

---

[7] The argument in this section is largely derived from Csibra and Gergely's treatment of this issue. See the following for additional detail: Gergely & Csibra, 2005, 2006; Csibra & Gergely, 2006; Gergely, Egyed, & Kiraly, 2007; Csibra, 2007.

Csibra and Gergely have referred to this aspect of chimpanzee tool use as *simple teleology* (Csibra & Gergely, 2006; Gergely & Csibra, 2005, 2006). They argue that nonhuman primates functional representations of objects 'tend to be transient and local, involving only short periods of functional insight about objects as potential tools that is likely to be forgotten as soon as the goal is satisfied or abandoned, or the goal object is lost sight of' (Gergely & Csibra, 2006, p. 2). In other words, though nonhuman primates can functionally appraise an object's affordances in relation to an immediate, visible goal (e.g., a nut needing to be cracked), they do not stably represent the objects that they use as tools with particular, fixed uses.

Gergely and Csibra speculate that the uniqueness of human tool use traces back to our ancestors abandonment of this simple teleological mode of representation (Csibra & Gergely, 2006; Gergely & Csibra, 2005, 2006). To summarize, their conjecture is that early hominids succeeded in inverting nonhuman primates teleological perspective, and began to reason about the *goals* that a particular object could facilitate (including goals that were not currently applicable) rather than only about the *object* needed to accomplish a particular, immediate goal. This *inverse teleology* allowed early hominids to stably represent tools as such—as objects with a fixed functional purpose. The stable representation of tools in turn enabled *recursive teleology*, or the ability to conceptualize objects as tools for making other tools. It is this recursive teleology, and the immense space of possible tools that it opens, that is argued to be the key feature differentiating our own tool use from that of other primates.

Though inverse and recursive teleology were great boons for the evolution of human tool use, their arrival also created a unique challenge for our ancestor's observational learning. To see why this is so, recall that a defining feature of simple teleology is that tools are only ever created and utilized in the immediate presence of relevant eliciting goals; chimpanzees only strip twigs for termite fishing, for example, when there is a termite hill available for immediate exploitation. This proximity between tool use and goal state has several salutary consequences for the individual attempting to learn about a tool by observing a conspecific's behavior. First, the goal state that the tool enables is easy to identify. Not only will low-level social cueing processes such as stimulus enhancement naturally direct the observer's attention to the relevant portion of the environment, but successful tool use will always be quickly and reliably followed by the availability of a primary reinforcer (Gergely & Csibra, 2005). Second, once this goal state has been identified, it becomes possible for the observer to attend more selectively to future examples of the tool's preparation and use, and thus to acquire the relevant behaviors more quickly. With inverse and recursive teleology, however, none of these favorable learning conditions apply. Since tools can be created in anticipation of future use rather than simply in response to immediate environmental cues, it will often be impossible to unambiguously identify the goal that a tool is intended to subserve. The ambiguity of the final goal in turn means that it is not possible for an observer to confidently differentiate relevant aspects of a tool's creation and use behavior from those aspects that are irrelevant. Csibra and Gergely (2006) illustrate this point nicely with the example of a child watching an adult carving small bits of wood off of a larger piece. The goal of such behavior could be to give the large piece of wood a particular shape, but it could equally be to obtain lots of tiny bits of wood (e.g., as tinder for starting a fire). If the child does not know which of these ends applies, it will be impossible to differentiate the idiosyncratic components of the display from those components that are essential to the goal; it is the goal itself that defines which features of the demonstration that are most relevant (a point that we will return to in due course). Furthermore, recursive tool use only magnifies these fundamental difficulties. When multiple instances of tool use separate a model's current behavior from an identifiable reward state, the relevance of each element in an observed action sequence becomes still more opaque.

In terms of our earlier vocabulary then, the human capacity for inverse and recursive teleology means that the informational significance of tool-mediated action is exceedingly difficult to estimate accurately. Especially when novel tools and artifacts are involved, there is often no safe way to ascertain where in an observed action hierarchy imitation becomes unnecessary and emulation will suffice. This conclusion brings us back to where we began: to the comparison between children's precise imitation in the hands-free condition of Gergely et al.'s (2002) lightbox study, and their selective imitation in the houses-present condition of the Carpenter et al. (2005) mouse study. The previously advanced argument, that the disparity in children's responses to these two situations can be accounted for by the differing domains in which they are situated, now seems more plausible. Our intuition that actions directed toward novel artifacts generally bear careful imitation is not something that can only be accounted for in terms of extensive experience with complex devices like computers and remote controls. Rather, given the observational learning difficulties that inverse and recursive tool use would have presented for our early hominid ancestors, it is plausible that this 'tool bias' in imitation is simply part of our species' social learning toolkit. Such an innate bias could explain how it is that even 14-month-olds appear to deploy different imitative strategies in situations where unfamiliar tools and artifacts are involved; they may be responding to a social learning mechanism that arose to address the learnability problems of advanced tool use.

6 Overimitation Revisited

We have covered a good deal of ground since the start of this chapter, so let us briefly recap. We began with the challenge of articulating a unified conceptual framework for understanding human imitative behavior. A particularly significant barrier to this conceptual unification was the puzzle of how children can imitate with such remarkable rational selectivity in some contexts, yet overimitate in such a seemingly blind way in others. Our attempts to reconcile this dichotomy began with formalizing the operations involved in imitating selectively. Locating the optimum 'imitation cut-off' within an observed behavior hierarchy—that is, picking the point in the hierarchy at which to switch from precise imitation to less constrained emulation—was identified as the key computation underlying selective imitation. We argued that children perform this calculation by estimating the informational significance of each level of an observed hierarchy, essentially attempting to infer the threshold separating signal from noise in an incoming behavior stream. We discussed children's ability to perform this calculation of informational significance in an impressively nuanced way, accounting for the intentions of the actor as well as for the domain of the model's task. With regard to this final aspect of children's calculation, we argued that the domain of tools and artifacts—particularly of novel tools and artifacts—is one that children treat with special care. Reasoning from the particular case of Gergely et al.'s (2002) lightbox study, children were argued to have a sensible bias against emulation in cases where unfamiliar tools and artifacts are involved. This bias was argued to be both rational and plausible in view of the observational learning pressures that likely accompanied our species' acquisition of sophisticated tool use abilities, particularly our ability to conceptualize tool function from both inverse and recursive teleological perspectives.

All of which brings us back to where we began: to the puzzle of reconciling selective imitation and overimitation. With the preceding points thus assembled, we can now see exactly why the apparent dichotomy between these two imitative phenomena is a false one. Specifically, overimitation is supported by exactly the same rational social learning mechanisms that give rise to selective imitation; the difference between the two is simply a contrast between the rational strategies that different domains demand. As we noted at the outset, overimitation is a phenomenon that appears to be quite specific to the domain of artifacts. All reported instances of overimitation have occurred in situations where children were engaged in an observational learning task that involved acting with or on a novel tool or artifact. Because of this, overimitation sits on much the same rational turf as children's use of their forehead to activate Gergely et al.'s lightbox. Both behaviors can be understood as a rational response to the opacity that advanced tool use creates.

Before concluding this discussion, there is one important aspect of overimitation that bears explicit highlighting. Specifically, why is it that overimitation persists even in situations where the causal irrelevance of the adult's unnecessary actions *should* be extremely obvious to children? Why do children overimitate even when the novel objects that they see the adult act on are transparent, rendering the causal significance (or lack thereof) of each action component visible? Actually, the answer to this question is already on the table, but it bears highlighting. To see it, we need to recognize that the learnability challenges posed by inverse and recursive teleology are *independent* of the degree to which a model's actions are causally opaque or transparent. Recall the previously discussed example of a child watching an adult carving small bits of wood off of a larger chunk of wood (Csibra & Gergely, 2006). The difficulty of trying to learn from the adult's actions in this situation stems not from the causal opacity of the adult's behavior; all of the causal relationships involved in the whittling process itself, a sharp implement combined with a reasonably pliant substrate, are visible. Instead, as we have previously noted, the difficulty is that the ultimate goal of all these individually transparent actions is opaque. In a recursively tool using species, the actions that the adult performs could be embedded in an unlimited range of goal-directed sequences, each of which may implicate different components of the demonstration as causally central. For example, if the goal of the adult's carving is to create tinder for a fire, the size and thinness of the bits of wood carved from the main block will later be causally important. On the other hand, if the adult's goal is to do something with the main block itself—make a hand tool for example—than the shape of the bits carved off has no causal significance; instead, it is the shape of the remaining wood itself that carries future causal implications. Thus, the lack of certain knowledge about the model's ultimate goal makes it impossible to safely discard any component of the demonstration; actions that appear to be causally irrelevant at the present moment, or for a particular inferred goal, may well turn out to have significant causal implications at some point in the future. This same sort of analysis applies to situations in which children overimitate on causally transparent objects. Even when there is good evidence to suggest that an action component is causally irrelevant for the sub-goal that is inferred to be active at the present moment, that action component may well bear causal significance at a later stage. Precisely reproducing all of the observed actions is thus the most prudent course of action. The rationality of overimitation is independent of the causal transparency of the target objects.

7 Conclusions

There can be little doubt that imitation is among our species' most important cognitive faculties. It is therefore quite easy to understand exactly why mirror neurons have captured so much interest amongst the scientific community and the general public alike. These neurons, with their common coding of action perception and action performance, have been a tantalizing target for those interested in the neural underpinnings of our imitative capacity. However, while there have been initial suggestions as to the role that the human mirror system might play in imitation, such theorizing has been arguably weakened by an oversimplified view of imitative phenomena. The full complexity of imitation cannot be captured by resonance metaphors that imply a more-or-less direct connection between perceptual input and motor output. Instead, any model of human imitation needs to account for the considerable degree of representation, interpretation, and inference making that accompanies imitation—even from the first months of life. This chapter has been intended to serve as a starting point for constructing this sort of empirically grounded model. We have seen that two seemingly opposite poles of imitative behavior—selective imitation and overimitation—can be unified into a single overarching continuum of rational social learning. It has been argued that both phenomena reflect a common underlying computation: that of inferring the optimal point in an observed behavior hierarchy to switch from copying a model's means to simply emulating the modeled ends. It is hoped that this conceptual framework, unifying a full spectrum of empirical results, will provide a much more useful foundation for neurophysiological theories of imitation than prior resonance metaphors. The cognitive view of imitation presented here may be far more complicated, but any theory about the ultimate relationship between the mirror system and human imitation will ultimately need to address it.

Acknowledgments The author wishes to thank Frank Keil, Laurie Santos, and Deena Skolnick Weisberg for their helpful comments on this work. The author was supported by a National Defense Science and Engineering Graduate Research Fellowship.

References

Bekkering, H., Wohlschlager, A., & Gattis, M. (2000). Imitation of gestures in children is goal-directed. *Quarterly Journal of Experimental Psychology, 53A*, 153–164.

Boesch, C., & Boesch, H. (1990). Tool use and tool making in wild chimpanzees. *Folia Primatologica, 54*, 86–99.

Byrne, R. W., & Russon, A. E. (1998). Learning by imitation: A hierarchical approach. *Behavioral and Brain Sciences, 21*, 667–721.

Call, J., & Carpenter, M. (2002). Three sources of information in social learning. In K. Dautenhahn & C. L. Nehaniv (Eds.) *Imitation in Animals and Artifacts* (pp. 211–228). Cambridge: MIT Press.

Call, J., Carpenter, M., & Tomasello, M. (2005). Copying results and copying actions in the process of social learning: Chimpanzees (*Pan troglodytes*) and human children (*Homo sapiens*). *Animal Cognition, 8*, 151–163.

Carpenter, M., Call, J., & Tomasello, M. (2002). Understanding "prior intentions" enables two-year-olds to imitatively learn a complex task. *Child Development, 73*(5), 1431–41.

Carpenter, M., Call, J., & Tomasello, M. (2005). Twelve- and 18-month-olds copy actions in terms of goals. *Developmental Science, 8*, F1–F8.

Carpenter, M., Nagell, K., & Tomasello, M. (1998). Social cognition, joint attention, and communicative competence from 9 to 15 months of age. *Monographs of the Society for Research in Child Development, 63*(4), 1–143.

Csibra, G. (2007). Action mirroring and action interpretation: An alternative account. In P. Haggard, Y. Rosetti, & M. Kawato (Eds.) *Sensorimotor foundations of higher cognition. Attention and performance XXII*. Oxford: Oxford University Press.

Csibra, G., & Gergely, G. (2006). Social learning and social cognition: The case for pedagogy. In Y. Munakata & M. H. Johnson (Eds.) *Processes of change in brain and cognitive development. Attention and performance, XXI*. Oxford: Oxford University Press.

Csibra, G., Gergely, G., Bíró, S., Koós, O., & Brockbank, M. (1999). Goal attribution without agency cues: The perception of 'pure reason' in infancy. *Cognition, 72*, 237–267.

Gallese, V., Fadiga, L., Fogassi, L., & Rizzolatti, G. (1996). Action recognition in the premotor cortex. *Brain, 119*, 593–609.

Gergely, G., Bekkering, H., & Király, I. (2002). Rational imitation in preverbal infants. *Nature, 415*, 755.

Gergely, G., & Csibra, G. (2005). The social construction of the cultural mind: Imitative learning as a mechanism of human pedagogy. *Interaction Studies, 6*, 463–481.

Gergely, G., & Csibra, G. (2006). Sylvia's recipe: The role of imitation and pedagogy in the transmission of cultural knowledge. In N. J. Enfield & S. C. Levenson (Eds.) *Roots of human sociality: Culture, cognition, and human interaction*. Oxford: Berg Publishers.

Gergely, G., Egyed, K., & Ildiko, K. (2007). On pedagogy. *Developmental Science, 10*, 139–146.

Gleissner, B., Meltzoff, A. N., & Bekkering, H. (2000). Children's coding of human action: Cognitive factors influencing imitation in 3-year-olds. *Developmental Science, 4*, 405–414.

Goodall, J. (1986). *The chimpanzees of gombe: Patterns of behavior*. Cambridge, MA: Harvard University Press.

Gordon, H. (1923). Hand and ear tests. *British Journal of Psychology, 13*, 283–300.

Head, H. (1920). Aphasia and kindred disorders of speech. *Brain, 43*, 87–165.

Horner, V., & Whiten, A. (2005). Causal knowledge and imitation/emulation switching in chimpanzees (*Pan troglodytes*) and children (*Homo sapiens*). *Animal Cognition, 8*, 164–181.

Kamewari, K., Kato, M., Kanda, T., Ishiguro, H., & Hiraki, K. (2005). Six-and-a-half-month-old children positively attribute goals to human action and to humanoid-robot motion. *Cognitive Development, 20*, 303–320.

Kephart, N. C. (1971). *The Slow Learner in the Classroom*. Columbus, OH: Charles Merrill.

Lyons, D. E., & Santos, L. R. (2006). Ecology, domain specificity, and the origins of theory of mind: Is competition the catalyst? *Philosophy Compass, 1*, 1–12.

Lyons, D. E., Santos, L. R., & Keil, F. C. (2006). Reflections of other minds: How primate social cognition can inform the function of mirror neurons. *Current Opinion in Neurobiology, 16*, 1–5.

Lyons, D. E., Young, A. G., & Keil, F. C. (2007). The hidden structure of overimitation. Proc Natl Acad Sci., 104, 19751–19756.

McGuigan, N., Whiten, A., Flynn, E., & Horner, V. (2007). Imitation of causally opaque versus causally transparent tool use by 3- and 5-year-ld children. *Cognitive Development* 2007, 22, 353–364.

Meltzoff, A. N. (1988). Infant imitation after a 1-week delay: Long-term memory for novel acts and multiple stimuli. *Developmental Psychology, 24*, 470–476.

Meltzoff, A. N. (1995a). What infant memory tells us about infantile amnesia: Long-term recall and deferred imitation. *Journal of Experimental Child Psychology, 59*, 497–515.

Meltzoff, A. N. (1995b). Understanding the intentions of others: Re-enactment of intended acts by 18-month-old children. *Developmental Psychology, 31*, 838–850.

Nagell, K., Olguin, K., & Tomasello, M. (1993). Processes of social learning in the tool use of chimpanzees (*Pan troglodytes*) and human children (*Homo sapiens*). *Journal of Comparative Psychology, 107*, 174–186.

Nielsen, M. (2006). Copying actions and copying outcomes: Social learning through the second year. *Developmental Psychology, 42*, 555–565.

Phillips, A. T., & Wellman, H. M. (2005). Infants' understanding of object-directed action. *Cognition, 98*, 137–155.

Rizzolatti, G., & Craighero, L. (2004). The mirror-neuron system. *Annual Review of Neuroscience, 27*, 169–192.

Rizzolatti, G., Fadiga, L., Fogassi, L., & Gallese, V. (1996). Premotor cortex and the recognition of motor actions. *Brain Research, Cognitive Brain Research, 3*, 131–141.

Rizzolatti, G., Fogassi, L., & Gallese, V. (2001). Neurophysiological mechanisms underlying the understanding and imitation of action. *Nature Reviews, Neuroscience, 2*, 661–670.

Schofield, W. N. (1976). Do children find movements which cross the body midline difficult? *Quarterly Journal of Experimental Psychology, 28*, 571–582.

Schwier, C., van Maanen, C., Carpenter, M., & Tomasello, M. (2006). Rational imitation in 12- month-old infants. *Infancy*, 10(3), 303–311.

Sodian, B., Schoeppner, B., & Metz, U. (2004). Do infants apply the principle of rational action to human agents? *Infant Behavior and Development*, *27*, 31–41.
Swanson, R., & Benton, A. L. (1955). Some aspects of the genetic development of right-left discrimination. *Child Development*, *26*, 123–133.
Tomasello, M. (1996). Do apes ape? In C. M. Heyes & B. G. Galef (Eds.) *Social learning in animals: The roots of culture* (pp. 319–346). New York: Academic Press.
Tomasello, M., & Call, J. (1997). *Primate cognition*. New York: Oxford University Press.
Tomasello, M., Carpenter, M., Call, J., Behne, T., & Moll, H. (2004). Understanding and sharing intentions: The origins of cultural cognition. *Behavioral and Brain Sciences*, *28*, 675–691.
Uzgiris, I. C. (1981). Two functions of imitation during infancy. *International Journal of Behavioral Development*, *4*, 1–12.
Want, S. C., & Harris, P. L. (2002). How do children ape? Applying concepts from the study of non-human primates to the developmental study of 'imitation' in children. *Developmental Science*, *5*, 1–41.
Wapner, S., & Cirillo, L. (1968). Imitation of a model's hand movements: Age changes in transposition of left-right relations. *Child Development*, *39*, 887–894.
Whiten, A. (2005). The second inheritance system of chimpanzees and humans. *Nature*, *437*, 52–55.
Whiten, A., Custance, D. M., Gomez, J-C., Teixidor, P., & Bard, K. A. (1996). Imitative learning of artificial fruit processing in children (*Homo sapiens*) and chimpanzees (*Pan troglodytes*). *Journal of Comparative Psychology*, *110*, 3–14.
Woodward, A. L. (1998). Infants selectively encode the goal object of an actor's reach. *Cognition*, *69*, 1–34.

Part III
Neural Basis

From Embodied Representation to Co-regulation

Gün R. Semin and John T. Cacioppo

Abstract A central feature of the research paradigm, namely the nearly exclusive focus on subjects as passive observers and how this particular vision informs theorizing in this field is discussed along with some of its implications. In advancing a novel model of social cognition, we place the function of a possible mirror neuron system within a general framework ranging from perception to co-action. The proposed social cognition model relies on a unit of analysis that is minimally dyadic. In concluding we draw out some of the implications of the proposed model for research on mirror neurons.

Keywords Social cognition · Co-regulation · Reception · Reproduction · Representation

1 Introduction

The discovery of the mirror neuron system (MNS) has conquered the imagination of the scientific community at large with a promise of furnishing the embodied grounding of a number of age-old puzzles that have been central to understanding the human condition. The discovery has been seen as providing the answer to a range of social behaviors ranging from the evolution of language (Arbib, 2005; Rizzolatti & Arbib, 1998) to imitation and empathy (Iacoboni, 2005).

The common denominator to all these intellectual challenges is the possibility of an answer to what mechanisms bridge the gap between individuals, namely how is intersubjectivity achieved (Gallese, 2005b). These are issues that have occupied major sociological figures such as Max Weber, Alfred Schütz, and the entire ethnomethodological tradition initiated by Garfinkel

G.R. Semin
Department of Social Psychology, Free University Amsterdam, van der Boechorststr. 1, 1081 BT Amsterdam, The Netherlands
e-mail: GR.Semin@psy.vu.nl

J.A. Pineda (ed.), *Mirror Neuron Systems*, DOI: 10.1007/978-1-59745-479-7_5,

and colleagues. The same questions were at the heart of 19th century psychology, which attempted to overcome an individual centered analysis, with the notion of *Völkerpsychologie* which originated in the 1850ies (Lazarus, 1861; Lazarus & Steinhal, 1860) culminating in Willem Wundt's work and the diverse origins of the emerging field of modern social psychology in the early 20th century with notions of 'group mind' and 'instinct,' associated with the classic contributions by Durkheim, LeBon, Ross, Tarde, and Wundt arguing in different voices for collective representations, group mind, collective mind, collective consciousness, or *Völkerpsychologie*, along with specific psychological mechanisms such as 'Einfühlung' introduced by Lipps (1903).

The mirror neuron system is currently seen as holding the promise of a neuroscientific bridge that would lead the way to answers to age-old questions that have been elusive intellectual problems for philosophers, psychologists, linguists, and anthropologists. The discovery that reinvigorated the imagination began with the demonstration of a particular class of visuomotor neurons (in the F5 area of the macaque monkey premotor cortex) that discharge when the monkey engages in a particular action (e.g. grasping a peanut) *and* when it observes another monkey engaging in the same action (e.g. di Pellegrino, Fadiga, Fogassi, Gallese, & Rizzolatti, 1992; Gallese, Fadiga, Fogassi, & Rizzolatti, 1996; Rizzolatti, Fadiga, Gallese, & Fogassi, 1996). Rizzolatti, Fadiga, Gallese, and Fogassi (1996), for a review see Rizzolatti and Craighero (2004).

Our chapter is structured in four parts. We begin with a brief overview of research on the MNS with a view of highlighting the central features of the research paradigm that is used in this field. In the second section, we examine some of the implications of the nearly exclusively focus on subjects as passive observers and this particular vision informs theorizing in this field. In the third, we advance a model of social cognition that attempts to capture the processes taking place from perception to co-action and relies on a unit of analysis that is minimally dyadic. In the concluding section, we shall draw out some of the implications of the proposed model for research on mirror neurons.

2 The Representation Paradigm

Research on the neural notation of the perception of intentional action gained momentum with the discovery of mirror neurons in area F5 of monkey premotor cortex (e.g. di Pellegrino, Fadiga, Fogassi, Gallese, & Rizzolatti, 1992; Gallese, Fadiga, Fogassi, & Rizzolatti, 1996; Rizzolatti, Fadiga, Gallese, & Fogassi, 1996). Specifically, Rizzolatti et al. (1996) demonstrated the existence of a particular class of visuomotor neurons (in the F5 area of the macaque monkey premotor cortex) that discharge when the monkey engages in a particular action (e.g. grasping a peanut) and when it observes another monkey engaging in the same action (for a review see Rizzolatti & Craighero, 2004). A subset of these neurons termed 'mirror neurons' – also become active even when the final part of the action (e.g. gasping the peanut) is hidden (Umiltà et al., 2001).

In subsequent research, specific populations of neurons ('audiovisual mirror neurons') have been identified in the ventral premotor cortex of the monkey that discharge not only when a monkey performs a specific action but also when it *sees or hears* another monkey perform the same action (Keysers et al., 2003; Kohler et al., 2002). These neurons therefore represent actions independently of whether these actions are performed, heard, or seen. Moreover, single neurons in the premotor cortex resonate not only to the actions that the other is executing, but also to the 'goal' of the action represented and inferred in different modalities.

An emerging body of evidence supports the existence of a MNS in humans. For instance, Buccino et al. (2001) using functional magnetic resonance imaging (fMRI) were able to localize areas of the brain that were active during the observation of movement by another individual. The areas activated in the premotor cortex corresponded to the regions that would be active were the individual to have executed the observed actions. Other evidence has revealed that observing movements of finger, hand, arm, mouth, or foot leads to the activation of motor-related areas of cortex (e.g. Grafton, Arbib, Fadiga, & Rizzolatti, 1996; Iacoboni et al., 1999; Manthey, Shubotz, & von Cramon, 2003; Stevens, Fonlupt, Shiffrar, & Decety, 2000). Notably, this seems to be the case for movements that are biologically possible and not for impossible movements (Stevens et al., 2000). Other research has shown that listening to speech sounds activates premotor areas in the brain overlapping with areas that are responsible for the production of speech (Wilson, Saygin, Sereno, & Iacoboni, 2004).

In 1909, Edward Tichner argued that people could never know what another felt by reasoning, that they could only know by feeling themselves into the other's feelings. Since that time, a large literature in human and nonhuman animals has accrued demonstrating mimicry and contagion effects, typically without any conscious awareness of control by the individuals involved. However, the neurophysiological mechanism by which these effects could occur was not specified. Converging evidence for a neuronal mirroring of emotional behavior is accumulating rapidly, and with it the purview of the putative human MNS. Briefly, it has been shown that observing facial expressions of disgust and feelings of disgust activated very similar sites in the anterior insula and anterior cingulate cortex (e.g. Wicker et al., 2003) and are involved in inducing what they term 'empathy' (e.g. Decety, 2005; Jabbi, Swart, & Keysers, 2007). Indeed, single neuron recording experiments with humans have shown that the observation of pain and its experience activate the same neurons (Hutchison, Davis, Lozano, Tasker, & Dostrovsky, 1999). The argument developed by a number of authors (e.g. Adolphs, 2006; Carr, Iacoboni, Dubeau, Mazziotta, & Lenzi, 2003; Decety, & Grèzes, 2006) is akin to an insight offered by Lipps in the beginning of the 20th century (1903). The studies on emotional 'mirroring' suggest that the observer experiences 'Einfühlung' or, as the researchers suggest, 'empathy.' Cumulatively, the research suggests that mirroring for action is also applicable to emotion (cf. Adolphs, 2006).

Using fMRI, Carr, Iacoboni, Dubeau, Mazziotta, & Lenzi, (2003) found that the brain region is important for action representation and imitation, such as the superior temporal sulcus (STS), are functionally connected to the insula and amygdala – regions in the limbic lobe that are involved in emotions. Because of the role of the insula in interoception, there is evidence for an association between activation of the insula and awareness of one's own bodily state (Critchley, Wiens, Rotshtein, Oehman, & Dolan, 2004). Recent research suggests that the observation of another person's emotional state recruits structures like the insula (Jackson, Meltzoff, & Decety, 2005; Singer et al., 2004), a part of the cortex, which is also involved not only in interoception but also in the representation of our own somatic states. Indeed, observations of emotional states (e.g. a fearful body) not only induces increased activity in brain areas related to emotional processes but also areas that are responsible for the representation of action, suggesting that the synchronization process goes beyond establishing symmetry in experience, but puts the organism in a state of readiness for action (e.g. De Gelder, Snyder, Greve, Gerard, & Hadijkhani, 2004).

Together, these studies suggest that the architecture of the human perceptual-motor system is specifically designed for the reproduction of movements of conspecifics in a privileged way. Indeed, if mapping human beings and their bodily movements does constitute a special or distinctive case, then it should be species specific. Recent evidence suggests that this is the case (cf. Buccino et al., 2004). These findings are also consistent with the notion that the MNS entails a type of sensory neural representation that has an entirely different ontological status than knowledge about the world in general.

In sum, our brief review suggests an impressive cumulative pattern about how multimodal actions resonate in an observer and indicate a mechanism with potentially innate bases (Meltzoff & Moore, 1997, 1999; for reviews) that provide a view to how "the epistemic gulf separating single individuals can be overcome" (Gallese, 2006, p. 16). In the next section, we focus on a generic feature of the empirical paradigm driving this research. We shall argue that this feature introduces specific limitations that have to be overcome if work on the MNS is deemed to be the royal road to solving the puzzle of *social* cognition.

3 From the Passive Observer to Reception, Reproduction, and Representation

The empirical paradigm on which the above reviewed research relies employs the neural circuitry recruited by a *passive* observer. It therefore maybe limited in what it reveals because cognition evolved for the control of adaptive action and social cognition evolved for the control of adaptive interaction.

The fundamental evolutionary demands on cognition are the organism's survival and reproduction, which (for humans) always takes place in a social context (Caporael, 1997; Fiske, 1992). Reliance on an observation paradigm limits the view of the operational field of social cognition, namely the co-regulation of action and its dynamic and adaptive functions. Thus, watching a tennis forehand stroke and examining the regions that are activated in the execution and observation of the stroke are informative but only a part of the story. Unless one is only hitting a ball against a practice wall, a tennis forehand stroke also involves the co-regulation of action as an integral part of social interaction, and thus also the defensive backhand stroke that is immediately activated upon the observation of the forehand. Seen that way, the research and theorizing about the MNS is in fact about one specific component function of what is neurally recruited, namely the *monitoring* of the action of the other, but this research is typically less revealing about another important component of *social* cognition. The complexity of the social environment and the adaptiveness required of social cognition is not only to continuously monitor but also to *selectively respond to significant features of a dynamic social environment* by setting goals for action. It is important to address how adaptive action is induced in response to 'significant' stimuli. Thus, the passive observer paradigm buys into a reproduction perspective rather than an interaction capable of promoting the co-regulation of two or more individuals' behavior. One possible consequence of the current emphasis on a passive observer paradigm is that both research and theory is about an individual rather than a dyad, triad, or group. The co-regulation of individuals is about jointly recruited processes and the resulting emergent cognition and behavior.

Relying on a passive observational paradigm described above introduces implicit assumptions that inform theory formation. First and foremost cognition becomes *reception*, namely the construction of inner neural representations based on observed behavior. Second, (specifically in the context of MNS) research and theory derived from an observational paradigm narrows social cognition down with the use of a *reproduction metaphor* (e.g. empathy, resonance, imitation, shared representations, social cognition), namely reproduction that takes place at a neural level. Finally, since the primary focus is at the neural level, the development of models that attempt to bridge the existential gap between individuals and provide an informed answer to how intersubjectivity achieved remain at a purely representational level neglecting the reciprocal nature and co-regulation of social behavior.

The three Rs (reception, reproduction, and representation) are conceptual consequences of relying on an observer paradigm that has left its mark on the diverse accounts addressing the meaning and function of the findings reviewed above (e.g. Firth & Firth, 2006; Grèzes & Decety, 2001; Decety & Grèzes, 2006; Iacoboni, 2005; Jeannerod, 1999; Goldman, 2002, 2005; Keysers & Gazzola, 2006). For space reasons, we shall illustrate the conceptual concerns raised by the three Rs in the case of one prominent account, namely Gallese's conceptualization of 'shared manifold' (e.g. Gallese, 2001,

2003, 2005a), namely a shared meaningful interpersonal space (e.g. Gallese, 2001). Gallese regards the 'shared manifold' as grounding the understanding of action or emotion. The shared manifold is achieved in a process in which the perception of an action or emotion induces a motor 'simulation.' This embodied simulation process is direct, automatic and escapes conscious access, and leads the observer to achieve a neural coupling with an agent thereby reconstituting the other's emotion or intentional action and creating a shared bodily state. In Gallese's view, this 'experienced state' constitutes the basis of mutual or direct understanding. "By means of embodied simulation, we do not just 'see' an action, an emotion, or a sensation. Side by side with the sensory description of the observed social stimuli, internal representations of the body states associated with these actions, emotions, and sensations are evoked in the observer, 'as if' he/she would be doing a similar action or experiencing a similar emotion or sensation" (Gallese, 2006). The neural correlate of such an embodied simulation mechanism is to be found in MNS and – as the above quote implies – in the shared neural state – the bodies of the agent and observer follow the same functional rules in which the other becomes another 'self.' Mutual understanding in this view is non-propositional. The shared bodily state achieved through embodied simulation is at a basic level of understanding and does not entail declarative representation. Most importantly, in this view the function of mutual understanding is not merely modeling agents, their actions and emotions, but also the successful *prediction* of upcoming events.

Admittedly, social cognition does not consist of merely passive experiences of observing another perform an action or express a particular affective state. Moreover, social cognition does not consist of internal or intra individual processes alone, even if these processes constitute a genuinely social state, as in the case of a 'shared manifold' to which Gallese refers. Nor is social cognition equivalent to the "becoming another self". Social cognition consists of dynamic, unfolding processes that take place between two or more agents, who are engaged in action and reaction, and the shape of this interaction takes a multitude of forms ranging from a wide range of different non-verbal to verbal exchanges. What does the embodied simulation model represent in the context of a broader conception of social cognition, namely as distributed processes taking place between two or more individuals?

What is at issue here is not the simulation process with which we are in agreement, but the interpretation of the *function* and the *reach* of the account advanced by Gallese. Gallese's model of social cognition is informed by a research paradigm involving a passive observer rather than two agents in a *continuously unfolding dynamic* monitoring process of co-regulation. That is, the unit of analysis is the individual. What is more, the 'shared manifold' alone does not shape unfolding interaction. As we shall argue in the next section, social cognition is about more than reception, reproduction, and representation. (cf. Semin & Cacioppo, 2007).

4 Co-regulation – The Social-Cognition Model

Social cognition is for the adaptive regulation of the behavior of another person (e.g. issuing instructions to another person) and adaptive co-regulation (e.g. the regulation of social interaction). Understanding what social cognition entails not only explicating that it is *distributed across brains* in a distinctive manner but also the processes by which the regulation of others' behavior and the co-regulation of social interaction is achieved. Figure 1 below captures the model we have precisely developed for this purpose (Semin & Cacioppo, 2007). The first point advanced in this model is that:

> The observation of the action of another person serves as a stimulus whose effects depend upon its goal-relevance.

This is Point 1 in Figure 1. Any action independent of whether it activates a significant goal or not also activates an implicit *monitoring* processes that we refer to as:

> Monitoring Synchronization.

The multimodal neurophysiological sensorimotor processes involved in the execution of any real (or imagined) action give rise to synchronization of neurophysiological sensorimotor processes in the observer of human action (point 2a in Fig. 1). The process of synchronization is time-locked to the observed stimulus. Specifically, we define synchronization as *jointly and simultaneously recruited sensory motor processes* that are evident in a neurophysiological mirroring of the producer by the perceiver. These synchronization processes enable organisms to continuously monitor and adaptively respond

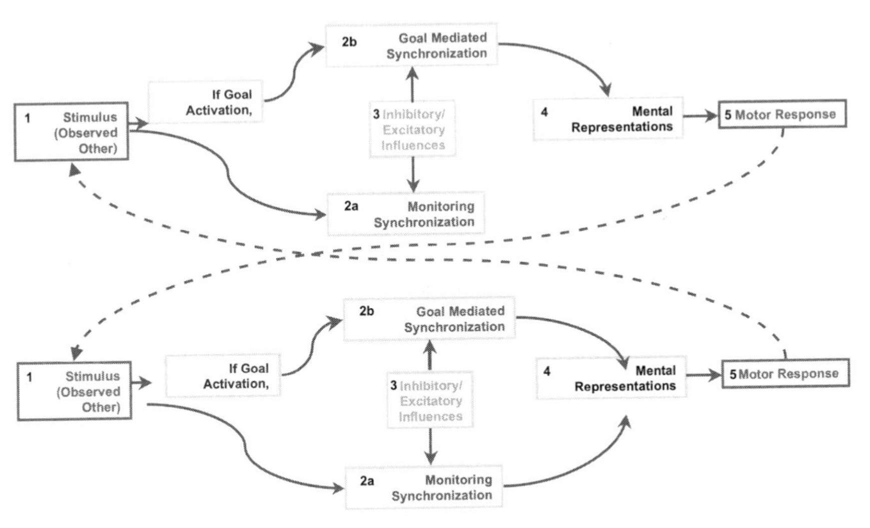

Fig. 1 The Social Cognition (SC) model (from Semin & Cacioppo, in press)

to their social environment by means of *discrete*, *recurrent*, and *short-lived temporal intervals* (Andrews & Coppola, 1999; Andrews & Purves, 2005; van Rullen & Koch, 2003); that is, these recurrent processes maintain the sensitivity of the organism to changes in the environment, serve a mutual monitoring function, and facilitate dynamic adaptation.

Aside from their monitoring function, these processes link two or more human agents putting them on a similar footing. It is *jointly recruited processes* with *overlapping* 'identities' that facilitates understanding (co-cogitation) and adaptive co-action (co-regulation) between two or more individuals. In other words, such mechanisms facilitate reaching a state of correspondence between the individuals. What counts for the one member may not initially have counted for the other but through interaction these two become synchronized to approach being on the same page – that is, for what counts for one individual also counting for the other. Such synchronization can occur without the presence of explicit intent or goals (see Fig. 1).

Does one become 'the other self'?

If synchronization *alone* were to lead to complete equivalence between the sensorimotor processes of a producer and a perceiver then there would be confusion between producer and perceiver, self and other (e.g. Adolphs & Spezio, 2007). Moreover, complete identity could lead to a never-ending loop of continuously performing the very same actions. Therefore, synchronization promotes partial, not full, correspondence between producer and perceiver (see Fig. 1, point 2a).

The complexity of the social environment and the adaptiveness required of social cognition is not only to continuously monitor but also to *selectively respond to significant features of a dynamic social environment* by setting goals for action. It is important to address how adaptive action is induced in response to 'significant' stimuli. Visualize a table tennis game, in which your partner makes a backhand smash. While monitoring that action is critical, it will not win you the game. You have to take effective *complementary* actions. Thus, if the stimulus is goal-relevant, then higher-level cognitive processes are recruited which entail goal mediated synchronization of neurophysiological activations whose specific form need not mirror the actions that are observed. While the monitoring process is critical, adaptive action is not merely a product of monitoring processes. We therefore propose that the identification of significant stimuli recruits goal-driven higher-level decision processes that can run in addition to continuous monitoring processes (see Fig. 1, Point 2b).

The higher order goal-mediated synchronization and adaptive action depicted in Figure 1, Point 2b are separable functionally from the monitoring synchronization depicted in Figure 1, Point 2a. One reason for this functional distinction is that the processes depicted in 2a are thought to be automatic whereas those depicted in 2b involve automatic and controlled processes. In terms of the table tennis example, this means that the backhand smash of the opponent is monitored – a process that is achieved by the nonconscious synchronization of neurophysiological activations corresponding to that motor movement. Synchronization as a continuous monitoring process serves a predictive function by detecting changes in the environment. Yet, the table tennis player has to make a decision about how to counter the backhand and defend himself or herself. This decision to counter the backhand leads to the requisite neurophysiological activation for the execution of the counter move rather than simply the mirror image of the observed action. Although specific aspects of goal-mediated synchronization can be automated, the act as a whole is not.

> Synchronization and higher-level processes are dissociated, but jointly shape the mental representation of the stimulus and are both subject to inhibitory and excitatory influences (Fig. 1, Point 3) before they shape these mental representations (Fig. 1, Point 4) that is then translated to action in the form of a motor response.

Both goal-mediated synchronization processes and monitoring synchronization recruit neurophysiological activity and jointly contribute to the shape of the representations that are translated into motor action (Point 4 in Fig. 1). In the absence of goal-mediated synchronization, the neurophysiological activation from monitoring synchronization shapes the representation. However, this does not necessarily mean the automatic execution of the observed action because other factors can contribute to the representation. Moreover, both excitatory and inhibitory factors can modulate the contributions of the processes depicted in Figure 1, Points 2a and 2b to the representation. For instance, inhibitory processes can contribute to blocking continuous automatic reproduction of what is observed (e.g. Baldissera, Cavallari, Craighero, & Fadiga, 2001).

> The structural features of the environment or task will contribute to the type of co-regulation processes that will emerge.

The type of task environment requiring two or more persons, which can be socially shaped (e.g. dancing, playing tennis, conversation) or by the physical characteristics of a joint task (e.g. carrying a large and heavy object, etc.) presents distinctive affordances that shape the co-regulation of the social interaction.

> The behavior of one individual becomes the stimulus for the other when the individuals are actively interacting, with each iteratively producing co-action effects. Consequently, synchronization and adaptive co-regulation of behaviors are detectible when the unit of analysis is at the social rather than the individual level.

This yields the emergent social cognition as represented in Figure 1 with direct loops from 5 to 1, namely where the behavior of a member of the dyad provides information pickup for the other member in an iterative process.

5 Conclusions

The Social Cognition (SC) model outlined above furnishes a framework, which locates the function of the MNS in the perception-action – co-regulation complex, by specifying it as a monitoring synchronization process. The monitoring synchronization process while critical to checking the moment-by-moment unfolding of the dynamic behavior co-regulation process is only a supportive component of social cognition. Social cognition is the product of agents who engage in joint action by means of jointly recruited processes and emergent behaviors. The SC model not only specifies the functional significance of the MNS, but also provides a way of systematizing our understanding of the behavioral research as well. There is a wide range of research that runs under diverse labels ranging from entrainment (e.g. Condon & Ogston, 1966, 1967), mimicry (e.g. Bavelas, Black, Chovil, Lemery, & Mullett, 1988; Bernieri, 1988; Bernieri, Reznick, & Rosenthal, 1988; Bernieri, Davis, Rosenthal, & Knee, 1994; Condon & Ogston, 1966, 1967; Kendon, 1970; Tickle-Degnen & Rosenthal, 1987), and contagion (e.g. Hatfield et al., 1994) – a focus that was rediscovered at the end of the last century (e.g. Chartrand & Bargh, 1999); and coordination which has been studied in diverse forms, in particular in communicative contexts and dialogue (e.g. Sachs, Schegloff, & Jefferson, 1974), but also analyzes of co-action from an ecological perspective (e.g. Marsh, Richardson, Baron, & Schmidt, 2006). The model we have described provides an overarching formulation for understanding and investigating these diverse observations.

We have distinguished between three qualitatively different forms of co-action as instances of co-regulation (Semin & Cacioppo, 2007). The first is *entrainment* and is exemplified by periodic co-action and occurs in cycles. This can be illustrated with the example of rhythmic clapping. The second form is when co-action a non-periodic, as in the case of *mimicry* (e.g. Chartrand & Bargh, 1999). The third case is exemplified in situations that involve a number of people who have to interface with each other's actions in the course of performing a complex task (e.g. open-heart surgery, playing tennis). This is the functional effect of co-action, namely *coordination* that entails the execution of complementary actions in the pursuit of accomplishing the task (e.g. successful surgery, winning in tennis). These different forms of the co-regulation of behavior can obviously all occur simultaneously and to different degrees.

It is effectively possible to systematically differentiate the relative contributions of monitoring synchronization, higher order goal-mediated synchronization, and the structural features of the environment or task that contribute to the shape of co-regulation (see Semin & Cacioppo, 2007 for detail). For instance, *entrainment* is likely to occur under circumstances either when: (i) a higher order goal is promoted by the explicit synchronization of rhythmic outputs (choral singing, chanting), or (ii) higher order goal-mediated synchronization is not operative and the monitoring synchronization process is therefore not inhibited. When explicit instructions are given experimentally to entrain, it is interesting to note how, the extent to which, when, and for how long entrainment occurs (e.g. Shockley, et al., 2003; Shockley, Baker, Richardson, & Fowler, 2007).

Although entrainment, mimicry, and coordination can be segmented for experimental purposes, they may all occur during a normal social interaction. Take for instance a dialogue. Any dialogue features a variety of instances of multimodal coordination, entrainment, and mimicry. A dialogue can simultaneously manifest *coordination* as in the case of turn taking in a conversation (e.g. Sachs, Schegloff, & Jefferson, 1974), or introducing a new topic, at a syntactic level (e.g. syntactic priming, Bock, 1986, 1989; Bock & Loebell, 1990), affective level (e.g. mood contagion, Neumann & Strack, 2000). Simultaneously, it is possible to see cyclically occurring instances of affective facial expressions (e.g. Dimberg, Thunberg, & Elmehed,2000), and breathing movements (e.g. Furuyama, Hayashi, & Mishima, 2005). Coordination and entrainment can converge when joint behavior is goal driven (e.g. playing tennis versus choral singing) be consciously accessible or escape conscious access (two people moving a heavy object versus emotional contagion) or a combination of both.

In the SC model we are suggesting that the goal of joint action along with the type of task environment (social or physical features of the task) contributes to the shape and nature of social cognition and joint action. The nature of coordination, mimicry, and entrainment in co-regulation will obviously vary depending on whether two people are attempting to move forward a heavy stone, a large table, an abstract idea, play tennis, or unfold a concrete business plan. To understand fully the neurophysiological and information processing bases of emergent social cognition and behavior, we may need to take into consideration units of analysis at the level of interacting dyads and beyond.

References

Adolphs, R. (2006). How do we know the minds of others? Domain specificity, simulation, and enactive social cognition. *Brain Research, 1079*, 25–35.

Adolphs, R., & Spezio, M. (2007). The neural basis of affective and social behavior. In Cacioppo, J. T., Tassinary, L. G., & Berntson, G. G. (Eds.), *Handbook of psychophysiology* (3rd ed., pp. 540–554). New York: Cambridge University Press.

Andrews, J. T., & Purves, D. (2005). The wagon-wheel illusion in continuous light. *Trends in Cognitive Sciences. 9*, 261–263.

Andrews, T. J., & Coppola, D. M. (1999). Idiosyncratic characteristics of saccadic eye movements when viewing different visual environments. *Vision Research, 39*, 2947–2953.

Arbib, M.A. (2005). From monkey-like action recognition to human language: An evolutionary framework for neurolinguistics. *Behavioral and Brain Sciences, 28*(2), 105–124.

Caporael, L. (1997). The evolution of truly social cognition: The core configurations model. *Personality and Social Psychology Review, 1*, 276–298.

Decety, J., & Grèzes, J. (2006). The power of simulation: Imagining one's own and other's behavior. *Brain Research, 1079*(1), 4–14.

Fiske, A. P. (1992). *Structures of social life*. New York: Free Press.

Iacoboni, M. (2005). Understanding Others: Imitation, Language, and Empathy. In S. Hurley & N. Chater (Eds.), *Perspectives on imitation: From neuroscience to social science: Vol. 1: Mechanisms of imitation and imitation in animals*. (pp. 77–99). Cambridge, MA, US: MIT Press.

Baldissera, F., Cavallari, P., Craighero, L., & Fadiga, L (2001). Modulation of spinal excitability during observation of hand actions in humans, *European Journal of Neuroscience, 13*, 190–194.

Bavelas, J. B., Black, A., Chovil, N., Lemery, C. R., & Mullett, L. (1988). Form and function in motor mimicry: Topographic evidence that the primary function is communicative. *Human Communication Research, 14*, 275–299.

Bernieri, F. J. (1988). Coordinated movement and rapport in teacher student interactions. *Journal of Nonverbal Behavior, 12*, 120–138.

Bernieri, F. J., Davis, J. M., Rosenthal, R., & Knee, C. R. (1994). Interactional synchrony and rapport: Measuring synchrony in displays devoid of sound and facial affect. *Personality and Social Psychology Bulletin, 20*, 303–311.

Bernieri, F. J., Reznick, J. S., & Rosenthal, R. (1988). Synchrony, pseudosynchrony, and dissynchrony: Measuring the entrainment process in mother infant interactions. *Journal of Personality and Social Psychology, 54*, 243–253.

Bock, L. J., & Loeblell, H. (1990). Framing sentences. *Cognition, 35*, 1–39.

Bock, L. J. (1986). Syntactic priming in language production. *Cognitive Psychology, 18*, 355–387.

Bock, L. J. (1989). Close class immanence in sentence production, *Cognition, 31*, 163–189.

Buccino, G., Binkofski, F., Fink, G. R., Fadiga, L., Fogassi, L., Gallese, V., et al. (2001) Action observation activates premotor and parietal areas in a somatotopic manner: An fMRI study. *European Journal of Neuroscience, 13*, 400–404.

Buccino, G., Lui, F., Canessa, N., Patteri, I., Lagravinese, G., Benuzzi, F., et al. (2004). Neural circuits involved in the recognition of actions performed by nonconspecifics: An fMRI study. *Journal of Cognitive Neuroscience. 16*, 114–126.

Carr, L., Iacoboni, M., Dubeau, M.-C., Mazziotta, J. C., & Lenzi, G. L. (2003). Neural mechanisms of empathy in humans: A relay from neural systems for imitation to limbic areas. *Proceedings of the National Academy of Sciences of the United States of America, 100*, 5497–5502.

Chartrand, T. L., & Bargh, J. A. (1999). The Chameleon effect: The perception-behavior link and social interaction, *Journal of Personality and Social Psychology, 76*, 893–910.

Condon, W. S., & Ogston, W. D. (1966). Sound film analysis of normal and pathological behavior patterns. *Journal of Nervous and Mental Diseases, 143*, 338–457.

Condon, W. S., & Ogston, W. D. (1967). A segmentation of behavior. *Journal of Psychiatric Research, 5*, 221–235.

Critchley, H. D., Wiens, S., Rotshtein, P., Oehman, A., & Dolan, R. J. (2004). Neural systems supporting interoceptive awareness, *Nature Neuroscience, 7*, 189–195.

De Gelder, J., Snyder, D., Greve, G. Gerard, & N. Hadijkhani, (2004). Fear fosters flight: A mechanism for fear contagion when perceiving emotion expressed by a whole body. *Proceedings of the National Academy of Sciences of the United States of America*, 47, 16701–16706.

Decety, L. (2005) Perspective taking as the royal avenue to empathy. In: B.F. Malle & S.D. Hodges (Eds.), *Other minds: How humans bridge the divide between self and other* (pp. 135–149), New York: Guilford Publications.

Decety, J., & Grèzes, J. (2006) Multiple perspectives on the psychological and neural bases of understanding other people's behavior. *Brain Research, 1079*, 4–14.

di Pellegrino, G., Fadiga, L., Fogassi, L., Gallese, V., & Rizzolatti, G. (1992). Understanding motor events: A neurophysiological study. *Experimental Brain Research, 91*, 176–180.

Dimberg, U., Thunberg, M., & Elmehed, K. (2000). Unconscious facial reactions to emotional facial expressions. *Psychological Science, 11*, 86–89.

Firth, C. D., & Firth, U. (2006). How we predict what other people are going to do. *Brain Research, 1079*, 36–46.

Furuyama, N., Hayashi, K., & Mishima, H. (2005). Interpersonal coordination among articulations, gesticulations, and breathing movements: A case of articulation of /a/ and

flexion of the wrist. In H. Heft & K. L. Marsh (Eds.). *Studies in perception and action* (pp. 45–48) Mahwah, NJ: Erlbaum.
Gallese, V. (2001). The "shared manifold" hypothesis: From mirror neurons to empathy. *Journal of Consciousness Studies*, *8*(5–7), 33–50.
Gallese, V., (2003). The manifold nature of interpersonal relations: The quest for a common mechanism. *Philosophical Transactions of the Royal Society London, Series B Biological Sciences*, *358*, 517–528.
Gallese, V. (2005a). Embodied simulation: From neurons to phenomenal experience. *Phenomenology and Cognitive Science*, *4*, 23–48.
Gallese, V. (2005b). From mirror neurons to the shared manifold hypothesis: A neurophysiological account of intersubjectivity. In S. T. Parker, J. Langer, & C. Milbrath (Eds.), *Biology and knowledge revisited: From neurogenesis to psychogenesis. The Jean Piaget symposium series* (pp. 179–203). Mahwah, NJ, US: Lawrence Erlbaum Associates Publishers.
Gallese, V. (2006). Intentional attunement: A neurophysiological perspective on social cognition and its disruption in autism. *Brain Research*, *1079*, 15–24.
Gallese, V., Fadiga, L., Fogassi, L., & Rizzolatti, G. (1996). Action recognition in the premotor cortex. *Brain*, *119*, 593–609.
Goldman, A. I. (2002). Simulation theory and mental concepts. In: J. Dokic & J. Proust, (Eds.), *Simulation and knowledge of action* (pp. 1–19), Amsterdam: John Benjamins Publishing Company.
Goldman, A. I. (2005). Imitation, mind reading, and simulation. In: S. Hurley & N. Chater (Eds.), *Perspective on imitation, from neuroscience to social science Vol. 2* (pp. 79–93). Cambridge: MIT Press.
Grafton, S. T., Arbib, M. A., Fadiga, L., & Rizzolatti, G. (1996). Localization of grasp representations in humans by PET: 2. Observation compared with imagination. *Experimental Brain Research*, *112*, 103–111.
Grèzes, J., & Decety, J. (2001). Functional anatomy of execution, mental simulation, observation, and verb generation of actions: A meta-analysis. *Human Brain Mapping*, *12*, 1–19.
Hutchison, W. D., Davis, K. D., Lozano, A. M., Tasker, R. R., & Dostrovsky, J. O. (1999). Pain-related neurons in the human cingulate cortex. *Nature-Neuroscience*, *2*, 403–405.
Iacoboni, M. (2005). Neural mechanisms of imitation. *Current Opinion in Neurobiology*, *15*, 632–637.
Iacoboni, M., Woods, R. P., Brass, M., Bekkering, H., Mazziotta, J. C., & Rizzolatti, G. (1999). Cortical mechanisms of human imitation. *Science*, *286*, 2526–2528.
Jabbi, M., Swart, M., & Keysers, C. (2007). Empathy for positive and negative emotions in the gustatory cortex. *Neuroimage*, *34*, 1744–1753.
Jackson, P.L., Meltzoff, A.N., & Decety, J., (2005). How do we perceive the pain of others: A window into the neural processes involved in empathy. *Neuroimage*, *24*, 771–779.
Jeannerod, M. (1999). The 25th Bartlett Lecture. To act or not to act: Perspectives on the representation of actions. *Quarterly Journal of Experimental Psychology: Human Experimental Psychology*, *52*, 1–29A.
Kendon, A. (1970). Movement coordination in social interaction: Some examples described. *Acta Psychologica*, *32*, 100–125.
Keysers, C., & Gazzola, V. (2006). Towards a unifying neural theory of social cognition. *Progress in Brain Research,* 156, 379–401.
Keysers, C., Kohler, E., Umiltà, M. A., Nanetti, L., Fogassi, L., & Gallese, V. (2003). Audiovisual mirror neurons and action recognition. *Experimental Brain Research*, *153*, 628–636.
Kohler, E., Keysers, C., Umiltà, M.A., Fogassi, L., Gallese, V., & Rizzolatti, G. (2002). Hearing sounds, understanding actions: Action representation in mirror neurons. *Science*, *297*, 846–848.
Lazarus, M. (1861). Über das Verhältnis des Einzelnen zur Gesamtheit. *Zeitschrift für Völkerpsychologie und Sprachwissenschaft*, *2*, 393–453.

Lazarus, M., & Steinhal, H. (1860). Einleitende Gedanken über Völkerpsychologie als Einladung zu für Völkerpsychologie und Sprachwissenschaft. *Zeitschrift für Völkerpsychologie und Sprachwissenschaft*, *1*, 1–73.

Lipps, T. (1903). Einfühlung, innere Nachahmung, und Organempfindungen. *Archiv für die gesamte Psychologiue*, *2*, 185–204.

Manthey, S., Schubotz, R. L., & von Cramon, D. Y. (2003). Premotor cortex in observing erroneous action: An fMRI study. *Cognitive Brain Research*, *15*, 296–307.

Marsh, K. L., Richardson, M. J., Baron, R. M., & Schmidt ,R.C. (2006). Contrasting approaches to perceiving and acting with others. *Ecological Psychology*, *18*, 1–38.

Meltzoff, A.N., & Moore, M.K. (1997). Explaining facial imitation: A theoretical model. *Early Development and Parenting, 6*, 179–192.

Meltzoff, A.N., & Moore, M.K. (1999). Persons and representation: Why infant imitation is important for theories of human development. In J.B.G. Nadel (Ed.), *Imitation in infancy: Cambridge Studies in Cognitive Perceptual Development* (pp. 9–39). New York: Cambridge University Press.

Neumann, R., & Strack, F. (2000). "Mood contagion": The automatic transfer of mood between persons. *Journal of Personality and Social Psychology*, *79*, 211–223.

Rizzolatti, G., & Arbib, M. A. (1998) Language within our grasp. Trends Neurosci., 21(5), 188–194.

Rizzolatti, G., & Craighero, L. (2004). The mirror-neuron system. *Annual Review of Neuroscience*, *27*, 169–192;

Rizzolatti, G., Fadiga, L., Gallese, V., & Fogassi, L. (1996). Premotor cortex and the recognition of motor actions. *Cognitive Brain Research*, *3*, 131–141.

Sachs, H. A, Schegloff, E. A., & Jefferson, G. (1974). A simplest systematics for the organization of turn-taking for conversation. *Language*, *50*, 696–735.

Semin, G. R., & Cacioppo, J. T. (in press). Grounding social cognition: Synchronization, entrainment, and coordination. In G.R. Semin & E.R. Smith (Eds.), *Embodied grounding: Social, cognitive, affective, and neuroscientific approaches*. New York: Cambridge University Press.

Shockley, K., Santana, M. V., & Fowler, C.A. (2003). Mutual interpersonal postural constraints are involved in cooperative conversation. *Journal of Experimental Psychology: Human Perception and Performance*, *29*, 326–332.

Shockley, K., Baker, A. A. Richardson, M. J., & Fowler, C. A. (2007). Articulatory Constraints on Interpersonal Postural Coordination. *Journal of Experimental Psychology: Human Perception and Performance*, *33*, 201–208.

Singer, T., Seymour, B., O'Doherty, J., Kaube, H., Dolan, R.J., & Frith, C. F. (2004). Empathy for pain involves the affective but not the sensory components of pain. *Science*, *303*, 1157–1162.

Stevens, J. A., Fonlupt, P., Shiffrar, M., & Decety, J. (2000). New aspects of motion perception: Selective neural encoding of apparent human movements. *NeuroReport*, *11*, 109–115.

Tickle-Degnen, L., & Rosenthal, R. (1987). Group rapport and nonverbal behavior. *Review of Personality and Social Psychology*, *9*, 113–136.

Umilta, M.A., Kohler, E., Gallese, E., Fogassi, L., Keysers, C., & Rizzolatti, G. (2001). I know what you are doing: a neurophysiological study. *Neuron, 31*(1), 155–65.

VanRullen, R., & Koch, C. (2003). Is perception discrete or continuous? *Trends in Cognitive Sciences*, *7*, 207–213.

Wicker, B., Keysers, C., Plailly, J., Royet, J.-P., Gallese, V., & Rizzolatti, G., (2003). Both of us disgusted in my insula: The common neural basis of seeing and feeling disgust. *Neuron*, *40*, 655–664.

Wilson, S. M., Saygin, A. P., Sereno, M. I., & Iacoboni, M. (2004). Listening to speech activates motor areas involved in speech production. *Nature Neuroscience*, *7*, 701–702.

The Problem of Other Minds Is Not a Problem: Mirror Neurons and Intersubjectivity

Marco Iacoboni

Abstract This chapter discusses evidence from single unit recordings in monkeys and brain imaging in humans suggesting that neural mechanisms of mirroring allow the sharing of mental states between individuals. Such sharing solves the old philosophical problem of other minds. Correlations between mirroring and empathy also suggest that relatively simple neurobiological mechanisms of mirroring maybe at the basis of a secular morality which is built upon biological predispositions we have inherited from our evolutionary ancestors.

Keywords Mirror neurons · Theory of mind · Simulation theory · Sociality · Imitation

1 Introduction

The things we take for granted, the things that go unnoticed, are the things we consider ordinary, the ones we do not give a second thought. Some of these things, however, are absolutely extraordinary, and they have baffled philosophers and psychologists for centuries. What I have in mind, in particular, is our effortless, automatic ability to understand the mental states of our fellows. Let me give you an example. Suppose I am at a social event that kicks off a scientific meeting. Suppose I am amiably talking with a colleague. It turns out that this particular colleague is a die-hard supporter of a scientific theory that is in complete opposition with the theory that seems strongly supported by my experimental work. For instance, this particular colleague is a believer in the theory-theory account of mindreading (more on this in the next section of the chapter). Now the colleague grasps a glass of wine. How do I know he is not

---

M. Iacoboni
Ahmanson-Lovelace Brain Mapping Center, Department of Psychiatry and Biobehavioral Sciences, Semel Institute for Neuroscience and Human Behavior, Brain Research Institute, David Geffen School of Medicine at UCLA, Los Angeles, CA, USA
e-mail: iacoboni@ucla.edu

J.A. Pineda (ed.), *Mirror Neuron Systems*, DOI: 10.1007/978-1-59745-479-7_6,

going to throw the glass at me? Or, when I see a research assistant in the lab getting a tool, how do I know that he is going to work on the Transcranial Magnetic Stimulator, rather than using it against me?

We literally make tons of these decisions every day of our life. We maybe wrong once in a while, but we tend to predict our fellows' behavior and to understand their mental states pretty well, continuously, and with no effort. How do we do it? The main thesis of this chapter is that we use relatively simple neural mechanisms of mirroring when we interpret the intentions associated with everyday actions of our fellows that we happen to observe. The understanding of the intentions associated with those actions allows us to anticipate what our fellow will do next, allows the prediction of the actions we have not seen yet. Before we dive into the neuroscience details of these mirroring mechanisms, I believe we should briefly review the philosophical and theoretical arguments made over the centuries on how we understand the minds of other people.

2 The Problem of Other Minds

The most dominant view in thinking about the mind – at least in Western culture – originates from a position that most thinkers believe going back to the French philosopher Descartes. This position looks at the starting point of the mind as the solitary, private, individual act of thinking, the famous *cogito* of *cogito, ergo sum*. If one accepts these premises, all sorts of problems arise. A famous one, is called the *problem of other minds*. Several philosophers, among them Wittgenstein, existential phenomenologists, and some Japanese philosophers, have considered the problem of other minds not really a problem, but actually a pseudo-problem. But, first of all, what is the problem of other minds?

Intersubjectivity, the sharing of meaning between people, has always been perceived as a problem in classical cognitivism. The problem of intersubjectivity, or the problem of other minds, is as follows: if I have only access to my own mind, which is a very private entity I can only access, how can I possibly understand the minds of other people? How can I possibly share the world with others, how can people possibly share their own mental states?

One classical solution to this problem has been provided by what is called the *argument from analogy*. The argument from analogy goes as follows: I only have access to one mind, which is my own mind. My access to the minds of other people must be mediated by what they do. The behavior of other people must somehow be the clue to understand what goes on in their minds. But how can it possibly work? Well, maybe the use of an *analogy* can help. If I analyze my own mind and its activity in relation with my own body and its actions, I realize that there are some links between my mind and my body. If I am nervous, I may sweat even though it is not hot. If I am in pain, I may scream. I then look at the other person and find an *analogy* between the body of the other person and my own body. If there is an analogy between my own body and somebody else's body, there may also be an analogy between the other person's body and the other person's mind. So, if I see the other person sweating when it is not hot, I may conclude that the other person is nervous. If I see the other person screaming, I may conclude that the other person is in pain. Although this kind of analogy does not allow me to be completely positive about the mental states of other people, and it does not allow me to share with them their feelings and experiences, it certainly does allow me to conclude with reasonable certainty that people have minds like my own.

The argument from analogy has been heavily criticized by some past thinkers. One of the most common criticisms is that this kind of reasoning is way too complex for something we seem to accomplish so naturally, effortlessly, and quickly. In more recent years, philosophers of mind have been joined by cognitive scientists, psychologists, and recently even neuroscientists in the attempt to understand how we 'read the mind' of other people. As of today, I would say that there are two main theoretical positions. One, called *theory theory*, assumes that we understand the mental states of other people by using a set of causal laws linking together human behavior and internal mental states. People would understand the mental states of others using an inferential approach not dissimilar to the one advocated by the argument from analogy, but more like a scientist that tries to understand a physical phenomenon on the basis of some data and some well-known physical laws.

The main theoretical camp opposite to theory theory is called *simulation theory*. According to simulation theory, we understand the mental states of other people by literally putting ourselves in their shoes, by imagining what we would do if we were in their situation. No knowledge of a set of causal laws linking behavior and mental states and not a lot of inferential thinking are required. There are obviously more nuanced positions within the two opposing camps of theory theory and simulation theory (see for instance the recent book by Alvin Goldman, Simulating Minds) (Goldman, 2006). However, the two main theoretical positions are captured well enough, for our purposes, by these simple descriptions.

Theory theory was the dominant view in the eighties and early nineties. At some point, however, things changed. More and more scholars became convinced that simulation theory was best equipped to deal with empirical findings from the neuroscience, especially from the discovery of mirror neurons in macaques and of a mirror neuron system in humans. The next sections of the chapter will discuss these discoveries.

3 Mirror Neurons and Hidden Actions

As also discussed in other chapters of the book, mirror neurons are cells in the ventral premotor cortex and inferior parietal cortex of the macaque brain that fire when the monkey performs goal-oriented hand and mouth actions (grasping, holding, biting, sucking, etc.) and communicative mouth actions (such as lipsmacking), but also when the monkey observes somebody else performing those very same actions (Gallese, Fadiga, Fogassi, & Rizzolatti, 1996; Ferrari, Gallese, Rizzolatti, & Fogassi, 2003; Rizzolatti & Craighero, 2004; Iacoboni & Dapretto, 2006). The hypothesis that these cells were key neural mechanisms for understanding not only the actions of other people, but also the mental states associated with those actions, was first proposed by Gallese and Goldman, that obviously mapped the neural properties of mirror neurons onto some form of simulation of the actions and mental states of others (Gallese & Goldman, 1998).

There are at least two major experimental attempts to link mirror neuron activity in monkeys with the coding of the intentions of other individuals. The first experiment was led by Alessandra Umiltà – now a faculty at the University of Parma – when she was a graduate student in the lab of Giacomo Rizzolatti. In one experimental condition, the monkey observes a human experimenter grasping an object. As expected, mirror neurons do fire at the sight of a grasping action. In the other condition, the monkey observes a human experimenter pantomiming a grasping action, in absence of a real object to be grasped. In this case, the sight of a pantomime does not trigger a discharge in mirror neurons. In a third experimental condition, the monkey sees an object (say, an orange) on the table. Subsequently, a screen is placed in front of the orange such that the monkey cannot see the orange anymore. After the screen has occluded the sight of the orange, a human experimenter reaches with her right hand behind the screen, where the orange is located. However, because the screen occludes the sight of the monkey, the grasping action is not visible. The question here is: do mirror neurons fire when the grasping action is actually hidden? The answer is yes; approximately 50% of the mirror neurons tested by Alessandra Umiltà and her colleagues did fire even though the monkey had not seen the completion of the grasping action, occluded by the screen.

In the final experimental condition, the monkey sees that there is no object (the orange) on the table. Subsequently, the screen is moved as in the previous condition, such that it occludes the sight of the table. After this, a human experimenter reaches with her right hand behind the screen. Note that – from a visual standpoint – at this point this experimental condition is identical to the previous one. The only difference between the two conditions is the prior knowledge about the presence of an object (the orange) on the table. The question here is: do mirror neurons fire, as if the human experimenter is actually grasping something behind the screen, or do mirror neurons remain silent, as if the human experimenter is only pantomiming a grasp behind the screen? The answer is: mirror neurons do not fire. The prior knowledge of the *absence* of an object on the table is sufficient for the cell to consider the hidden action – visually identical to the previous hidden action that generated the firing of the mirror neuron – not a hidden grasping action but simply a hidden pantomime (Umiltà et al., 2001).

This experiment clearly shows that mirror neurons do not simply form a neural system matching performed actions and observed ones. Mirror neurons can provide a nuanced coding of the actions of others, using prior information to differentiate the meaning of partially occluded actions that are visually identical. Is this sufficient evidence in support of the idea that mirror neurons code the intentions of the person grasping the object? Probably not. After all, to differentiate between the two hidden actions one needs only to take into account *prior information* such that under one condition the human action is categorized as 'grasping,' whereas in the other condition the human action is categorized as 'not grasping.' This experiment, then, does not fully address the fundamental question of whether mirror neurons can differentiate between, say, grasping to eat versus grasping to place the food in the refrigerator. This is why Leo Fogassi led – some years after Alessandra Umiltà's experiment – another experiment directly investigating the role of mirror neurons in understanding the intentions of other people.

4 Mirror Neurons and Intentions

The surprising properties of mirror neurons suggest that the brain of primates has neurons in its motor regions that are helpful for recognizing the actions of other people. The neurons performing the recognition of the actions of others are the same ones we use to perform those actions ourselves. This suggests that the action recognition process implemented by mirror neurons is some sort of simulation or *internal imitation* of the actions of other people. That is, the same neurons I use in my brain to send signals to my muscles when I grasp a cup, are firing when I simply see you grasping a cup. Given that our actions are almost invariably associated with very specific intentions, the activation in my brain of the same neurons I use to perform my own actions when I see other people performing their own actions, may also allow me to understand the intentions of other people while I watch their actions. However, there is a problem, because the same action can be associated with different intentions. How do mirror neurons differentiate between the same action associated with different intentions?

A recent experiment led by Leo Fogassi addressed this question quite directly (Fogassi et al., 2005). Here, monkeys' neural activity was measured during a variety of grasping execution and grasping observation conditions. In one of the grasping execution conditions, the monkey – starting from a fixed position – reached for and grasped a piece of food located in front of the animal and brought the food to the mouth to eat it. In another grasping execution condition, the monkey reached for an object – located where the food was located in the previous experimental condition. After grasping it, the monkey placed the object in a container. In this condition, several trials were performed while the container was located close to the mouth of the animal, such that the arm and hand movements of the grasping-to-eat and grasping-to-place conditions would be closely matched. The main question in comparing these two conditions was whether cells that fire while the monkey is grasping, fire differently if the grasping action will lead to eating food as opposed to placing the object in the container. The grasping action is the same action, but the intentions associated with the actions are quite different. Note also that after completing the placing-in-the-container trials, monkeys were rewarded, such that the amount of reward for grasping-to-eat and grasping-to-place was identical.

Between a third and a fourth of the recorded neurons fired equivalently during grasping-to-eat and grasping-to-place. The remaining neurons, however, that is, the majority of the recorded cells, fired differently for the same grasping action associated with different intentions. About 75% of these cells discharged more vigorously during grasping actions followed by bringing the food to the mouth in order to eat, compared to grasping actions followed by placing the object in the container. The remaining 25% of cells demonstrated the opposite pattern, with a more robust discharge for grasping actions associated with the intention to place, compared to grasping actions associated with the intention to eat.

Was it possible that the differential discharge was due simply to the fact that in one condition the monkey was grasping food, whereas in the other condition the animal was grasping an object? To test for this possibility, the neural activity of the monkey was also tested under the condition of 'food placing' and this condition was compared to the condition of 'food eating,' as described above. In the 'food placing' conditions, monkeys were trained to grasp the food and put it in the container. This was achieved by showing the monkey one of her favorite pieces of food before the trial started, so that the monkey was motivated to place the food in the container in order to receive her favorite food as reward for completing the trial. This new experimental condition, when compared to the 'food eating' condition, demonstrated that the differential responses previously observed between motor neurons coding grasping actions were not due to differences in the type of object grasped. The results previously observed were confirmed again. The majority of cells discharged preferentially during grasping for eating, a minority preferred grasping for placing, and the cells that had not shown any 'intention' effect still discharged equivalently during grasping for eating and grasping for placing.

The experimenters then proceeded to test the visual properties of these cells. Here, one of the experimenters was seated in front of the monkey and was performing grasping actions as the ones the monkey had previously performed. There were two main experimental conditions: When a container was present and visible to the monkey, the experimenter grasped the food and placed it in the container. When the container was not present, the experimenter grasped the food, brought it to the mouth, and ate it. Thus, the presence of the container acted as a visual clue allowing the monkey to predict the next movement of the experimenter. The empirical question was whether the discharge of the mirror neurons occurring while the monkey observed the experimenter grasping the food was similar or different in the two conditions of observing grasping-to-eat or observing grasping-to-place. The results demonstrated that even when the monkey was completely still and was simply observing the experimenter grasping the food, at the moment of the grasping action mirror neurons fired differently at the sight of the same grasping action associated with different intentions. In fact, the pattern of neuronal firing during grasping observation *mirrored* the pattern of neuronal firing measured during grasping execution. That is, if a cell discharged more vigorously while the monkey was grasping the food in order to eat, that very same cell discharged more vigorously also while the monkey was observing the human experimenter grasping the food in order to eat it. Also, if a cell discharged more vigorously while the monkey was grasping the food in order to place it in the contained, that very same cell discharged more vigorously while the monkey was observing the human experimenter grasping the food in order to place it in the container. Finally, if a cell discharged equally while the monkey was grasping-to-eat and grasping-to-place, that very same cell discharged equally also while observing the human experimenter grasping the food in order to eat it or in order to place it in the container.

The results of Fogassi's experiment demonstrate that the coding of the actions of other people provided by mirror neurons is much more sophisticated than initially thought. Although Vittorio Gallese and Alvin Goldman had speculated quite early on – after the initial discovery of mirror neurons – that these cells may provide a key neural mechanism for understanding the mental states of others, until Fogassi's experiment there was much more support in the scientific community for a more parsimonious account of the functions of mirror neurons, such that these cells would simply provide a form of action recognition. The experiment by Fogassi and colleagues clearly supports the initial intuition of Gallese and Goldman. Mirror neurons let us understand the intentions of other people.

Leo Fogassi's study provides robust empirical evidence in support of the hypothesis that we understand the mental states of others by simulating them in our brain. The empirical study of intention understanding has always been considered almost impossible to achieve, intentions typically being considered 'too mental' to be studied with empirical tools. The research on mirror neurons is giving the neuroscience community the opportunity to tackle experimentally these slippery issues for the first time. The fact that mirror neurons code differently the same grasping action associated with different intentions both when we do it and when we see somebody else doing it, suggests that at the fine-grained level of a single cell our brains are capable of mirroring the deepest aspects of the minds of others.

5 The Tea Party Experiment

But what about humans? Obviously, in humans we cannot easily study single-cell activity, although this is now possible when patients with epilepsy are implanted with depth electrodes in order to localize the epileptic focus of activity that needs to be surgically removed when the seizures are not well controlled pharmacologically. Typically, however, what is used in humans to study the mirror neuron system is functional magnetic resonance imaging (fMRI), Transcranial Magnetic Stimulation (TMS), magnetoencephalography (MEG), and electronecephalography (EEG). In my lab, we have used both TMS and fMRI to study the human mirror neuron system and its relations with imitation and a variety of aspects of social cognition (Iacoboni & Dapretto, 2006; Iacoboni et al., 1999; Koski et al., 2002; Heiser, Iacoboni, Maeda, Marcus, & Mazziotta, 2003; Koski, Iacoboni, Dubeau, Woods, & Mazziotta, 2003; Carr, Iacoboni, Dubeau, Mazziotta, & Lenzi, 2003; Iacoboni, 2005; Uddin, Kaplan, Molnar-Szakacs, Zaidel, & Iacoboni, 2005; Aziz-Zadeh, Wilson, Rizzolatti, & Iacoboni, 2006; Aziz-Zadeh, Koski, Zaidel, Mazziotta, & Iacoboni, 2006; Dapretto et al., 2006). The way we investigated with a brain imaging experiment those slippery things called intentions is conceptually similar to the way it was done in the experiment performed in monkeys by Fogassi and colleagues. Indeed, our brain imaging experiment on understanding intentions with mirror neurons (Iacoboni et al., 2005) preceded and inspired Fogassi's experiment in monkeys on intention understanding. This experiment has become quite famous, and it is typically described almost invariably when people talk about mirror neurons. I heard people calling it the Tea Party experiment. Why? Our starting point was that the same action can be associated with different intentions. One can grasp a cup for many reasons. Among these, the two most common ones are probably to drink, or to put the cup in the dishwasher, to clean up. The context in which the action occurs often gives clues to the observer with regard to the intention associated with the action. For instance, if we are having a tea party at my house that had just started, and I see my wife reaching for a cup of tea, it is likely that she will drink from it. However, if all the guests are gone and the party is over, when I see my wife reaching for the cup, it is likely that she will put it in the dishwasher. True, she could have a last sip. However, this outcome is less likely than putting the cup in the dishwasher, on the basis of the context in which my wife's action is embedded. Indeed, the outcome of putting the cup in the dishwasher in the other scenario, when the tea party had just started, is *highly* unlikely.

If mirror neurons respond only to the grasping action, it does not really matter which context surrounds them. Indeed, it does not really matter at all that there is a context surrounding the action. A grasp is a grasp, with or without context. If mirror neurons, on the other hand, respond to the intention associate with the observed action – which basically means that they can predict the unseen actions that follow the observed one, that is, bringing the cup to the mouth for the drinking intention and putting the cup in the dishwasher for the cleaning intention – the presence of the context, and the *type* of context, should matter and should influence the activity of mirror neurons. Following this logic, we designed a brain imaging experiment in which subjects saw a series of videoclips (see Fig. 1). One type of videoclip, that we called among ourselves *Context* (we did not name the videoclips in front of our subjects), had a scene

Fig. 1 The experimental conditions of the Tea Party experiment. During the Action and the Intention condition, both precision grips (here shown in the *bottom row*) and whole hand prehension (*top row*) were presented in a counterbalanced fashion. From Iacoboni et al. (2005)

with some objects in it: a teapot, cookies, a mug, and so on. We made these clips with a little bit of Italian touch, and included also Nutella – the tasty Italian hazelnut-based spread – among the various objects. Subjects saw two different kinds of *Context* clips. In one case, everything was neatly organized, suggesting that somebody was going to have a tea. In the other case, the scene was quite messy, there were cookie crumbs, and a dirty napkin, and so on, thus suggesting that somebody already had tea.

A second type of clip, that we called *Action*, simply displayed a hand grasping a cup, in absence of any context. Different kinds of grasping actions were presented, but what happened after the hand grasped the cup was not shown. In the third type of clip, that we called *Intention*, we put together the elements forming the first two types of videoclips. That is, subjects watched a hand grasping a cup, as in the *Action* clip, but this time the grasping action was embedded in a scene that suggested a context, as in the *Context* clips. One type of context clearly suggested drinking (the neat one), and the other type of context suggested cleaning up (the messy context). The predictions related to this experiment were relatively simple: if mirror neurons simply code the observed grasping action, activity in mirror neuron areas should be equivalent for the *Action* clip and for the *Intention* clips. If mirror neurons, on the other hand, code the intentions associated with the observed action, thus predicting the actions that would follow the observed one, activity in mirror neuron areas should be different between the *Action* clip and the *Intention* clips, and possibly it should also be different between the two *Intention* clips.

When we looked at the brain imaging data, we immediately realized that the results supported the hypothesis that mirror neurons code the intentions associated with the observed action, not simply the action. Indeed, there was higher activity in the frontal mirror neuron area depicted in Fig. 2 while subjects observed the grasping actions embedded in the two contexts, compared to observing the grasping action without any context. There was also higher activity in the frontal mirror neuron area while subjects observed the grasping action embedded in the context that suggested drinking, compared to the grasping action embedded in the context that suggested cleaning up. This result also made sense, since drinking is a much more primary intention than cleaning up.

The fact that the brain cells we use to achieve our own intentions by grasping and subsequently drinking or cleaning, also fire up when we discern between different intentions associated with the actions of other people clearly supports the simulation model of our ability of understanding the intentions of other people. The form of simulation supported by mirror neurons is likely the automatic, effortless form of simulation. Indeed, mirror neurons are cells located in the part of the bran that is important for motor behavior, close to the primary motor cortex sending electric signals directly to our muscles. This type of cell seems to have no business with a deliberate, effortful, and cognitive pretense of being in somebody else's shoes. Indeed, we found identical activity in the frontal mirror neuron area coding intention in subjects that were explicitly asked to understand the intentions associated with the observed grasping actions and in subjects that were not explicitly asked to do so (Iacoboni et al., 2005).

But how do mirror neurons actually predict the action that follows the observed one, to let us understand the intention associated with it? Some mirror

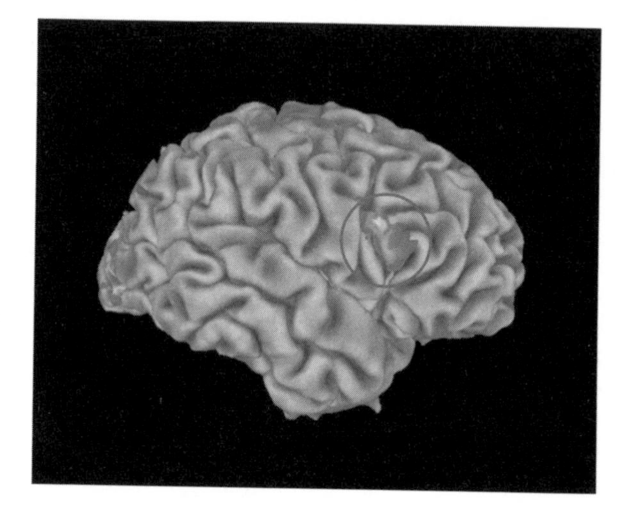

Fig. 2 The right frontal mirror neuron area (located in the dorsal sector of pars opercularis of the inferior frontal gyrus) showing activity compatible with intention coding. From Iacoboni et al. (2005)

neurons, called *logically related mirror neurons*, fire not for the same observed and performed action, but for logically related ones, such as 'grasping with the hand' and 'bringing to the mouth' (di Pellegrino, Fadiga, Fogassi, Gallese, & Rizzolatti, 1992). Logically related mirror neurons are likely key neuronal elements for understanding the intentions associated with the observed action. By means of the activation of a neuronal chain, these cells can simulate a whole sequence of simpler actions. The sequential activation of mirror neurons coding these simpler actions – reaching for the cup, grasping it, bringing it to the mouth – is the simulation in our brain of the intention of the fellow human we are watching. Mirror neurons help us re-enacting in our brain the intentions of other people, thus allowing a profound understanding of others. Are mirror neuron areas a bio-marker of our ability to empathize with others? The next section of the chapter addresses this question directly.

6 Grasping Intentions and Empathy

In my lab, we have performed a series of experiments on how the mirror neuron system interfaces with limbic areas to provide a simulation-based form of understanding the emotional states of others expressed through facial emotional expressions (Carr et al., 2003; Dapretto et al., 2006). Recently, however, we have asked a simpler question. Does mirror neuron activity evoked by the observation of simple, non emotional, everyday actions such as grasping a cup, map onto the ability of individuals to empathize with others?

Jonas Kaplan led this recent fMRI experiment (Kaplan & Iacoboni, 2006), using once again the stimuli depicted in Fig. 1. Jonas also measured subjects' empathic abilities with a well tested scale, the Interpersonal Reactivity Index (Davis, 1983). This empathy scale is composed of four subscales, two of them measuring cognitive empathy and the other two subscales measuring emotional empathy. The cognitive empathy scales assessed the ability to imagine another person's perspective and the tendency to imagine oneself in the place of fictional characters. The emotional empathy scales assessed the tendency to be concerned for the emotions of others and the emotional response the observer experiences when watching somebody else feeling strong emotions. While three out of four subscales demonstrated correlations with mirror neuron activity, the most cognitive of the subscales, the one that measures the ability to explicitly imagine another person's perspective failed to show any correlation with mirror neuron areas. This result suggests that the kind of empathy supported by mirror neuron activity is likely pre-reflective and automatic. This result also implicitly suggests that the kind of simulation process implemented by mirror neurons that help us understanding the intention of others, is also likely pre-reflective and automatic. In his experiment, Jonas also addressed an issue that was left unanswered by the previous study on understanding intentions. The way we grasp a cup, can reveal at least in part our intentions.

Especially when we grasp to drink, we tend to grasp the cup from the handle using the index finger and the thumb, in the type of grasping action called precision grip. The original Tea Party experiment was performed with a blocked design, in which precision grips and whole hand prehension (grasping the body of the cup with the whole hand) were counterbalanced across experimental blocks. In his follow up experiment, Jonas Kaplan used a single trial design that allowed him to compare activity in mirror neuron areas while subjects were watching the cup grasped with a precision grip or with a whole hand prehension in the drinking context. As predicted by a strong simulation account of the coding of the actions of other people, the mirror neuron area depicted in Figure 2 responded more strongly while subjects observed the cup grasped with precision grip in the drinking context.

7 Conclusions

In this chapter, I reviewed some recent evidence that suggests that neural mechanisms of mirroring likely implement a simulation-based form of intention understanding. Mirror neurons were likely selected during the evolutionary process because they provide the adaptive advantage of understanding the mental states of other people in an effortless and automatic way. We have also seen that the human mirror neuron system is correlated with the tendency to empathize with other people. This neural system seems to be the cornerstone of social competence, and may represent one of the first neuroscience evidence in support of the concept of a secular morality, a morality built bottom up from relatively simple neurobiological mechanisms for sensory-motor behavior that we have inherited from our evolutionary ancestors.

Acknowledgments For generous support the author wish to thank the Brain Mapping Medical Research Organization, Brain Mapping Support Foundation, Pierson-Lovelace Foundation, The Ahmanson Foundation, William M. and Linda R. Dietel Philanthropic Fund at the Northern Piedmont Community Foundation, Tamkin Foundation, Jennifer Jones-Simon Foundation, Capital Group Companies Charitable Foundation, Robson Family and Northstar Fund.

References

Aziz-Zadeh, L., Koski, L., Zaidel, E., Mazziotta, J., & Iacoboni, M. (2006). Lateralization of the human mirror neuron system. *Journal of Neuroscience, 26*, 2964–2970.

Aziz-Zadeh, L., Wilson, S. M., Rizzolatti, G., & Iacoboni, M. (2006). Congruent embodied representations for visually presented actions and linguistic phrases describing actions. *Current Biology, 16*, 1818–1823.

Carr, L., Iacoboni, M., Dubeau, M. C., Mazziotta, J. C., & Lenzi, G. L. (2003). Neural mechanisms of empathy in humans: A relay from neural systems for imitation to limbic areas. *Proceedings of National Academy of Science U S A, 100*, 5497–5502.

Dapretto, M., Davies, M. S., Pfeifer, J. H., Scott, A. A., Sigman, M., Bookheimer, S. Y., et al. (2006). Understanding emotions in others: Mirror neuron dysfunction in children with autism spectrum disorders. *Nature Neuroscience, 9*, 28–30.

Davis, M. H. (1983). Measuring individual differences in empathy: Evidence for a multidimensional approach. *Journal of Personality & Social Psychology, 44*, 113–126.

di Pellegrino, G., Fadiga, L., Fogassi, L., Gallese, V., & Rizzolatti, G. (1992). Understanding motor events: A neurophysiological study. *Experimental Brain Research, 91*, 176–180.

Ferrari, P. F., Gallese, V., Rizzolatti, G., & Fogassi, L. (2003). Mirror neurons responding to the observation of ingestive and communicative mouth actions in the monkey ventral premotor cortex. *European Journal of Neuroscience, 17*, 1703–1714.

Fogassi, L., Ferrari, P. F., Gesierich, B., Rozzi, S., Chersi, F., & Rizzolatti, G. (2005). Parietal lobe: From action organization to intention understanding. *Science, 308*, 662–667.

Gallese, V., Fadiga, L., Fogassi, L., & Rizzolatti, G. (1996). Action recognition in the premotor cortex. *Brain 119*(Pt 2), 593–609.

Gallese, V., & Goldman, A. (1998). Mirror neurons and the simulation theory of mind-reading. *Trends in Cognitive Science, 2*, 493–501.

Goldman, A. (2006). *Simulating minds: The philosophy, psychology, and neuroscience of mind-reading*. New York, NY: Oxford University Press.

Heiser, M., Iacoboni, M., Maeda, F., Marcus, J., & Mazziotta, J. C. (2003). The essential role of Broca's area in imitation. *European Journal of Neuroscience, 17*, 1123–1128.

Iacoboni, M. (2005). Neural mechanisms of imitation. *Current Opinion in Neurobiology, 15*, 632–637.

Iacoboni, M., & Dapretto, M. (2006). The mirror neuron system and the consequences of its dysfunction. *Nature Review Neuroscience, 7*, 942–951.

Iacoboni, M., Molnar-Szakacs, I., Gallese, V., Buccino, G., Mazziotta, J. C., & Rizzolatti G. (2005). Grasping the intentions of others with one's own mirror neuron system. *PLoS Biology, 3*, e79.

Iacoboni, M., Woods, R. P., Brass, M., Bekkering, H., Mazziotta, J. C., & Rizzolatti G. (1999). Cortical mechanisms of human imitation. *Science, 286*, 2526–2528.

Kaplan, J. T., & Iacoboni, M. (2006). Getting a grip on other minds: Mirror neurons, intention understanding and cognitive empathy. *Society for Neuroscience, 1*, 175–183.

Koski, L., Iacoboni, M., Dubeau, M. C., Woods, R. P., & Mazziotta, J. C. (2003). Modulation of cortical activity during different imitative behaviors. *Journal of Neurophysiology, 89*, 460–471.

Koski, L., Wohlschläger, A., Bekkering, H., Woods, R. P., Dubeau, M. C., Mazziotta, J. C., et al. (2002). Modulation of motor and premotor activity during imitation of target-directed actions. *Cerebral Cortex, 12*, 847–855.

Rizzolatti, G., & Craighero, L. (2004). The mirror-neuron system. *Annual Review of Neuroscience, 27*, 169–192.

Uddin, L. Q., Kaplan, J. T., Molnar-Szakacs, I., Zaidel E., & Iacoboni, M. (2005). Self-face recognition activates a frontoparietal "mirror" network in the right hemisphere: An event-related fMRI study. *Neuroimage, 25*, 926–935.

Umiltà, M. A., Kohler, E., Gallese, V., Fogassi, L., Fadiga, L., Keysers, C., et al. (2001). I know what you are doing. A neurophysiological study. *Neuron, 31*, 155–165.

Hierarchically Organized Mirroring Processes in Social Cognition: The Functional Neuroanatomy of Empathy

Jaime A. Pineda, A. Roxanne Moore, Hanie Elfenbeinand, and Roy Cox

Abstract Mirroring-like processes occur at all levels of information processing in the central nervous system, producing a gradient of faculties that vary in complexity from stimulus enhancement, response facilitation, emotional contagion, mimicry, simulation, and emulation to imitation, empathy, and theory of mind. These processes reflect the range of mechanisms for organizing the hierarchical representations of information. Many of these processes occur in other species and thus there is evidence for phylogenetic continuity. The organizational structure of mirroring is more functional than anatomical in that it is dynamic and modifiable at the level of the individual. Furthermore, the increased interconnectivity and complexity of the human brain provides the substrate for more complex forms of mirroring, such as empathy. Hence, in the pathways from perception to action, mirroring provides the foundational basis for social cognition.

Keywords Phylogenetic continuity · Theory of mind · Insula · Cingulate cortices · Somatosensory cortices · Amygdala · Mirror neuron system · Self-other

1 Introduction

Many investigators have recognized that humans are an especially prosocial species (Batson et al., 1988; Silk et al., 2005; Henrich et al., 2005) exhibiting behaviors such as empathy and altruism, the roots of which have been difficult to explain. Empathy is the glue that makes human society possible. Furthermore, it is the glue that arises not just from our rational organs but from foundational mechanisms at every level of information processing activated by our social

J.A. Pineda
Department of Cognitive Science and Neurosciences Program, University of California, San Diego, La Jolla, CA 92093-0515, USA
e-mail: pineda@cogsci.ucsd.edu

J.A. Pineda (ed.), *Mirror Neuron Systems*, DOI: 10.1007/978-1-59745-479-7_7,

interactions. Many attempts to define empathy have put emphasis on behavioral and/or affective responses. For example, empathy has been thought of as "an affective response more appropriate to someone else's situation than to one's own" (Hoffman, 2000), or "a sense of similarity between the feelings one experiences and those expressed by others" (Decety & Jackson, 2004), or "a complex form of psychological inference in which observation, memory, knowledge, and reasoning are combined to yield insights into the thoughts and feelings of others" (Ickes, 1997). Although the construct of empathy has been difficult to define in the past, the discovery of systems that provide a neural substrate for mirroring processes has allowed for substantial progress to be made by suggesting a plausible unifying mechanism for inferring and experiencing what another feels by simulating it through a shared self-other representation.

Current definitions of empathy can now be grounded in neurobiology, making the construct more tractable and testable. A strong trend in this regard has been toward viewing empathy as a monolithic response involving higher-order cognition. In a recent review (Blair, 2005), it was argued that this is a misleading perspective and that empathy should be viewed "not as a unitary system but rather as a loose collection of partially dissociable neurocognitive systems." Blair makes the argument for the existence of at least three such systems, namely cognitive empathy, motor empathy, and emotional empathy. This is an argument based primarily on the empathic dysfunctions experienced by two groups of individuals – those diagnosed with autism and those with psychopathy. Blair's view differs from accounts in social psychology and comparative ethology that offer a more nuanced and fractured system. For example, Preston and de Waal view empathy as "a super-ordinate category that includes all sub-classes of phenomena that share the same mechanism and includes emotional contagion, sympathy, cognitive empathy, helping behavior, etc." (Preston & de Waal, 2002). The view taken in this chapter is a combination of these ideas. Empathy is seen as a super-ordinate construct incorporating cognitive, motor, and emotive subcomponents in a loose collection of partially dissociable systems that evolve out of a rich repertoire of more fundamental processes, such as emotional contagion and imitation (see Box 1 for definitions). Empathy itself may provide some of the foundation for even more complex processes, such as theory of mind.

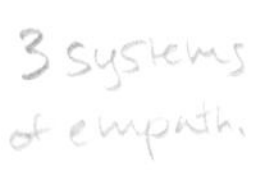

BOX 1. Definition of terms

emotional contagion:	the involuntary adoption of another's posture/mannerism/emotion.
emulation:	the increased ability to reproduce an observed goal (not the kinematics of the movement).
imitation:	reproduction of the exact goal and precise kinematics of movement, based on priming, such that a significant elevation in the frequency of an observed action occurs.

mimicry:	adopting the same behavior or emotion that is observed.
mirroring:	process that allows direct understanding of the meaning of actions and emotions of others by internally replicating or simulating them without requiring explicit reflective mediation.
priming:	increased accessibility to a representation based on prior activation of that representation.
response facilitation:	the increased likelihood of voluntarily or involuntarily performing the same action following observation of that action.
simulation:	same as mirroring but applicable when 'observed' object is not present.
stimulus enhancement:	the increased tendency to pay attention to a place or object after observing actions at that place or in conjunction with those objects.
social contagion:	an increase in the motivational homogeneity of a group increasing the tendency to perform the same behaviors at the same time.
theory of mind:	ability to represent one's thoughts, beliefs, knowledge, and internal processes, as well as those of others; often depends on empathic processes.

2 Characteristics of Empathy

Frans de Waal and Stephanie Preston have argued persuasively that the roots of empathy can be traced at least to our closest living relatives, the chimpanzees. Indeed, a number of different studies show that humans and other animals respond empathically in a similar way to a conspecific they have encountered previously and thus to the perceived overlap (e.g., familiarity or similarity) between the construct of self and other (Preston & de Waal, 2002). One problem with this viewpoint is that it does not account for the limited and qualitatively different processes seen in chimps compared to humans. More precisely, although nonhuman primates exhibit empathy, they do not exhibit the range of empathy seen in humans. Thus, an alternative perspective assumes fundamental differences between humans and other primates in their empathic responses. Among these emergent properties in humans and human societies are the capacity for imitation and a well-developed theory of mind (Tomasello, Call & Hare, 2003; Bowles & Gintis, 2004; Povinelli & Eddy, 1996).

Attempts to reconcile phylogenetically continuous versus discontinuous explanations of empathy are rare. Nonetheless, answers to the following questions might provide a beginning: (1) how are human empathy, theory of mind, imitation, mimicry, and related processes linked? and (2) what are the constitutive components of empathy? These issues motivate the discussion in the rest of the chapter.

2.1 Phylogenetic Continuity

One clue that may provide an organizational principle, and therefore a link, among empathy, theory of mind, imitation, mimicry, and related processes is their Presumed phylogenetically continuous characteristic (Decety, Chaminade, Grezes, & Meltzoff, 2002). It can be argued that successful integration into a social group requires the need for not only effective but affective communication, something that characterizes communication in a variety of animals (Preston & de Waal, 2002). In environments with ever increasing social complexity, as seen in many large primate groups, it becomes important to be able to communicate information as well as the emotional context of that information with other members of the group. Empathy appears to enhance social functioning by tuning our response to the emotions of another (Smith, 2006). Hence, knowing the emotional state of another, and making decisions based on that understanding, facilitates the most advantageous response to the situation, whether fleeing or joining in a fight.

Before the evolution of language, nonverbal communication was the method of transmission of information, as evidenced by the continued importance of 'body language.' One often cited example is the expression of taste aversion whereby the expressed disgust of another upon eating some food substance leads to avoidance of that food in observing individuals (Blair, 2005). Evidence for this learned attraction and aversion comes from the many species of social eaters (for a review, see Visalberghi, 1994). Internalizing the approach and avoidance behaviors of others would have served as a strong driving force for the increasing state of social competence in humans. Early in social development, orienting to the signs of distress of conspecifics would have been highly advantageous, if not utterly necessary (Donald, 1991).

Emotion is a universal human experience, though its expression varies widely among cultures (Ekman, 1972). Six emotions with universal facial expressions have been recognized in humans: anger, disgust, fear, happiness, sadness, and surprise (Ekman, 1972, 1973). These expressions are homologous to phylogenetically older facial expressions, such as the wide-eyed fear grimace in monkeys and in humans (Van Hooff, 1973). Chimpanzees are able to discriminate Ekman's basic categories, a finding that argues for the phylogenetic continuity of these facial expressions and for their importance to our closest primate relative (Parr, Hopkins, & de Waal, 1998). In work by Bernston, Boysen, Bauer, & Torello, (1989), it was found that chimpanzees' heart rates change based on the emotional vocalizations of others. In that study, chimpanzees were exposed to recordings of both play sounds and distress calls. In the former, heart rates increased while in the latter there was a deceleration, suggesting that these apes attend to the emotional state of others and modulate their own state in response. More recently, chimpanzees have been shown to exhibit an autonomic sympathetic nervous system response to viewing both static emotional expressions and video clips of emotionally negative situations involving conspecifics (Parr, 2001;

Parr & Hopkins, 2000). These studies support the existence in apes of social understanding of others and provide clues for the emergence of empathy in humans. A recent study reported that rhesus monkeys require extended training to be able to discriminate the emotional expressions of conspecifics (Parr & Hopkins, 2000), thus their difficulties with recognizing expressions correlate with the greater phylogenetic distance between monkeys and humans.

Much of the literature on the evolution of empathy focuses on the importance of the complex social environment in which it evolved (see for example Smith, 2006; Decety et al., 2002; Rizzolatti & Arbib, 1998) and on the close link with the development of the fundamental human capacity to understand and make inferences about the mental states of others, or *theory of mind* (Williams, Whiten, Suddendorf, & Perrett, 2001). Arguably, theory of mind is one of the most complex thought processes available to humans and critical for social communication. It involves understanding not only of oneself but also of others, the abilities to keep those distinctions separate as well as survey past situations to draw conclusions about the current one. Terms such as 'mentalizing' and 'self-reflection' refer to similar cognitive functions – that is, the ability to represent ones thoughts, beliefs, knowledge, and internal processes as well as those of others (Frith & Frith, 2003; Premack & Woodruff, 1978). This ability, combined with the human drive to help others, may be the cognitive consequence of the emergence of empathy (Williams et al., 2001).

Cognitive empathy, a term closely associated with theory of mind – as argued by Blair (2005) – may also have evolved through its role in deception and manipulation. If we can understand the perspective of another, we can use it to our advantage by conveying false information and knowing when others are doing the same (Smith, 2006). However, the more important evolutionary role for empathy as a superordinate construct that includes cognitive, motor, and affective components may be its promotion of social cohesion, feelings of group membership, and inclusive fitness (Farrow et al., 2001; Sturmer, Snyder, Kropp, & Siem, 2006; Preston & de Waal, 2002). Additionally, emotional awareness facilitates the formation of long-lasting relationships and reconciliation after disputes (Parr, 2001) and strengthens the mother-infant bond, in itself a powerful enough application to warrant its rapid evolution (Smith, 2006; Preston & de Waal, 2002).

The more fundamental abilities from which empathy might have arisen include social contagion, response facilitation, and emulation (for a review, see Tomasello & Call, 1997). All of these are considered categories of social learning – from observational learning to active teaching. Some of the most salient emotional responses are acquired through observing the behaviors of others, especially those involved in avoidance and withdrawal, such as taste aversion and predator fear (Tomasello & Call, 1997). Naive rhesus monkeys, for example, have been found to adopt the fear state of another in response to the presence of a snake (Mineka & Cook, 1993). The great advantage is that a young rhesus monkey can learn to avoid a snake or other predators without having its own negative encounter.

Thus, from an evolutionary perspective, empathy has survival value in that it helps individuals gather and hunt for food, detect predators, enhance courtship, and ensure reproductive success (Plutchik, 1980). Empathy could be considered one of the most important social skills in humans because it not only provides for strong relationships but affects social learning as well. One of the problems in making such an evolutionary argument, however, is that it is difficult to see how natural selection works at the level of superordinate categories, such as empathy. It is more logical to suggest that it is simpler processes that are selected and that those in turn are combined to produce complex responses. But, just what are these constitutive components of empathy?

To a large extent we agree with Decety and Jackson (2004) that the basic macrocomponents of empathy are mediated by specific neural systems. Furthermore, they have suggested that these include shared neural representations, self-awareness, mental flexibility, and emotion regulation. We agree that mirroring processes, by which it's meant that the shared neural representation of information about self and other along multiple levels and across multiple modalities of the central nervous system, are critical components. We believe that natural selection for mirroring produced the abilities to mimic and imitate, which in turn generated the ability to learn through such mechanisms. Learning through mirroring may have facilitated the evolution of the abilities to resonate and respond to the emotions of others (empathy) as well as to represent the beliefs and intentions in other minds (a theory of mind). Empathy and theory of mind, plus the underlying functional infrastructure, likely allowed for social learning and the faithful transmission of culture (McElreath, 2003; Gintis, 2003).

2.2 Foundational Mechanisms

In cognitive-developmental psychology, it has been recognized for some time that complex behavior is constructed by combining and coordinating low level components (e.g., mental, perceptual, or motor schemes) into novel sequences. For de Waal and colleagues (2002) and for us, empathy comes in many forms rather than an all-or-nothing phenomenon, as described by Decety and Jackson (2004). There are clearly different levels of learning in the human brain, from priming to classical and operant conditioning to imitation. All such learning involves manipulations of stored internal representations that flow along hierarchically organized networks exhibiting high degrees of interconnectivity. Byrne and Russon (1998) have argued that hierarchical organizations of control "are easier than linear ones to repair when they fail, allow the economy of multiple access to common subroutines, and combine efficient local action at low hierarchical levels while maintaining the guidance of an overall structure." We assume that mirroring representations occur at all levels of this information flow (see Fig. 1) producing a gradient of faculties that vary in complexity. We

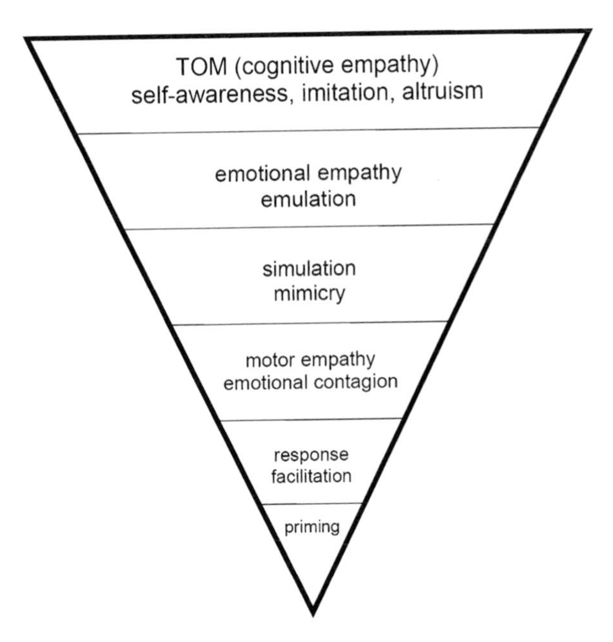

Fig. 1 The hierarchy of 'mirroring' and social cognition. In a framework that conceptualizes social cognition as hierarchically organized mirroring functions, no single neural structure is necessary and sufficient for it. Damage to any layer will result in an alteration of social cognition but the lower the level of malfunctioning the more fundamental the alteration. Full-fledged social cognition (e.g., empathy) requires the engagement of newly evolved structures, specifically the frontal cortex. Adapted from Dietrich (Dietrich, 2003)

thus associate priming with the simplest form of mirroring, and processes such as stimulus enhancement, response facilitation, emotional contagion, mimicry, simulation, emulation, imitation, empathy, and theory of mind as progressively more complex forms of mirroring. These mirroring processes simply reflect the range of mechanisms for manipulating the hierarchical representations of information at different levels of organization (Byrne & Russon, 1998).

Several corollaries from this broader view of mirroring processes can be deduced. First, mirroring occurs in other species and thus there is evidence for phylogenetic continuity. Second, mirroring is the outcome of more primitive functions that can be reorganized to produce different levels with slightly different functionalities and gradations. Byrne and Russon (1998), for example, have distinguished 'action level' imitation from 'program level' imitation. The latter refers to a high-level, constructive mechanism, adapted for the efficient learning of complex skills, while the former is a detailed and linear specification of sequential acts. Third, there are gradations in terms of voluntary control over such a hierarchy, with greater degrees of freedom in the higher levels of processing. Fourth, the organizational structure of mirroring is more functional than anatomical in the sense that it is dynamic and modifiable at the level of the individual. That is, processes of self-awareness, self-regulation of emotions, and language play critical roles in modifying behavior. Finally, the increased interconnectivity and complexity in the human brain may provide the substrate for qualitatively different forms of 'mirroring,' such as perspective-taking, self-awareness, and emotion reappraisal.

2.3 *Interdependencies and Dissociability*

The interdependent nature of mirroring in the conceptual model we are developing makes it difficult, though theoretically not impossible, to dissociate them. It has been unclear, for example, to what extent theory of mind and empathy are overlapping or distinct entities. Empathizing refers to the process which allows one to infer and experience what another person's experience of an emotional state feels like. Theory of mind has a more explicit nature than empathic processes, which tend to be more involuntary and unconscious (Singer, 2006). It could be argued that we always model another person's perspective when empathizing. However, in a review of imaging studies on the neural basis of empathy and theory of mind, Singer (2006) concludes that the two are in fact distinct abilities that rely on different neuronal structures. She argues that empathy-related processes rely on phylogenetically older limbic and paralimbic areas, while theory of mind processing is more dependent on evolutionarily younger neocortical regions. Furthermore, developmental studies suggest this pattern may be reflected in ontogeny. For example, children show mastery of false belief theory of mind tasks by about age 4, whereas newborns already show evidence of emotional contagion in the form of contagious crying. Thus, Singer suggests that the ability to reflect on others and the ability to share their feelings should have different and testable ontogenetic trajectories due to reliance on different underlying structures. Mentalizing structures are activated during attention to one's own mental state as well as to another's, and it may be the case that these regions generally represent beliefs about the world that are decoupled from the actual state of the world and so potentially false. In contrast, empathy involves structures that allow us to sense and represent our own internal feeling states and to evaluate the affective outcome of an event for ourselves and others. Unlike humans, monkeys show no sign of theory of mind and apes struggle with it, yet both show signs of empathy, as previously discussed with reference to autonomic system changes. These findings in nonhuman primates support the relative independence of these processes.

Another argument for the dissociability of empathy and theory of mind stems from psychopathological populations. It has been known for some time that a key problem in psychopathy is a reduced ability for empathy and an inability to respond to emotional cues (Cleckley, 1941). Psychopaths show less aversive conditioning than normal subjects and their electrodermal skin conductance response, a measure of emotional response, is diminished or absent when presented with punishment-related or socially important stimuli such as facial expressions (e.g., Birbaumer et al., 2005; for an overview see Herpertz & Sass, 2000). Interestingly, psychopaths do not seem to have any impairment in theory of mind (see Blair, 2005). In fact, one of the defining characteristics of psychopathy is the tendency to manipulate others in order to obtain one's goals, which arguably relies on the ability to take others' perspectives into account. Autism, unlike psychopathy, is marked by serious deficits in the ability to model another individual's point of view. Blair (2005) even hints at a double dissociation when he suggests that autistics have no emotion processing impairment, unless the emotion involved is complex and cognitive. In any event, it seems clear that theory of mind and affect sharing are partly dissociable both in neural and psychopathological terms, though we would argue not in typical functioning.

3 Functional Neuroanatomy of Empathy

The central thesis of this chapter is that empathy is made possible by mirroring mechanisms that allow us to infer and simulate the feelings, beliefs and intentions of others. This mirroring includes both the simulation of externally observable motor behaviors such as bodily gestures, facial expressions and goal directed hand actions, and the simulation of subtler, internal bodily states and feelings that are not so clearly expressed to others. In this section, the functional neuroanatomy involved both in motor mirroring, which allows us to infer intentions and goals from behaviors, and in affect mirroring, which allows us to color those inferences with the feelings of others, are discussed. The interoceptive system is considered particularly important for the mirroring of affect, while the mirror neuron system is considered particularly important for the mirroring of meaningful actions.

3.1 Interoception (Insular and Cingulate Cortices)

Because empathy refers to the sharing of how another feels, a general theory of the nature of feeling states will ultimately play a role in our understanding of empathy. The James-Lange theory of emotion, over 100 years old but still well appreciated, suggests that perception of a salient stimulus produces changes in the body, and the evaluation of the state of the body in response to the stimulus is the basis of emotional feeling (Carlson, 2003). Recent findings have improved our understanding of how this evaluation of the state of the body takes place.

A critical process of somatic evaluation is known as interoception. Interoception relies on afferent A-delta and C fibers that innervate essentially all the tissues of the body, and which carry a broad array of information regarding temperature, itch, pain, tickle, sensual touch, muscular and visceral sensations, anxiety, hunger and thirst (Craig, 2003). A-delta and C fibers serve to maintain the body's homeostatic balance through continual monitoring of the body's state. Craig proposes the term "homeostatic emotions" to describe the positively and negatively valenced sensations and motivations that are the eventual consequences of these inputs.

The Lamina I pathway is the afferent pathway that carries this A-delta and C fiber information, critical for affective feeling, to the cortex. Lamina 1 of the spinal dorsal horn is the only neural region that receives monosynaptic input from A-delta and C afferent fibers. It relays this information to the thalamic posterior ventromedial nucleus, a dedicated thalamocortical relay nucleus, which projects topographically to the mid/posterior dorsal insula, "interoceptive cortex," in primates (Craig, 2002). In humans, this somatic information is re-represented in the anterior insula, where it appears to be organized in a lateralized fashion with parasympathetic information mainly in the left hemisphere and sympathetic information mainly in the right hemisphere. Hence the insular cortices serve to represent "homeostatic emotions", pleasant and unpleasant feelings and drives that reflect the survival needs of the body.

As the insula's role in interoception plus interoception's role in emotion processing predict, functional imaging studies in humans have shown that a large number of tasks involving emotion activate the anterior insula (Singer, 2007). Recall-generated sadness, anger, anticipatory anxiety and pain, feelings of panic, disgust, sexual arousal, perceived trustworthiness of faces, affiliative touch, and even positive and negative responses to music have all been associated with anterior insula activation.

Furthermore, according to Damasio's somatic marker hypothesis (Damasio, 1996), the multiple representations of the body's state found in the insular cortices could allow for simulation of future actions, in order to use the feelings generated by the simulation to guide decision making. And according to Damasio and other theorists, the explanatory power of anterior insular re-representations of the body could extend beyond feeling states and decision making, and even empathy and knowing other minds, to an understanding of subjective awareness itself. What Damasio calls 'the mental self', which is characterized by a sense that the self is constant despite the reality of continual change, may arise from these continuously maintained cortical representations of the homeostatic state of the body (Damasio 2003). Damasio also asserts that having a feeling differs from feeling a feeling (consciously having a feeling) in that the latter relies on secondary representations (re-representations), such as those believed to exist in the anterior insula. These ideas are supported by empirical work showing that activation in right anterior insular cortex (along with OFC) when experiencing graded cooling stimulation is correlated with reports of the subjective experience of coolness, while activation in posterior insular cortex, "interoceptive" cortex, is correlated with the objective temperatures applied (Craig 2000).

Because empathy, we believe, relies on mirroring processes, the insular cortices are likely to play a significant role in empathy, by subserving simulations of feeling states. Functional imaging studies have not necessarily recognized the important role the insular cortices play in emotional processing through interoception, and have sometimes attributed insular activation in their imaging studies to narrower functions such as relaying information (Carr, Iacoboni, Dubeau, Mazziotta, & Lenzi, 2003). According to this reasoning, the insular cortices are activated during observed emotion processing because of their anatomical connectivity to the action imitation circuit and the limbic system, including the amygdala, such that they act as a relay for signaling between the two. It is also quite plausible that this insular activation reflects a fundamental role in interoceptive emotion processing relevant to knowledge of one's emotions and, through simulation, of the emotions of others.

A number of functional imaging studies of empathy, some of which will be described below, have confirmed that the insular cortices, and in particular the anterior insula, are involved in representing feeling states during empathy (Jackson, Meltzoff, & Decety, 2005; Singer et al., 2004; Wicker et al., 2003a). Disgust and pain are two domains of interoceptive feeling and emotion that have recently been explored with reference to mirroring and empathy.

Disgust. Wicker and colleagues (2003a, b) explored whether the brain areas used to understand the emotions of others are those same areas used to experience one's own emotions. Subjects passively viewed movies of individuals sniffing a substance in a container and then reacting to the odor with a smile, a disgusted grimace, or a neutral expression. Later, subjects actually inhaled pleasant odorants and disgusting odorants themselves. Both disgusting and pleasant smells activated the amygdala and insular cortices. The authors reported clusters of overlap during the viewing of disgusted faces and during the experience of disgust, which localized to the left anterior insula extending to just caudal of the inferior frontal gyrus (IFG), as well as to the right anterior cingulate cortex (ACC). The amygdala was not found to be activated during the viewing condition. This study supports the notion that observing the facial expression for disgust and actually experiencing the emotion involve shared neural representations in the left anterior insula and right ACC.

Pain. The insular cortices, as well as the ACC and in some cases somatosensory cortices and the thalamus, typically show increased activation during pain processing (Peyron, Laurent, &Garcia-Larrea, 2000). If empathy for pain involves a mirroring mechanism, one would predict that these areas would also be activated when observing another in pain and feeling for another's suffering. Jackson and colleagues (2005) confirmed that a network involving the insular cortices as well as the ACC is activated by the observation of photos of pain inflicted on hands and feet. Subjects rated the pain intensity they imagined each photo to represent, and these ratings correlated significantly with ACC activity when viewing the photos. While significant ACC and insula activation was reported, the researchers did not find significant amygdala activation, which they interpreted as implying that observing the pain of others does not simply induce a general state of emotional distress, but rather activates the structures we use to feel pain ourselves, an instance of mirroring.

The literature on pain processing distinguishes between the sensory components of pain, involving processing in primary and secondary somatosensory cortices, and the emotional or motivational components of pain, involving processing in the insula and ACC (Peyron et al., 2000). In an fMRI study involving observing a signal that indicated when a loved one was receiving a pain stimulus, Singer and colleagues (2004) found that empathy for pain involves simulating the aversive qualities of the pain (the motivational significance of pain) but not its precise somatic characteristics. However, a transcranial magnetic stimulation study has shown that watching pain inflicted on a person's hand at the site of a specific muscle reduces the motor excitability of that muscle in observers, and in the same way as self-performance reduces the motor excitability (Singer & Frith, 2005). Further, the hand muscle evoked potential is unaffected by viewing a needle stabbing another's foot or even a different site on the hand, implying that the simulation mechanism involved is somatotopic with a high degree of precision. The implication of this finding is that sensorimotor aspects of pain are also involved in empathy, in addition to the motivational aspects of pain revealed by neuroimaging.

In an fMRI examination of how humans detect pain in another's face, Saarela and Hari (Saarela et al., 2007) used photos of facial expressions from chronic pain sufferers which varied in the intensity of depicted suffering. Not only were bilateral anterior insula, left anterior cingulate, and left inferior parietal lobe activated, but the amount of these activations correlated with subjects' estimates of the intensity of observed pain. Furthermore, bilateral IFG plus anterior insula, bilateral supplementary motor area, left ACC, left premotor cortex, and left inferior parietal lobe all showed more activation for intense, provoked pain than for chronic pain expressions. Finally, the strength of activation in the left anterior insula and left IFG when observing intensified pain correlated with subjects' self-rated pain. Since we know that anterior insula activation increases proportionally with one's own experienced pain intensity, this constitutes evidence of a mirroring mechanism, wherein empathy for another's pain involves simulating it with the representation that creates the sensation of pain in oneself.

3.2 *Exteroception (Somatosensory Cortex)*

The body map found in somatosensory cortex also plays a role in allowing emotions to be simulated in the act of empathizing. Researcher Ralph Adolphs examined 108 subjects with focal brain lesions in different areas throughout the cortex, and tested their ability to recognize emotions expressed by faces (Adolphs, Damasio, Tranel, Cooper, & Damasio, 2000). He found that lesions to right somatosensory cortex were the most detrimental to the ability to recognize facial emotion. These patients also showed preliminary evidence of a correlation between degree of impairment in facial emotion recognition and impairment in their own somatosensation, again suggesting shared representations for one's own somatosensory experience and for comprehension of the experience of another.

3.3 Amygdala

The amygdala has long been implicated in emotional processing, and fear processing in particular. Visual emotional stimuli can reach the amygdala via cortical and subcortical routes (LeDoux, 1996). While the cortical route is required for conscious identification of the stimulus, the subcortical pathway seems to be involved in more automatic and reflexive processing. The subcortical route to the amygdala can lead an individual to experience the autonomic aspects of an emotional response to facial expressions in the absence of awareness and this allows for fast, automatic responses to emotionally significant stimuli, in the same sense as emotional contagion. Two important aspects of emotional contagion are a matching of states between the observed and the observer, and the absence of a distinction between the self and other (Preston & de Waal, 2002). Both of these aspects are functional characteristics of this subcortical processing route, as it induces emotional resonance on an unconscious level. The amygdala does not only respond to unconsciously presented fearful stimuli, but also to masked expressions of happiness (Killgore & Yurgelun-Todd, 2004), anger (Nomura et al., 2004), and sadness (Dannlowski et al., 2007). The notion that emotional stimuli can lead to activation of an automatic response in the absence of awareness is robust. For example, when subjects fearful of either snakes or spiders are presented with brief images of these stimuli immediately followed by a masking picture they show larger skin conductance responses to feared than to non-feared or neutral stimuli. However, they were still unable to identify the masked picture above chance level (Ohman & Soares, 1994). In an extracellular recording paradigm (Rolls, 2004) such masking led to a decrease of spiking activity and informational content in neurons of the inferior temporal cortex of monkeys. This degraded cortical signal is likely to influence the amygdala only in a very crude manner, if at all. Evidence that unconscious emotional processing is mediated by a subcortical route to the amygdala was provided by Vuilleumier, Armony, Driver, and Dolan (2003). Using a functional magnetic resonance imaging paradigm, they showed that the amygdala is co-active with other structures belonging to the proposed subcortical route. This structure is well-situated to activate the autonomic and hormonal systems leading to characteristic emotional responses.

3.4 Mirror Neuron System

Although we have argued that theory of mind (or cognitive empathy) and emotional empathy are to a certain extent dissociable, it is evident that in the aggregate response there is interactivity, that is, we explicitly take the perspective of another in order to understand what they feel, and, in fact, feel what they feel. A number of researchers have proposed that a distinct mirror neuron system is involved in action understanding and imitation (Rizzolatti &

Craighero, 2004; Lyons, Santos, & Keil, 2006). It seems reasonable to posit that the action understanding properties of the mirror neuron system may serve as a foundational cornerstone for the higher order perspective taking that humans engage in during empathy. Adopting someone else's view minimally requires that the other's actions be understood; else no accurate prediction of the other's behavior can be made. Such understanding or imitative behavior, however, does not necessarily have to be conscious. Indeed, the mimicry of another individual's postures, facial expressions, vocalizations, movements, and mannerisms is often executed in the absence of awareness. This effect is known as the chameleon effect, motor empathy, motor contagion, or emotional contagion (Leslie, Johnson-Frey, & Grafton, 2004).

The mirror neuron system is the canonical mirroring system in the brain. The extended mirror neuron loop in rhesus monkeys (Rizzolatti & Gallese, 2006; Rizzolatti & Craighero, 2004) begins in the rostral part of the superior temporal sulcus, which contains neurons that respond to viewing biologically relevant actions, such as head, body, and eye movements as well as to static pictures merely implying biological motion (Jellema & Perrett, 2005). Information is then thought to flow to area PF on the rostral cortical convexity of the inferior parietal lobule. A subset of the cells in this area has mirror properties: that is, they discharge both when the monkey observes an action and executes it. Area PF, in turn, sends projections to area F5 on the ventral premotor cortex, which also contains mirror neurons. An important characteristic of mirror neurons in macaque monkeys is their activation by the interaction between a biological agent and a target object (Rizzolatti & Craighero, 2004). Moreover, there is congruence between the actions that activate a mirror neuron when observed, and the actions that activate the same neuron when executed. For example, a neuron may show an increase in firing activity when the monkey reaches for a particular object, as well as when the monkey observes another individual (whether monkey or human) reaching for same object. Such congruence can be strict, when the observed and performed actions have to be virtually identical to lead to a neuronal response, or broad, as when the observed and performed actions are similar or functionally related. Both inferior parietal lobe and ventral premotor cortex send information back ('efference copies') to the superior temporal sulcus (STS), which is thought to match the sensory consequences of the proposed motor plan with the observed action. These latter connections are thought to be of importance especially during imitation (Iacoboni, 2005).

Evidence for a human equivalent of the mirror neuron system exists, although such evidence is necessarily less direct, stemming mostly from behavioral, electroencephalographic (EEG), and functional magnetic resonance imaging (fMRI) studies (Iacoboni & Dapretto, 2006). As shown in Fig. 2, converging evidence suggests that a network consisting of the rostral inferior parietal lobule, *pars opercularis* of the inferior frontal gyrus (Broca's area) and other premotor areas respond to both viewing an action and executing or imitating it (Rizzolatti & Craighero, 2004; Carr et al., 2003; Leslie et al., 2004). The parietal mirror area is thought to receive input reflecting biological

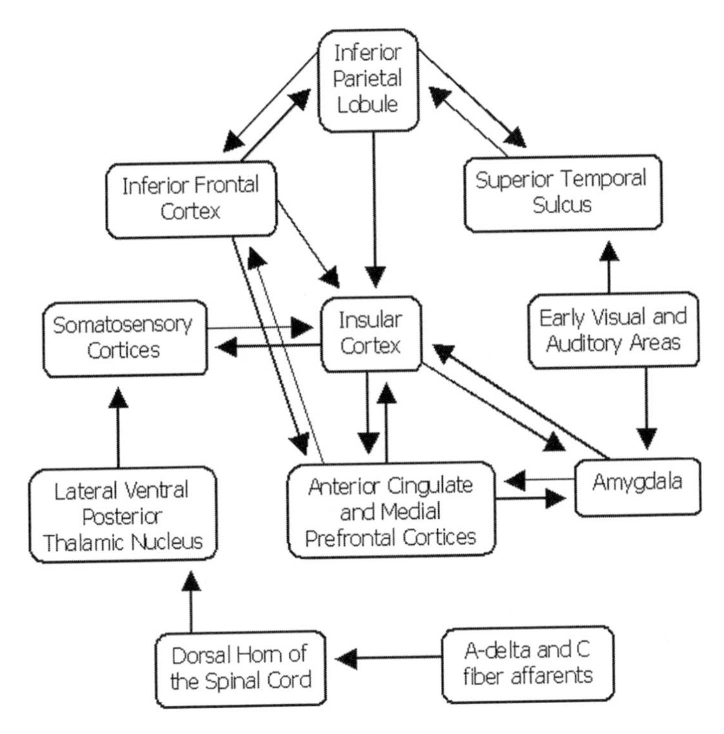

Fig. 2 Neuroanatomical Circuits involved in Empathy

action from the posterior superior temporal sulcus. Using fMRI, Leslie and colleagues (2004) examined participants who watched short movies of a model smiling or frowning. Subjects were asked to view or imitate the facial expression, or to smile or frown irrespective of what the model was doing. Results indicated that imitation of a facial expression leads to more widespread pattern of brain activation than passive viewing. However, both conditions led to significantly increased activation in the right ventral premotor cortex, superior and inferior temporal gyri, and fusiform gyrus compared to the control condition. In another neuroimaging study, Carr and colleagues (2003) had subjects viewing pictures of human faces displaying the prototypical six emotions discussed earlier. They were asked to observe or imitate the emotion, that is, internally generate the target emotion. An overlapping network consisting of the *pars opercularis* of the inferior frontal gyrus, premotor face area, superior temporal sulcus, amygdala, and anterior insula was active during both conditions. Moreover, all these regions were more active during the imitation condition than during the mere observation condition. However, it is not yet entirely clear to what degree the human mirror neuron system maps onto the one extensively described in the monkey. Rizzolatti and Craighero (2004) suggest that the inferior parietal cortex in humans corresponds to monkey area PF, and that the *pars opercularis* of the inferior frontal gyrus is the human homologue of monkey area F5.

The relevant question vis-à-vis empathy is whether and how the mirror neuron system enables an organism to understand actions, imitate the actions of others, allow an individual to take the perspective of another, and develop a theory of mind. It has been suggested that activation of the mirror neuron system by merely observing an action automatically engages the organism's own motor representations (Rizzolatti & Craighero, 2004). Arguably, the reason that this system is more active during action execution than during observation (and even more so during imitation, which involves both observation *and* execution), is because performing an action requires a higher degree of activation of the involved motor representations. Thus, the only requirement for understanding an action is the presence of a preexisting motor representation of that action. This is assumed to allow an organism the ability to infer a conspecific's goal or intention, as long as it is familiar with using the observed pattern of motor actions for the observed (or inferred) goal.

These complex forms of mirroring still rely, ultimately, on the more basic forms of mirroring. Many instances of 'real' imitation (learning by copying), for example, can be explained by simpler mechanisms, such as stimulus enhancement, response facilitation, and emulation (Byrne & Russon, 1998), although these are often indistinguishable from imitation based solely on behavioral outcome. These authors correctly note that these are, in effect, priming processes. Thus, nonhuman primates watching a human experimenter trying to open a box, might 'imitate' this behavior because of increased stimulus salience, priming of the motor representation of the observed action, or priming of the goal representation of the observed action. The latter two explanations would most likely rely on mirroring processes to activate representations based on perceptual input alone. Imitation thus likely relies on mirroring mechanisms for recruiting relevant motor representations (Iacoboni, 2005). Indeed, applying transcranial magnetic stimulation to the *pars opercularis* of the inferior frontal gyrus, an area involved in mirroring, disrupts subjects' ability to imitate simple movements (Heiser, Iacoboni, Maeda, Marcus, & Mazziotta, 2003).

It has been postulated that there are at least two types of neurons in the frontal mirror system – broadly and strictly congruent (Rizzolatti & Craighero, 2004). The distinction between them may map onto the notions of response facilitation and emulation, respectively. Strictly congruent mirror neurons fire to observed and performed actions only when these are identical; therefore, their priming could lead to execution of the same action that was observed, which is, in essence, response facilitation. Broadly congruent mirror neurons, on the other hand, show activity during performance of a specific action (e.g., grasping an object with the hand), and during observation of identical as well as related actions (e.g., holding or manipulating an object, or grasping it with the mouth). Like strictly congruent mirror neurons, their priming may lead to a heightened probability of executing the action they code. However, what primes these cells is not the observation of any particular action; it is the observation of a class of actions. In other words, watching a human manipulate a box may lead to activation of neurons that do not precisely reflect the observed actions, but that nevertheless code the intended goal ('open the box') quite well. Such neural facilitation could be the basis of emulation, where the goal is mimicked, without necessarily using the observed route of arriving at it.

Theory of mind seems to rely to a considerable degree on the medial prefrontal cortex (mPFC), temporo-parietal junction, and temporal pole (see Singer, 2006; Decety & Jackson, 2004). Vogeley and colleagues (2001) found an increase in activation with fMRI in the right anterior cingulate cortex and left temporal pole when subjects listened to a story that involved perspective taking, as compared to listening to a story that did not require perspective taking. Likewise, Mitchell, Macrae, and Banaji (2006), found increased activation in the mPFC when subjects were instructed to mentalize about other persons' opinions. Adopting the perspective of an individual with similar sociopolitical views led to a more ventral focus of activation, while considering dissimilar points of view resulted in more dorsal activation. Finally, Ohnishi and colleagues (2004) also found theory of mind-related activity in the mPFC, temporal pole, and temporo-parietal junction.

The enormous enlargement of the human PFC with respect to other primates would allow for more complex computations using mirror neuron input. Interestingly, there is evidence that the inferior parietal lobule has gone through a substantial expansion during hominid evolution as well (Simon, Mangin, Cohen, Le, & Dehaene, 2002). Thus, the combination of prefrontal and inferior parietal enlargement might well account for the improved capacities for action understanding and perspective taking in humans. Indeed, there certainly is evidence that theory of mind-related stimuli and action observation activate a partly overlapping network involving posterior superior temporal sulcus (Ohnishi et al., 2004). Thus, an intact mirroring circuit may be a requirement for the ability to mentalize. The widespread hypothesis that autistic patients have some dysfunction of the mirror neuron system, as evidenced by imitation impairments, might therefore lead to downstream effects as well. This would explain this population's difficulty with perspective taking.

In summary, the mirror neuron system is involved in producing action understanding by mapping others' actions onto those available in one's own motor repertoire. Assuming this is the case, it makes sense that premotor areas respond to mere action observation. This mechanism then sets the stage for imitation: the relevant motor representations are simply primed to carry out the same movements that are observed in another individual. As such, it can account for the motor aspects of emotional contagion: the often unconscious adoption of another person's stance and movements. It has been noted that displaying a facial expression may actually lead to autonomic changes and changes in subjective ratings of emotional experience, even in the absence of awareness (Soussignan, 2002). Thus, the mirror neuron system can induce the expression of a certain emotion in response to observation of this emotion in another individual. This in turn, can activate the autonomous state associated with the emotion. Of course, we do not engage in imitative behavior continuously. These actions may actually be inhibited by prefrontal areas. Indeed, it is well-known that damage to prefrontal cortex may lead to imitative tendencies (Brass, Derrfuss, Matthes-von, & von Cramon,, 2003). Disruption of this top-down control system may lead to 'motor release' effects via a disinhibition of the mirror neuron system. Yet, under normal circumstances this system exerts its influence very subtly, such as when we unintentionally mirror somebody's posture, or in an obvious manner, such as when we follow instructions to imitate.

A study by Keysers and colleagues (2004) examined mirroring mechanisms for the experience of touch, by performing fMRI while subjects were touched on the legs, or while they watched movies of others being touched on the legs. In both conditions they found certain areas of secondary somatosensory cortex in the left hemisphere activated, confirming that some of the representations that allow us to process the experience of being touched also allow processing of the observation of others being touched. However, even the sight of inanimate objects rather than legs being touched by an object produced this effect, suggesting that an abstracted notion of 'touch' is also processed like the experience of being touched. Subjects were later asked whether they imagined what the person whose legs were being touched in the movie was feeling and they reported they had not, suggesting that the simulation process is an automatic rather than a voluntary or conscious one.

4 Other Dimensions of Empathy

The functional neuroanatomy of empathy is as diverse and complex as the construct itself. Distinguishing among the various processes that give rise to empathy may help make sense of the multitude of neural structures involved. Although empathy and theory of mind may be dissociable as suggested by Singer (2006), an empathic response likely engages multiple neural systems. Thus, there appears to be some agreement in the psychological literature that an empathic response is multidimensional and involves not only an emotional dimension characterized by the sharing of the emotional experience of the other person, as well as a cognitive dimension characterized by the ability to take the perspective of the other person and understand their experience, but also a self-other dimension characterized by maintaining the distinction between the self and other.

4.1 Self-Other

Although some affect sharing properties of empathy may be unconscious and involuntary, when taking somebody else's perspective and empathizing we do not normally hold the other's feelings for our own. It is true we often overestimate what others know and assume that others' motivational states are closer to our own states then they are in reality, yet complete misattribution of our feelings does not usually occur. Such confusion would not be beneficial, as holding someone else's sadness for our own, for example, would not evoke comforting or consoling behavior. Instead, it would drive us away from whatever caused it, which clearly is contrary to the social goal of empathy. Thus, the neural mechanisms underlying high-level social cognition must account for this ability to distinguish between self and other.

Recent findings suggest that both self-related and other-related thought rely on networks overlapping with structures important for mentalizing. The fMRI study by Vogeley and colleagues (2001) investigated the neural basis of theory of mind and assessed the effects of self involvement during both the perspective taking, and the non-perspective taking conditions. Self involvement resulted in more activation in right motor/premotor areas, right inferior parietal lobule, bilateral anterior cingulate, and bilateral precuneus. In the absence of self involvement there was more activity in the left inferior parietal lobule. An fMRI study asking subjects to view food names and to decide if they themselves or a friend would like or dislike the food item found bilateral activation in medial prefrontal regions, medial parietal regions and the insular cortices in both conditions (Seger, Stone, & Keenan, 2004). Furthermore, deciding about your own compared to another's preferences led to increased activation in the precuneus. The opposite, that is, more activity in the other versus the self condition, was true for a more inferior and posterior site in the precuneus.

A different approach by Farrer and Frith (Farrer & Frith, 2002) assessed brain activation while subjects used a joystick to move a circle into a certain direction on a screen. In some of the trials they were in control of the circle, while in other trials they were not. The sense of agency these participants experienced corresponded to different patterns of brain activity. When subjects attributed the action to themselves the left anterior insula was active. In contrast, when they were not in control and likely imparting an agent of 'other' the right intraparietal sulcus, precuneus, and occipitoparietal fissure were active. A direct comparison between the two conditions yielded increased activation in bilateral insular cortices for the 'in control' condition and in both the angular gyrus and precuneus bilaterally when subjects did not control the action. Recent studies have indicated that the perception of agency also activates the right STS while self-evaluation involves activity in the ventromedial PFC and recognizing emotions expressed by another activates the right ventrolateral PFC (Tankersley, Stowe, & Huettel, 2007; Seitz, Nickel, & Azari, 2006).

These and other studies (see Decety & Jackson, 2004; Decety & Grezes, 2006) give rise to the idea that the inferior parietal lobule, medial frontal cortex, and medial parietal areas are important for keeping self and other separate. Although the exact manner in which these structures contribute to distinguishing self from other is not clear, recent findings suggest that neurodynamics may be important. Grezes, Frith, and Passingham (2004) witnessed activation in the intraparietal sulcus and premotor cortex when subjects viewed actions performed by others or by themselves videotaped eight months earlier. Interestingly, the latency of the hemodynamic signal was slightly but significantly larger in these regions when the observed action was performed by others. This hints at preferred neural processing when it concerns the self. The fact that these structures belong to the human mirror neuron system suggests that although motor representations may be shared for perceiving and acting, there may still be subtle differences in the way these are put to use in different situations. Moreover, the involvement of all the self/other-related structures in mentalizing and mirror processes attests to the idea that these faculties are to a considerable degree interrelated.

4.2 Mnemonic Processes

It has been suggested that it is useful to think of the self as one's memory for oneself (Klein, 2001). Whether perception-based or meaning-based, self-knowledge is represented in the individual's memory. Hence, self-other distinctions would appear to require the encoding and retrieval of memories. Encoding and retrieval of episodic, and especially autobiographical memories may provide the necessary background information (Stark & Squire, 2001), although other types of memories, such as working and implicit memories, play important roles. Autobiographical memories contain information about oneself and one's own experiences, including relevant emotions. Memory is likely to be involved in the formation of an internal script useful in solving problems and coping with novel situations by relying on similar ones from the past (Greicius, Krasnow, Reiss, & Menon, 2003). Similarly, memory enhances our ability to empathize by evoking similar situations. Recalling a similar situation allows one to feel how another person might feel in a similar situation by recalling one's own past emotional state (Preston & de Waal, 2002). On the other hand, when someone you trust shows emotional distress in response to a situation, empathy might allow us to also understand the association between the situation and emotional state through the actions of another person without having to experience it first hand (Blair, 2005). In this way, memory can facilitate empathy by serving as a foundation for understanding and feeling the current emotional environment. Reciprocally, our ability to empathize may make remembering a particular emotional situation unnecessary in that we can gather all the necessary information by adopting the state of another. Thus, empathy may involve not only monitoring the present state of another, but also the past through the retrieval and processing of memories both about one's self and the other in the current situation.

4.3 Hemispheric Asymmetries

Generally, the right hemisphere appears to be more involved in emotional tasks while more verbal and concrete tasks appear to primarily engage the left hemisphere (Rizzolatti & Arbib, 1998). Therefore, it would be expected that empathy might engage the right hemisphere more than the left. On the whole, this is true as evidenced by patients with specific brain lesions to either hemisphere (Perry et al., 2001). In a report of several carefully examined cases, patients with frontotemporal right hemisphere damage lost their affect, were unable to appreciate emotional salience, and could not respond with empathy to their family members. The patients with similar left hemisphere lesions showed extreme cognitive impairment though their performance on emotional tasks was preserved (Perry et al., 2001).

Conversely, studies suggest that imitation and movements for symbolic imitation requires the left inferior parietal area and lesions of this area result in imitation deficits (e.g., Makuuchi, 2005). Other functional imaging studies have shown that the act of imitation involves the left superior temporal sulcus and inferior parietal areas while observing the imitation of one's own action activates the right inferior parietal area (Decety et al., 2002). This left-hemisphere dominance for imitation holds true though the more emotional involvement required for a task the more areas of activation are found in the right hemisphere (Williams et al., 2006).

In a recent fMRI study by Iacoboni and colleagues, subjects were shown stimuli either in their left or right visual field. The stimuli were designed to activate the mirror neuron system. Several brain regions showed similar activation regardless of presentation visual field. These included the left parietal cortex, bilateral inferior frontal gyrus, and right STS (Aziz-Zadeh, Koski, Zaidel, Mazziotta, & Iacoboni,, 2006). The results suggest a circuit involved in the understanding of socially relevant information that is independent of where in the visual field the stimuli are. All this evidence suggests there is not a simple dichotomy between the hemispheres and that the component tasks of empathy determine left-right activation.

5 Conclusions

Matt Ridley has argued in his book 'The Origins of Virtue' (Ridley, 1996) that "Society was not invented by reasoning men. It evolved as part of our nature. It is as much a product of our genes as our bodies are. To understand it we must look inside our brains at the instincts for creating and exploiting social bonds that are there." It is our contention that mirroring mechanisms, in the pathways from perception to action, and especially those that engage affective processing, provide the foundational basis for social cognition, including empathy. Perhaps the most commonly accepted definition of mirroring refers to functions that allow us to directly understand the meaning of the actions and emotions of others by internally replicating or simulating them without any explicit reflective mediation (Gallese, Keysers, & Rizzolatti,, 2004). Mirroring is economical in that one representation serves multiple functions. The perspective taken in this chapter is that mirroring processes also *involve the re-representation of information – for example, from one level of the nervous system to another, from one modality to another (visual-auditory), a cross-mapping of information into two or more dimensions (perception-action), and a sharing of those representations across multiple neural systems.* These mirroring processes are assumed to be phylogenetically continuous with parallel, distributed, hierarchical, and interdependent properties. In the human brain, they have led to highly complex computations, such as empathy toward those around us. Decety and Jackson (2004) and Blakemore and Frith (2005) have made similar points in arguing that the study of the mirror neuron system provides a neurophysiological basis underlying the ability to understand the meaning of one's own and another person's actions. They view mirroring as fundamental to higher level social processes. Our view differs slightly in that we see mirroring processes as more universal and necessary for all levels of information processing relevant to social cognition.

References

Adolphs, R. (2006). How do we know the minds of others? Domain-specificity, simulation, and enactive social cognition. *Brain Research, 1079*, 25–35.

Adolphs, R., Damasio, H., Tranel, D., Cooper, G., & Damasio, A. R. (2000). A role for somatosensory cortices in the visual recognition of emotion as revealed by three-dimensional lesion mapping. *Journal of Neuroscience, 20*, 2683–2690.

Aziz-Zadeh, L., Koski, L., Zaidel, E., Mazziotta, J., & Iacoboni, M. (2006). Lateralization of the human mirror neuron system. *Journal of Neuroscience, 26*, 2964–2970.

Batson, C. D., Dyck, J. L., Brandt, J. R., Batson, J. G., Powell, A. L., McMaster, M. R., & Griffitt, C. (1988). Five studies testing two new egoistic alternatives to the empathy-altruism hypothesis. *Journal of Personality and Social Psychology, 55*, 52–77.

Bernston, G. G., Boysen, S. T., Bauer, H. R., & Torello, M. S. (1989). Conspecific screams and laughter: cardiac and behavioral reactions of infant chimpanzees. *Developmental Psychology, 22*, 771–787.

Birbaumer, N., Veit, R., Lotze, M., Erb, M., Hermann, C., Grodd, W., & Flor, H. (2005). Deficient fear conditioning in psychopathy: a functional magnetic resonance imaging study. *Archives of General Psychiatry, 62*, 799–805.

Blair, R. J. (2005). Responding to the emotions of others: dissociating forms of empathy through the study of typical and psychiatric populations. *Conscious and Cognition, 14*, 698–718.

Blakemore, S. J., & Frith, C. (2005). The role of motor contagion in the prediction of action. *Neuropsychologia, 43*, 260–267.

Bowles, S., & Gintis, H. (2004). The evolution of strong reciprocity: cooperation in heterogeneous populations. *Theoretical Population Biology, 65*, 17–28.

Brass, M., Derrfuss, J., Matthes-von, C. G., & von Cramon, D. Y. (2003). Imitative response tendencies in patients with frontal brain lesions. *Neuropsychology, 17*, 265–271.

Byrne, R. W., & Russon, A. E. (1998). Learning by imitation: a hierarchical approach. *The Behavioral and Brain Sciences, 21*, 667–684.

Carlson, N. R. (2003). *Physiology of behavior*. Allyn and Bacon: Boston.

Carr, L., Iacoboni, M., Dubeau, M. C., Mazziotta, J. C., & Lenzi, G. L. (2003). Neural mechanisms of empathy in humans: a relay from neural systems for imitation to limbic areas. *Proceedings of the National Academy of Sciences of the United States of America 100*, 5497–5502.

Cleckley, H. M. (1941). *The mask of sanity: an attempt to reinterpret the so-called psychopathic personality*. Mosby: St. Louis, MO.

Craig, A. D. (2002). How do you feel? Interoception: the sense of the physiological condition of the body. *Nature Reviews. Neuroscience, 3*, 655–666.
Craig, A. D. (2003). Interoception: the sense of the physiological condition of the body. *Current Opinion in Neurobiology, 13*, 500–505.
Damasio, A. R. (1996). The somatic marker hypothesis and the possible functions of the prefrontal cortex. *Philosophical Transactions of the Royal Society of London. Series B, Biological Sciences, 351*, 1413–1420.
Damasio, A. (2003). Mental self: the person within. *Nature, 423*, 227.
Dannlowski, U., Ohrmann, P., Bauer, J., Kugel, H., Arolt, V., Heindel, W., & Suslow, T. (2007). Amygdala reactivity predicts automatic negative evaluations for facial emotions. *Psychiatry Research, 154*, 13–20.
Decety, J., Chaminade, T., Grezes, J., & Meltzoff, A. N. (2002). A PET exploration of the neural mechanisms involved in reciprocal imitation. *Neuroimage, 15*, 265–272.
Decety, J., & Grezes, J. (2006). The power of simulation: imagining one's own and other's behavior. *Brain Research, 1079*, 4–14.
Decety, J., & Jackson, P. L. (2004). The functional architecture of human empathy. *Behavioral and Cognitive Neuroscience Reviews, 3*, 71–100.
Dietrich, A. (2003). Functional neuroanatomy of altered states of consciousness: the transient hypofrontality hypothesis. *Conscious and Cognition, 12*, 231–256.
Donald, M. (1991). *Origins of the modern mind: three stages in the evolution of culture and cognition*. Harvard University Press: Cambridge, MA.
Ekman, P. (1972). *Emotion in the human face*. Cambridge University Press: Cambridge.
Ekman, P. (1973). *Darwin and facial expressions*. Academic Press: New York.
Farrer, C., & Frith, C. D. (2002). Experiencing oneself vs another person as being the cause of an action: the neural correlates of the experience of agency. *Neuroimage, 15*, 596–603.
Farrow, T. F., Zheng, Y., Wilkinson, I. D., Spence, S. A., Deakin, J. F., Tarrier, N., Griffiths, P. D., & Woodruff, P. W. (2001). Investigating the functional anatomy of empathy and forgiveness. *Neuroreport, 12*, 2433–2438.
Frith, U., & Frith, C. D. (2003). Development and neurophysiology of mentalizing. *Philosophical Transactions of the Royal Society of London. Series B, Biological Sciences, 358*, 459–473.
Gallese, V., Keysers, C., & Rizzolatti, G. (2004). A unifying view of the basis of social cognition. *Trends in Cognitive Sciences, 8*, 396–403.
Gintis, H. (2003). The hitchhiker's guide to altruism: gene-culture coevolution, and the internalization of norms. *Journal of Theoretical Biology, 220*, 407–418.
Greicius, M. D., Krasnow, B., Reiss, A. L., & Menon, V. (2003). Functional connectivity in the resting brain: a network analysis of the default mode hypothesis. *Proceedings of the National Academy of Sciences of the United States of America, 100*, 253–258.
Grezes, J., Frith, C. D., & Passingham, R. E. (2004). Inferring false beliefs from the actions of oneself and others: an fMRI study. *Neuroimage, 21*, 744–750.
Heiser, M., Iacoboni, M., Maeda, F., Marcus, J., & Mazziotta, J. C. (2003). The essential role of Broca's area in imitation. *The European Journal of Neuroscience, 17*, 1123–1128.
Henrich, J., Boyd, R., Bowles, S., Camerer, C., Fehr, E., Gintis, H., McElreath, R., Alvard, M., Barr, A., Ensminger, J., Henrich, N. S., Hill, K., Gil-White, F., Gurven, M., Marlowe, F. W., Patton, J. Q., & Tracer, D. (2005). "Economic man" in cross-cultural perspective: behavioral experiments in 15 small-scale societies. *The Behavioral and Brain Sciences, 28*, 795–815.
Herpertz, S. C., & Sass, H. (2000). Emotional deficiency and psychopathy. *Behavioral Sciences & the Law, 18*, 567–580.
Hoffman, M. L. (2000). *Empathy and moral development*. Cambridge University Press: New York.
Hutchison, W. D., Davis, K. D., Lozano, A. M., Tasker, R. R., & Dostrovsky, J. O. (1999). Pain-related neurons in the human cingulate cortex. *Nature Neuroscience, 2*, 403–405.

Iacoboni, M. (2005). Neural mechanisms of imitation. *Current Opinion in Neurobiology, 15*, 632–637.
Iacoboni, M., & Dapretto, M. (2006). The mirror neuron system and the consequences of its dysfunction. *Nature Reviews. Neuroscience, 7*, 942–951.
Ickes, W. (1997). *Empathic accuracy*. Guilford: New York.
Jackson, P. L., Meltzoff, A. N., & Decety, J. (2005). How do we perceive the pain of others? A window into the neural processes involved in empathy. *Neuroimage, 24*, 771–779.
Jellema, T., & Perrett, D. I. (2005). Neural basis for the perception of goal-directed actions. In: A. Easton & N. J. Emery (Eds.),*The cognitive neuroscience of social behavior* (pp. 81–112). Psychology Press: Hove/New York.
Keysers, C., Wicker, B., Gazzola, V., Anton, J. L., Fogassi, L., & Gallese, V. (2004). A touching sight: SII/PV activation during the observation and experience of touch. *Neuron, 42*, 335–346.
Killgore, W. D., & Yurgelun-Todd, D. A. (2004). Activation of the amygdala and anterior cingulate during nonconscious processing of sad versus happy faces. *Neuroimage, 21*, 1215–1223.
Klein, S. B. (2001). A self to remember: a cognitive neuropsychological perspective on how self creates memory and memory creates self. *Individual self, relational self, collective self* (pp. 25–46). Psychology Press/Taylor & Francis: Philadelphia, PA.
LeDoux, J. E. (1996). *The emotional brain*. Simon & Schuster: New York.
Leslie, K. R., Johnson-Frey, S. H., & Grafton, S. T. (2004). Functional imaging of face and hand imitation: towards a motor theory of empathy. *Neuroimage, 21*, 601–607.
Lyons, D. E., Santos, L. R., & Keil, F. C. (2006). Reflections of other minds: how primate social cognition can inform the function of mirror neurons. *Current Opinion in Neurobiology, 16*, 230–234.
Makuuchi, M. (2005). Is Broca's area crucial for imitation? *Cerebral Cortex, 15*, 563–570.
McElreath, R. (2003). Reputation and the evolution of conflict. *Journal of Theoretical Biology, 220*, 345–357.
Mineka, S., & Cook, M. (1993). Mechanims involved in the observational conditioning of fear. *Journal of Experimental Psychology General, 122*, 23–38.
Mitchell, J. P., Macrae, C. N., & Banaji, M. R. (2006). Dissociable medial prefrontal contributions to judgments of similar and dissimilar others. *Neuron, 50*, 655–663.
Nomura, M., Ohira, H., Haneda, K., Iidaka, T., Sadato, N., Okada, T., & Yonekura, Y. (2004). Functional association of the amygdala and ventral prefrontal cortex during cognitive evaluation of facial expressions primed by masked angry faces: an event-related fMRI study. *Neuroimage, 21*, 352–363.
Ohnishi, T., Moriguchi, Y., Matsuda, H., Mori, T., Hirakata, M., Imabayashi, E., Hirao, K., Nemoto, K., Kaga, M., Inagaki, M., Yamada, M., & Uno, A. (2004). The neural network for the mirror system and mentalizing in normally developed children: an fMRI study. *Neuroreport, 15*, 1483–1487.
Ohman, A., & Soares, J. J. (1994). "Unconscious anxiety": phobic responses to masked stimuli. *Journal of Abnormal Psychology, 103*, 231–240.
Parr, L. A. (2001). Cognitive and physiological markers of emotional awareness in chimpanzees (*Pan troglodytes*). *Animal Cognition, 4*, 233–239.
Parr, L. A., & Hopkins, W. D. (2000). Brain temperature asymmetries and emotional perception in chimpanzees, Pan troglodytes. *Physiology & Behavior, 71*, 363–371.
Parr, L. A., Hopkins, W. D., & de Waal F. B. M. (1998). The perception of facial expressions in chimpanzees (*Pan troglodytes*). *Evolution of Communication, 2*, 1–23.
Perry, R. J., Rosen, H. R., Kramer, J. H., Beer, J. S., Levenson, R. L., & Miller, B. L. (2001). Hemispheric dominance for emotions, empathy and social behaviour: evidence from right and left handers with frontotemporal dementia. *Neurocase, 7*, 145–160.
Peyron, R., Laurent, B., & Garcia-Larrea, L. (2000) Functional imaging of brain responses to pain. A review and meta-analysis (2000). *Neurophysiologie Clinique-Clinical Neurophysiology, 30*, 263–288.

Plutchik, R. (1980). A general psychoevolutionary theory of emotion. In: R. Plutchik & H. Kellerman (Eds.), *Theories of emotion* (pp. 3–33). Academic Press: New York.
Povinelli, D. J., & Eddy, T. J. (1996). What young chimpanzees know about seeing. *Monographs of the Society for Research in Child Development, 61*, 1-152.
Premack, D., & Woodruff, G. (1978). Does the chimpanzee have a theory of mind? *Behavioral and Brain Sciences, 4*, 515–526.
Preston, S. D., & de Waal, F. B. (2002). Empathy: its ultimate and proximate bases. *The Behavioral and Brain Sciences, 25*, 1–20.
Ridley, M. (1996). *The origins of virtue: human instincts and the evolution of cooperation.* New York, Penguin Books. Ref Type: Generic
Rizzolatti, G., & Arbib, M. A. (1998). Language within our grasp. *Trends in Neurosciences, 21*, 188–194.
Rizzolatti, G., & Craighero, L. (2004). The mirror-neuron system. *Annual Review of Neuroscience, 27*, 169–192.
Rizzolatti, G., & Gallese, V. (2006). Do perception and action result from different brain circuits? The three visual systems hypothesis. In: J. L. Van Hemmen & T. J. Sejnowski (Eds.), *23 problems in systems neuroscience* (pp. 369–393). Oxford University Press: New York.
Rolls, E. T. (2004). The functions of the orbitofrontal cortex. *Brain Brain and Cognition, 55*, 11–29.
Saarela, M. V., Hlushchuk, Y., Williams, A. C., Schurmann, M., Kalso, E., & Hari, R. (2007). The compassionate brain: humans detect intensity of pain from another's face. *Cerebral Cortex, 17*, 230–237.
Seger, C. A., Stone, M., & Keenan, J. P. (2004). Cortical activations during judgments about the self and an other person. *Neuropsychologia, 42*, 1168–1177.
Seitz, R. J., Nickel, J., & Azari, N. P. (2006). Functional modularity of the medial prefrontal cortex: involvement in human empathy. *Neuropsychology, 20*, 743–751.
Silk, J. B., Brosnan, S. F., Vonk, J., Henrich, J., Povinelli, D. J., Richardson, A. S., Lambeth, S. P., Mascaro, J., & Schapiro, S. J. (2005). Chimpanzees are indifferent to the welfare of unrelated group members. *Nature, 437*, 1357–1359.
Simon, O., Mangin, J. F., Cohen, L., Le, B. D., & Dehaene, S. (2002). Topographical layout of hand, eye, calculation, and language-related areas in the human parietal lobe. *Neuron 33*, 475–487.
Singer, T. (2006). The neuronal basis and ontogeny of empathy and mind reading: review of literature and implications for future research. *Neuroscience and Biobehavioral Reviews, 30*, 855–863.
Singer, T. (2007). The neuronal basis of empathy and fairness. *Novartis Foundation Symposium, 278*, 20–30.
Singer, T., & Frith, C. (2005). The painful side of empathy. *Nature Neuroscience, 8*, 845–846.
Singer, T., Seymour, B., O'Doherty, J., Kaube, H., Dolan, R. J., & Frith, C. D. (2004). Empathy for pain involves the affective but not sensory components of pain. *Science, 303*, 1157–1162.
Smith, A. (2006). Cognitive empathy and emotional empahy in human behavior and evolution. *The Psychological Record, 56*, 3–21.
Soussignan, R. (2002). Duchenne smile, emotional experience, and autonomic reactivity: a test of the facial feedback hypothesis. *Emotion, 2*, 52–74.
Stark, C. E., & Squire, L. R. (2001). When zero is not zero: the problem of ambiguous baseline conditions in fMRI. *Proceedings of the National Academy of Sciences of the United States of America, 98*, 12760–12766.
Sturmer, S., Snyder, M., Kropp, A., & Siem, B. (2006). Empathy-motivated helping: the moderating role of group membership. *Personality and Social Psychology Bulletin, 32*, 943–956.

Tankersley, D., Stowe, C. J., & Huettel, S. A. (2007). Altruism is associated with an increased neural response to agency. *Nature Neuroscience, 10*, 150–151.

Tomasello, M., & Call, J. (1997). *Primate cognition*. Oxford University: New York.

Tomasello, M., Call, J., & Hare, B. (2003). Chimpanzees understand psychological states – the question is which ones and to what extent. *Trends in Cognitive Sciences, 7*, 153–156.

Van Hooff, J. (1973). A structural analysis of the social behavior of a semi-captive group of chimpanzees. In: M. Von Cranach & I. Vine (Eds.), *Social communication and movement* (pp. 75–162). Academic Press: London.

Visalberghi, E. (1994). Learning processes and feeding behavior in monkeys. In: B. G. Galef, M. Mainardi & P. Valsecchi (Eds.), *Behavioral aspects of feeding. Basic and applied research on mammals* (pp. 257–270). Harwood Academic: Chur, Switzerland.

Vogeley, K., Bussfeld, P., Newen, A., Herrmann, S., Happe, F., Falkai, P., Maier, W., Shah, N. J., Fink, G. R., & Zilles, K. (2001). Mind reading: neural mechanisms of theory of mind and self-perspective. *Neuroimage, 14*, 170–181.

Vuilleumier, P., Armony, J. L., Driver, J., & Dolan, R. J. (2003). Distinct spatial frequency sensitivities for processing faces and emotional expressions. *Nature Neuroscience, 6*, 624–631.

Wicker, B., Keysers, C., Plailly, J., Royet, J. P., Gallese, V., & Rizzolatti, G. (2003a). Both of us disgusted in My insula: the common neural basis of seeing and feeling disgust. *Neuron, 40*, 655–664.

Wicker, B., Keysers, C., Plailly, J., Royet, J. P., Gallese, V., & Rizzolatti, G. (2003b). Both of us disgusted in My Insula: the common neural basis of seeing and feeling disgust. *Neuron, 40*, 655–664.

Williams, J. H., Waiter, G. D., Gilchrist, A., Perrett, D. I., Murray, A. D., & Whiten, A. (2006). Neural mechanisms of imitation and 'mirror neuron' functioning in autistic spectrum disorder. *Neuropsychologia, 44*, 610–621.

Williams, J. H., Whiten, A., Suddendorf, T., & Perrett, D. I. (2001). Imitation, mirror neurons and autism. *Neuroscience and Biobehavioral Reviews, 25*, 287–295.

Part IV
Relationship to Cognitive Processes

Mirror Neurons and the Neural Exploitation Hypothesis: From Embodied Simulation to Social Cognition

Vittorio Gallese

Abstract The relevance of the discovery of mirror neurons in monkeys and of the mirror neuron system in humans to a neuroscientific account of primates' social cognition and its evolution is discussed. It is proposed that mirror neurons and the functional mechanism they underpin, embodied simulation, can ground within a unitary neurophysiological explanatory framework important aspects of human social cognition. A neurophysiological hypothesis – the "neural exploitation hypothesis" – is introduced to explain how key aspects of human social cognition are underpinned by brain mechanisms originally evolved for sensory-motor integration. It is proposed that these mechanisms were later on exapted as new neurofunctional architecture for thought and language, while retaining their original functions as well. By neural exploitation, social cognition and language can be linked to the experiential domain of action.

Keywords Action · Embodied simulation · Folk psychology · Intentions · Language · Mind reading · Mirror neurons · Monkey · Neural exploitation · Premotor cortex · Social cognition · Theory of mind

1 Introduction

The traditional view in the cognitive sciences holds that humans are able to understand the behavior of others in terms of their mental states – intentions, beliefs, and desires – by exploiting what is commonly designated as 'Folk Psychology.' According to a widely shared view, non-human primates, apes included, do not rely on mentally based accounts of each other's behavior. This view prefigures a sharp distinction between non-human species, confined to

V. Gallese
Department of Neuroscience, Section of Physiology, University of Parma,
43100 Parma, Italy
e-mail: vittorio.gallese@unipr.it

J.A. Pineda (ed.), *Mirror Neuron Systems*, DOI: 10.1007/978-1-59745-479-7_8,

behavior reading, and our species, whose social cognition makes use of a different level of explanation: mind reading. The neuroscientific investigation of the supposedly unique human capacity for mind reading, however, faces several problems, which I'll briefly enumerate.

Mind reading is today the topic of empirical investigation, particularly after the development of the powerful brain imaging technologies, enabling us to directly look at what happens inside of our brains when engaged in a variety of perceptual, executive, and cognitive tasks. We should be aware, though, of the risks inherent in a blind reliance upon the heuristic power of the sole brain imaging approach. Especially so, if the brain imaging evidence is blindly used to validate a preconceived notion of what the human mind supposedly is and how it works.

The automatic translation of the Folk Psychology-inspired 'flow charts' into encapsulated brain modules, specifically adapted to mind reading abilities, should be carefully scrutinized. Language can typically play ontological tricks by means of its 'constitutiveness', that is, its capacity to give an apparent ontological status to the concepts words embody (Bruner, 1986, p. 64). Space can provide a telling example of how our language-based definitions do not necessarily translate into real entities in the brain. Space, although unitary when introspectively examined, is not mapped in the brain as a single multipurpose central processing unit for space. Truth is that in the brain there are numerous spatial maps (see Rizzolatti, & Gallese, 1997). The same logic might apply to the employment of the cognitive tools of Folk Psychology, that is, our language-mediated definition of what does it mean to mind read.

It is by no means obvious that behavior reading and mind reading constitute two autonomous realms. In fact, during our social transactions we seldom engage in explicit interpretative acts. Most of the time, our understanding of others' behavior is immediate, automatic, and almost reflex-like. The claim that our capacity to reflect on propositional attitudes like intentions, beliefs, and desires as the causal determinants of others' behavior is *all there is* in social cognition, is arguable. It is even less obvious that, while understanding the intentions of others, we employ a cognitive strategy totally unrelated to predicting the consequences of their observed behavior (see below).

Social cognition is not only 'social metacognition,' that is, explicitly thinking about the contents of someone else's mind by means of symbols or other representations in propositional format. We can certainly 'explain' the behavior of others by using our most sophisticated mentalizing abilities, but the neural mechanisms underpinning such abilities are far from being understood. Chances that we will find boxes in our brain containing the neural correlates of beliefs, desires, and intentions *as such* probably amount next to zero. Such a search looks like an ill-suited form of reductionism leading us nowhere.

A growing criticism toward a blind faith in Folk Psychology as the sole characterization of social cognition is indeed surfacing within the field of philosophy of mind. It has been recently stressed that the use in social cognition of the belief-desire propositional attitudes of Folk Psychology is overstated (see

Hutto, 2004). As emphasized by Bruner (1993, p. 40), "When things *are as they should be*, the narratives of Folk Psychology are unnecessary."

The standard approach to the empirical investigation of mind reading and social cognition faces another problem, that of 'mereological fallacy' (see Bennett & Hacker, 2003), that is, the problem of attributing to parts of an organism attributes that are properties of the whole being. This problem is not only typical of representationalist cognitive science, but also of some quarters of cognitive neuroscience. Mentalizing, whatever it is, is a personal-level competence and therefore it cannot be fully reduced to the sub-personal activity of 'mind reading-specific' clusters of neurons within brain cortical areas. Neurons *are not* epistemic agents. Neurons only 'know' about ions passing through their membranes. To mentalize we need a person (see also Wittgenstein, 1953, § 281), that is, a properly wired brain-body system interacting with a specific environment populated by other brain-body systems. Neurons (mirror neurons included) or brain areas can at best be *necessary but not sufficient* conditions for mentalizing (Gallese, 2000, 2001).

A third problem concerns the typical solipsistic attitude of the standard cognitive representationalist account of the human mind and of social cognition. The dominant approach in cognitive science and in some quarters of philosophy of mind consists in clarifying the formal rules structuring a Cartesian rational and solipsistic mind. Much less investigated is what triggers the sense of social identification that we experience with the multiplicity of 'other selves' populating our social world.

A final problem is posed to the mainstream representationalist account of social cognition by the relationship between mind reading and linguistic competence. Recent evidence shows that 15 months-old infants understand false beliefs (Onishi and Baillargeon, 2005). Furthermore, several studies show that non-human primates (monkeys included) can understand others as intentional agents (see Hare, Call, Agnetta, & Tomasello, 2000; Santos, Nissen, & Ferrugia, 2006; Flombaum & Santos, 2006). These results suggest that typical constitutive aspects of what mentalizing is taken to be can be explained on the basis of low-level mechanisms, which develop well before full-blown linguistic competence, or even without requiring language.

How can we overcome the above-mentioned problems? Perhaps by informing our scientific investigation according to the following points: (1) We should adopt a bottom-up approach, by thoroughly investigating the non-metarepresentational aspects of human social cognition, so far unduly minimized or even neglected. Perhaps we should even abandon terms like 'mind reading' and 'Theory of Mind', because they introduce unwarranted preconceived biases; (2) Our scientific investigation of human social cognition should conform with an evolutionary perspective, hence should be complemented by the neurophysiological and psychological investigation of the functional mechanisms implicated in non-human primates' social cognition. The only way to go beyond a mere correlative use of brain imaging data acquired in humans consists in unveiling the neural mechanisms leading to the activation of different brain regions in different tasks; (3) By doing so, it might be established to which extent the apparently different social cognitive abilities and strategies adopted by different species of primates maybe produced by similar functional mechanisms, which in the course of evolution acquired increasing complexity.

The discovery of mirror neurons has changed our views on the relations among action, perception and cognition, and has boosted a renewed interest in the neuroscientific investigation of the social aspects of primate cognition. The main point of this paper is that most of the time we have a direct access to the world of others. Direct understanding does not require explanation. This particular dimension of social cognition is embodied in that it mediates between the multimodal experiential knowledge of our own lived body and the way we experience others.

I present here accounts of how mirroring mechanisms, described from a functional point of view as embodied simulation (Gallese, 2001, 2003a, 2003b, 2005a, 2005b), can underpin basic forms of social cognition like the capacity of directly understanding others' actions, intentions, emotions, and sensations. I also show that embodied simulation can play an explanatory role on more sophisticated aspects of social cognition, like in language. I conclude by introducing the 'neural exploitation hypothesis', according to which a single functional mechanism, embodied simulation, is likely at the basis of various and important aspects of social cognition.

2 Mirror Neurons

About 16 years ago, a new class of motor neurons was discovered in a sector of the macaque monkey's ventral premotor cortex, known as area F5. These neurons discharge not only when the monkey executes goal-related hand movements like grasping objects, but also when observing other individuals (monkeys or humans) executing similar acts. These neurons were called 'mirror neurons' (Gallese, Fadiga, Fogassi, & Rizzolatti, 1996; Rizzolatti, Fadiga, Gallese, & Fogassi, 1996). Neurons with similar properties were later on discovered in a sector of the posterior parietal cortex reciprocally connected with area F5 (Gallese, Fogassi, Fadiga, & Rizzolatti, 2002; Fogassi et al., 2005). Typically, mirror neurons in monkeys do not respond to the observation of an object alone, even when it is of interest to the monkey.

Action observation causes in the observer the automatic activation of the same neural mechanism triggered by action execution. The novelty of these findings resides in the fact that, for the first time, a neural mechanism allowing a direct matching between the sensory (visual and auditory) description of a motor act and its execution has been identified. This matching system provides a parsimonious solution to the problem of translating the results of the visual analysis of an observed movement – in principle, devoid of meaning for the observer – into something that the observer is able to understand to the extent that the observer 'experientially owns' it already. It was proposed that this 'direct-matching' mechanism could be at the basis of a direct form of action understanding (Gallese et al., 1996; Rizzolatti et al., 1996; Rizzolatti & Gallese, 1997). If mirror neurons really mediate action understanding, their activity should reflect the meaning of the observed act, not its visual features.

Accordingly, experiments by Umiltà et al. (2001) showed that F5 mirror neurons are activated also during the observation of partially hidden actions, when the monkey can predict the action outcome, even in the absence of the complete visual information about it. Macaque monkey's mirror neurons therefore respond to acts made by others not exclusively on the basis of their visual description, but on the basis of the anticipation of the final goal-state of the motor act, by means of the activation of its motor neural 'representation' in the observer's premotor cortex.

These data can hardly be reconciled with 'minimalist' interpretations of mirror neurons, as that proposed by Knoblich and Jordan (2002), according to which mirror neurons merely code "the perceived effect the action exerts on the object" (2002, p. 116). Furthermore, these data alone seem to contradict the notion that the functional mechanism at the basis of the activation of mirror neurons qualifies as a form of 'direct perception' of the acts of others (see Gallagher, this volume). It is obvious that there must be a system that visually analyzes and describes the acts of others. A 'direct perception' of the observed motor acts most likely describes the activation of extrastriate visual neurons sensitive to biological motion. However, the view that such 'pictorial' analysis is per se sufficient to provide an understanding of the observed act must be questioned. Without reference to the observer's internal 'motor knowledge', this description is devoid of experiential meaning for the observing individual.[1] It must be stressed that this internal motor knowledge is *procedural* and *not representational*. Furthermore, the activation of mirror neurons in the experiment by Umiltà et al. (2001) testifies something even more different from a 'direct perception': it exemplifies a form of simulation-driven 'motor inference.'

In other experiments, it has been shown that a particular class of F5 mirror neurons ('audio-visual mirror neurons') respond not only when the monkey *executes* and *observes* a given hand action, but also when it just *hears* the sound typically produced by the action (Kohler et al., 2002). These neurons respond to the sound of actions and discriminate between the sounds of different actions, but do not respond to other similarly interesting sounds such as arousing noises, or monkeys' and other animals' vocalizations. Events as different as sounds, images, or voluntary acts of the body, are nevertheless mapped by the same network of audio-visual mirror neurons. The presence within a

---

[1] Studies conducted on human infants have consistently revealed a tight relationship between motor experience and the capacity of evaluating the goal-relatedness of the actions of others (see Woodward 1998; Sommerville, Woodward, & Needham, 2005; Sommerville & Woodward, 2005; Falck-Ytter, Gredeback, & von Hofsten, 2006).

non-linguistic species of such neural matching system can be interpreted as the dawning of an embodied 'conceptualization' mechanism, that is, a mechanism that grounds meaning in the situated and experience-dependent systematic interaction with the world (see Gallese, 2003c; Gallese & Lakoff, 2005). The world becomes 'our world' to the extent that it evokes and is subjected to our actions.

In the most lateral part of area F5 a population of mirror neurons related to the execution/observation of mouth acts has been discovered (Ferrari, Gallese, Rizzolatti, & Fogassi, 2003). The majority of these neurons discharge when the monkey executes and observes transitive, object-related ingestive acts, such as grasping, biting, or licking. However, a small percentage of mouth-related mirror neurons discharge during the observation of intransitive, communicative facial gestures performed by the experimenter in front of the monkey ('communicative mirror neurons', Ferrari et al., 2003). Macaque monkeys seem to have an initial capacity to control and emit "voluntarily" facial social signals, mediated by the frontal lobe. Most interestingly, this capacity develops in a cortical area – area F5 – that in humans became Brodmann's area 44, a key area for verbal communication (Rizzolatti & Arbib, 1998; Nelissen, Luppino, Vanduffel, Rizzolatti, & Orban, 2005).

A major step forward in the research on the MNS consisted in the discovery that parietal mirror neurons not only code the goal of an executed/observed motor act, like grasping an object, but they also code the overall action intention (e.g., bringing the grasped object to the mouth or into a container, Fogassi et al., 2005). The MNS maps integrated sequences of goal-related motor acts (grasping, holding, bringing, placing, the different 'words' of a 'motor vocabulary', see Rizzolatti et al., 1988) so to obtain different and parallel intentional 'action sentences', that is, temporally chained sequences of motor acts properly assembled to accomplish a more distal goal-state. The 'motor vocabulary' of grasping–related neurons, by sequential chaining, reorganizes itself as to map the fulfillment of an action intention. The overall action intention (to eat, to place the food, or object) is the goal-state of the ultimate goal-related motor act of the chain.

These results seem to suggest – at least at such a basic level – that the 'prior intention' of eating or placing the food is also coded by parietal mirror neurons. Of course, this does not imply that monkeys *explicitly represent* prior intentions as such.

3 Mirroring Mechanisms in Humans

Several studies using different experimental methodologies and techniques have demonstrated also in the human brain the existence of a mechanism directly matching action perception and execution, defined as the Mirror Neuron System (MNS) (for review, see Rizzolatti, Fogassi, & Gallese, 2001;

Gallese 2003a, 2003b, 2006; Gallese, Keysers, & Rizzolatti, 2004; Rizzolatti & Craighero, 2004). Although direct evidence of the presence in the human brain of *single neurons* exemplifying the properties displayed by mirror neurons in the monkey brain is at present missing, I think it is fair to interpret the extant evidence as implying their existence.

During action observation, there is a strong activation of premotor and posterior parietal areas, the likely human homologue of the monkey areas in which mirror neurons were originally described. The mirroring mechanism for actions in humans is coarsely somatotopically organized, with distinct cortical regions within the premotor and posterior parietal cortices being activated by the observation/execution of mouth, hand, and foot related acts (Buccino et al., 2001; see also Aziz-Zadeh, Wilson, Rizzolatti, & Iacoboni, 2006). A recent study by Buxbaum, Kyle, & Menon, (2005) on posterior parietal neurological patients with Ideomotor Apraxia has shown that they were not only disproportionately impaired in the imitation of transitive as compared to intransitive gestures, but they also showed a strong correlation between imitation deficits and the incapacity of recognizing observed goal-related meaningful hand actions. These results show that the same motor 'representations' underpin both action production and action understanding.

The MNS in humans is directly involved in imitation of simple movements (Iacoboni et al., 1999), imitation learning of complex skills (Buccino et al., 2004a), in the perception of communicative actions (Buccino et al., 2004b), and in the detection of action intentions (Iacoboni et al., 2005). In this study, participants witnessed three kinds of videos portraying: grasping hand actions without a context, context only (a scene containing objects), and grasping hand actions embedded in contexts. In the latter condition, the context suggested the intention associated with the grasping action (either drinking or cleaning up). Actions embedded in contexts, compared with the other two conditions, yielded a significant signal increase in the posterior part of the inferior frontal gyrus and the adjacent sector of the ventral premotor cortex where hand actions are represented. Thus, premotor mirror areas – areas active during the execution and the observation of motor acts – previously thought to be involved only in action recognition are actually also involved in understanding the 'why' of action, that is, the intention promoting it.

Another result of this study lends further support to the motor hypothesis of intention detection. After the scanning session, all participants were debriefed about the actions they had witnessed. Independently from having been or not explicitly instructed to determine the intention of the observed actions of others, they all correctly identified the action intentions. It must be stressed that the different instructions in the two groups did not affect the activation of premotor mirror areas. This means that – at least for simple actions as those employed in this study – the detection of intentions occurs by default and it is underpinned by the mandatory activation of an embodied simulation mechanism.

Our results seem to suggest that that most of the time even we humans do not explicitly represent intentions as such when understanding them in others.

Action intentions are not *just* propositional contents. They are embedded within the intrinsic intentionality of action, its intrinsic relatedness to an end-state, a goal. My point is that most of the time we do not *ascribe intentions to others*, we simply detect them. By means of embodied simulation, when witnessing others' behaviors their motor intentional contents can be directly grasped without the need of representing them in propositional format.

Other mirroring mechanisms seem to be involved with our capacity to share emotions and sensations with others (for review, see Gallese, 2003a, 2003b, 2006; Gallese et al., 2004; de Vignemont and Singer, 2006). When we perceive others expressing a given basic emotion such as disgust, the same brain areas are activated as when we subjectively experience the same emotion (Wicker et al., 2003). Similar direct matching mechanisms have been described for the direct and indirect perception of pain (Hutchison, Davis, Lozano, Tasker, & Dostrovsky, 1999; Singer et al., 2004; Jackson, Meltzoff, & Decety, 2005; Botvinick et al., 2005) and touch (Keysers et al., 2004; Blakemore, Bristow, Bird, Frith, & Ward, 2005).

These results altogether suggest that our capacity to empathize with others is mediated by embodied simulation mechanisms, that is, by the activation of the same neural circuits underpinning our own motor, emotional, and sensory experiences (see Gallese, 2001, 2003a, 2003b, 2005a, 2005b, 2006, 2007; Gallese et al., 2004).

4 The Development of Mirroring Mechanisms and Social Identification

One crucial issue still not clarified is how the MNS develops in ontogeny. To which extent are the mirroring mechanisms described in this paper innate and how are they shaped, and modeled during development? The answer to these questions is that we simply do not know.

We know that motor skills mature much earlier on than previously thought. In a recent study (Zoia et al., 2007), the kinematics of fetal hand movements were measured. The results showed that the spatial and temporal characteristics of fetal movements were by no means uncoordinated or unpatterned. By 22 weeks of gestation fetal hand movements show kinematics patterns that depend on the goal of the different motor acts fetuses perform. This results led the authors of this study to argue that 22 weeks old fetuses show a surprisingly advanced level of motor planning, already compatible with the execution of 'intentional actions.'

Were these results eventually confirmed and extended, one might speculate that during prenatal development specific connections may develop between the motor centers controlling mouth and hand goal-directed behaviors and the brain regions that will become recipient of visual inputs after birth. Such connectivity could provide functional templates (e.g., specific spatio-temporal patterns of neural firing) to areas of the brain that, once reached by visual inputs after birth, would be ready to specifically respond to the observation of biological motion like hand or facial gestures. In other words, neonates and infants, by means of a specific connectivity developed during the late phase of gestation between motor and 'to-become-visual' regions of the brain, would be ready to mirror and imitate the gestures performed by adult caregivers in front of them, and would be endowed with the neural resources enabling the reciprocal behaviors that characterize our post-natal life since its very beginning (see Braten, 1988, 1992, 2007; Meltzoff & Moore, 1977, 1998; Meltzoff & Brooks, 2001; Stern, 1985; Trevarthen, 1979, 1993; Tronick, 1989).

The shared intersubjective 'we-centric' space mapped by mirroring mechanisms is likely crucial in bonding neonates and infants to the social world, but it progressively acquires a different role. It provides the self with the capacity to simultaneously entertain self-other identification and difference (see Uddin, Kaplan, Molnar-Szakacs, Zaidel, & Iacoboni, 2005; Uddin, Molnar-Szakacs, Zaidel, & Iacoboni, 2006). Once the crucial bonds with the world of others are established, this space carries over to the adult conceptual faculty of socially mapping sameness and difference ("I am a different self").

Social identification, the 'selfness' we readily attribute to others, the inner feeling of being 'like-me' (Meltzoff & Brooks, 2001; Meltzoff, 2007a, 2007b) triggered by our encounter with others, are the result of the shared we-centric space enabled by embodied simulation. Self-other physical and epistemic interactions are shaped and conditioned by the same body and environmental constraints. This common relational character is underpinned, at the sub-personal level of the brain, by shared mirroring neural networks. These shared neural mechanisms enable the shareable character of actions, emotions, and sensations, the earliest constituents of our social life.

5 Embodied Simulation and Intentional Attunement

Our capacity to conceive of the acting bodies of others as *selves* like us depends on the constitution of a 'we-centric' shared meaningful interpersonal space. This 'shared manifold' (see Gallese, 2001, 2003a, 2003b, 2005a, 2005b) can be characterized at the functional level as embodied simulation, a specific mechanism by means of which our brain/body system models its interactions with the world. I propose that the different mirroring mechanisms described in this article constitute the sub-personal instantiation of embodied simulation.

According to my model, when we witness the intentional behavior of others, embodied simulation generates a specific phenomenal state of 'intentional attunement.' This phenomenal state in turn generates a peculiar quality of identification with other individuals, produced by the collapse of the others' intentions into the observer's ones. By means of embodied simulation we do not just 'see' an action, an emotion, or a sensation. Side by side with the sensory description of the observed social stimuli, neural correlates of the body states associated with these actions, emotions, and sensations are activated in the observer. As previously proposed (Gallese, 2001, 2003b), recent studies suggest that some of these mechanisms could be altered in individuals affected by the Autistic Spectrum Disorder (for a review, see Gallese, 2006; Oberman & Ramachandran, 2007).

The roots of intentionality are to be found in the intrinsically relational nature of action. The mirroring mechanisms described here map the different intentional relations in a fashion that is neutral about the specific quality or identity of the agentive/subjective parameter. By means of a shared functional state realized in two different bodies obeying to the same functional rules, the 'objectual other' becomes 'another self.'

Of course, embodied simulation is not the only functional mechanism underpinning social cognition. Social stimuli can also be understood on the basis of the explicit cognitive elaboration of their contextual perceptual features, by exploiting previously acquired knowledge about relevant aspects of the situation to be analyzed. Our capacity of attributing true or false beliefs to others, our most sophisticated cognitive abilities, likely involve the activation of large regions of our brain, certainly larger than a putative and domain-specific Theory of Mind Module, be it located in the anterior paracingulate cortex (see Gallagher & Frith, 2003) –already almost out of fashion (see Bird, Castelli, Malik, Frith, & Husain, 2004) – or in the nowadays much celebrated temporo-parietal junction (TPJ, see Saxe, 2005).

Recently there have been attempts to reconcile within social cognition the evidence about mirroring mechanisms with the activation of frontal midline structures during mentalizing tasks. These attempts aim to an ecumenical or hybrid account of mentalization (see Uddin, Iacoboni, Lange, & Keenan, 2007; Keysers & Gazzola, 2007; Goldman, 2006 and this volume). I think these attempts are open to the same criticism previously raised against the mainstream standard cognitive account of the solipsistic representational mind. In fact, these attempts reify a Cartesian Self, supposedly the recipient of the outcome of the mentalizing process, by reducing it to the neural processing instantiated by a localized network of mesial cortical areas. Reductionism works if it is methodological, not if – as in these examples – becomes ontological.

The embodied simulation model is immune to these criticisms, because it postulates a situated Self that in virtue of the facticity of its pragmatic being-in-the-world, is constitutively "open to the other", to which is connected by means of the we-centric shared manifold. A self whose proper development depends on the possibility of *mirroring and being mirrored from* the praxis of the other. A Self that most of the time doesn't even 'attribute' intentions to others, because these intentions are grasped as already embedded in the behavior of others. The witnessed behavior of others triggers at the sub-personal level the activation of mirroring neural networks, henceforth activating – at the functional level of the description – embodied simulation.

6 The MNS and Its Relevance in the Evolution of Social Cognition

The presence of mirror neurons in different species of primates such as macaques and humans seem to favor a continuist view of the evolution of social cognition. However, it is also true that the very same evidence must be reconciled with the uniqueness of human (social) cognition. This is a major topic for future research.

It is likely that monkeys use their MNS to optimize their social interactions, by making sense of their conspecifics' behavior. The evidence collected so far suggests that the MNS is enough sophisticated to enable its exploitation for different social purposes among macaque monkey, like social facilitation (Ferrari, Maiolini, Addessi, Fogassi, & Visalberghi, 2005), and the recognition of being imitated (Paukner, Anderson, Borelli, Visalberghi, & Ferrari, 2005).

It must be noted that the apparent inability of non-human primates (chimps included) to understand others as intentional agents (see Povinelli & Eddy, 1996; see also Povinelli & Vonk, 2003), given the presence of the MNS in non-human primates, has led scholars to dispute and argue against a role of the MNS in providing "sufficient basis for agentive understanding" (see Pacherie & Dokic, 2006, p. 106), or as playing – by means of motor simulation – a major role in social cognition (Jacob & Jeannerod, 2004).

The chimps' inability to understand others as intentional agents, however, turned out to be only apparent in particular *cooperative* contexts. In fact, there is evidence that chimps when engaged in a *competitive* setting do deduce what others know on the basis of where they are looking at (Hare et al., 2000; Tomasello, Carpenter, Call, Behne, & Moll, 2005).

Even more importantly, it has been recently shown that *rhesus monkeys* can establish a cognitive link between seeing and knowing, by systematically choosing to steal food from the human competitor that could not see the food, while refraining from doing it when the human competitor could see it (Flombaum & Santos, 2006). Similarly, it has been shown that rhesus monkeys choose to obtain food silently only in situations in which silence is crucial to remain undetected by a human competitor (Santos et al., 2006).

These results show that non-human primates – including macaques – possess the ability to recognize others as intentional agents and to deduce what others know about the world on the basis of ostensive behavioral cues like the direction of gaze. To which extent this capacity depends on the MNS will require further investigations. However, it is clear that the so far misconceived and underrated social cognitive abilities of non-human primates cannot constitute an argument against the relevance of the MNS for social cognition.

Monkeys most likely do not explicitly attribute mental states to others as a causal explanation of others' behavior. Many posit that in order to do that one needs language. To thoroughly understand social cognition we must therefore investigate the neuro-functional mechanisms at the basis of the faculty of language. Before doing that, however, we should clarify what we refer to we when speaking about language. Human language for most of its history has been just spoken language. This suggests that human language likely evolved from dialogic speech in order to provide individuals with a more powerful and flexible social cognitive tool to share, communicate and exchange knowledge.

In the final sections I propose to look at the experiential groundings of human linguistic competence and emphasize the tight connection between meaning and action. I then show how the functional mechanism supposedly describing the function of the MNS – embodied simulation – can also ground several aspects of social cognition, language included.

7 A 'Neurophenomenological' Account of Language: Action, Experience, and Their Expression

Speech is comparable to a gesture because what it is charged with expressing will be in the same relation to it as the goal is to the gesture which intends it. (Merleau-Ponty, 1960/1964, p. 89)

The intimate nature of language and the evolutionary process producing are still matter of debate. This is partly due to the complexity and multidimensional nature of language. What do we refer to when we investigate the language faculty and its evolution? Barrett, Henzi, & Rendall, (2007) have recently argued that apparent cognitive complexity in the social domain emerges from the interaction of brain, body, and world, rather than being a mere outcome of the level of intrinsic cognitive complexity primate species possess.

Viewing social cognition as an embodied and situated enterprise (see Clark, 1997; Barsalou, 1999; Lakoff & Johnson, 1980, 1999; Anderson, 2003; Gallese, 2003; Barrett & Henzi, 2005; Niedenthal, Barsalou, Winkielman, Krauth-Gruber, & Ric, 2005) offers the possibility of a new neuroscientific approach to language. Let us see why and how by first briefly introducing the perspective of phenomenology, which provides stimulating perspectiveson the nature and structure of human experience and its connection to language, by putting bodily action at the center of the stage. By following the phenomenological perspective we learn that language is a social enterprise in which action plays a crucial role.

A caveat first. Neuroscience's goal is not to validate or confute philosophical theories. However, when neuroscience aims at understanding personal-level issues like language and meaning, it cannot escape a serious confrontation with a philosophy. By means of such multidisciplinary dialogue, a new philosophy of nature can emerge.

Probably one of the phenomenology's greatest merits is to have pointed out that every form of consciousness is intentional, that is, *consciousness of something*, and to have stressed how cognitive forms of intentionality are rooted in the aboutness and relatedness to the world of our bodily actions. The facticity of human experience is at the core of Martin Heidegger's speculation, with his notion of being-in-the-world (1927/1962). According to Heidegger, being and world are to be seen as a unitary phenomenon, intrinsically and ontologically connected. This perspective substantially blurs distinctions between subject and object and between internal and external realms.

In Heidegger's perspective, animals and humans profoundly differ with respect to their relationship with the world. Only humans fully possess a world, because only human existence has a true historical dimension, which, in turn, depends on language (Heidegger, 1929/1995). According to Heidegger, language is meaningful because it reveals and discloses possibilities of contextual actions (1927/1962). Meaning emerges from a peculiar historical world to which humans are connected through their daily interactions with it (Heidegger, 1925/1985). Hence language is ontologically of practical nature. Terms like 'concepts' and 'thoughts,' according to Heidegger are to be understood as originating from our practical experience of the world. This is what Heidegger implies by claiming that meaning has its roots in the ontology of being-in-the-world (1927/1962). Being-in-the-world precedes reflection.

Our understanding of the meaning of a word like 'table' does not stem from our use of a linguistic game, which, at best, can specify when to apply a given word as a tag to a given object in the world. The meaning of 'table' stems from its use, from what we can do with it, that is, from the multiple and interrelated possibilities for action it evokes.[2]

Neuroscience today shows that the scientific investigation of the 'Korper' (the brain-body system) can shed light on the 'Leib' (the lived body of experience), as the latter is the lived expression of the former. The neurophysiological aspects of action did not interest philosophers like Husserl and Heidegger, also because of the mechanistic views purported by neurobiology at the beginning of last century. The phenomenological approach, though, clearly shows that meaning does not inhabit a pre-given Platonic world of ideal and eternal truths to which mental representations connect and conform. Phenomenology thus entertains a perspective partly compatible with many empirical results of contemporary cognitive neuroscience: Meaning is the outcome of our situated interactions with the world.

With the advent of language, and even more with the 'discovery' of written language, meaning is amplified as it frees itself from being dependent upon specific instantiations of actual experience. Language connects all possible actions within a network expanding the meaning of individual situated experiences. Language evokes the totality of possibilities for action the world calls upon us, and structures action within a web of related meanings. By endorsing this perspective, it follows that if we confine language to its sole predicative use, we reify a consistent part of language's nature. Our understanding of linguistic expressions is not solely an epistemic attitude; it is a way of being. Our way of

[2] It is worth noting how long it took before a similar perspective emerged in the field of cognitive psychology (see Gibson 1979; Lakoff and Johnson 1980; Glenberg 1997; see also Gallese 2000, 2003c).

being, in turn, depends on what we act, how we do it, and how the world responds to us.

The relationship between language and body is also emphasized by Maurice Merleau-Ponty (1960/1964). According to Merleau-Ponty, signification arouses speech as the world arouses the body. For the speaking subject to express a meaning is to become fully aware of it. In other words, the signifying intention of the speaker can be conceived as a gap to be filled with words. When we speak, by means of the shared neural networks activated by embodied simulation, we experience the presence of others in ourselves and of ourselves in others. This mirroring likely helps filling the gap.

A further contribution in clarifying the relationship between language, action, and experience comes from Paul Ricoeur's hermeneutic phenomenology. According to Ricoeur, language is first and foremost *discourse*, and therefore the "mimetic bond between the act of saying and effective action is never completely severed" (1986/1991, p. XIV). The hermeneutic development of phenomenology in Ricoeur's approach connects intentionality to meaning: the logical sense of language must be grounded in the broader notion of meaning that is coextensive with the notion of intentionality (Ricoeur, 1986/1991, p. 40). In 'From Text to Action' (1985/1991), Ricoeur builds upon the historical dichotomy introduced in linguistics by de Saussure (1973/1974) and by Hjelmslev (1959) between *language* and *speech*, or between *schema* and *use*, respectively, and draws an important distinction between the formal language studied by structural linguistics and discourse, and in particular its original form, speech. According to Ricoeur, discourse is seen as an event, taking place in time to a speaker who speaks about something. By means of discourse, language acquires a situated world. It is in discourse that all meanings are transferred, hence "…discourse not only has a world but has an other, another person, an interlocutor to whom it is addressed" (Ricoeur, 1986/1991, p. 78).

The action-related account of language proposed by phenomenology and its intersubjective framing suggest that the neuroscientific investigation of what language is and how it works should begin from the domain of action. This investigation has already produced remarkable findings. The MNS provides a matching mechanism that seems to play an important role in social cognition, thus it looks like a very good candidate also for grounding the social nature of language.

Accumulating evidence shows that humans, when processing language, by means of embodied simulation activate the motor system at several of the levels traditionally describing language.[3] Two of these levels will be addressed here. The first level, defined as "embodied simulation at the vehicle level", pertains to phono-articulatory aspects of language. The second level, defined as "embodied simulation at the content level", concerns the semantic content of a word, verb, or proposition.

---

[3] To which extent these levels can be conceived as distinctly mapped in the brain is not so obvious yet.

8 Embodied Simulation and Language: Simulation at the Vehicle Level

It is now ascertained that Broca's region, formerly considered an exclusive speech-production area, contains neurons activated by the execution/observation/imitation of oro-facial gestures and of hand acts. This region is known to be part of the MNS (Bookheimer, 2002; Rizzolatti & Craighero, 2004; Nishitani, Schurmann, Amunts, & Hari, 2005). In an elegant transcranial magnetic stimulation (TMS) experiment, Fadiga, Craighero, Buccino, and Rizzolatti, (2002) showed that listening to phonemes induces an increase of the amplitude of motor-evoked potentials (MEPs) recorded from the tongue muscles involved in their execution. This result was interpreted as a motor resonance mechanism at the phonological level.

These findings have been complemented by a TMS study of Watkins, Strafella, Paus, (2003), showing that listening to and viewing speech gestures enhanced the amplitude of MEPs recorded from lip muscles. A recent fMRI study demonstrated the activation of motor areas devoted to speech production during passive listening to phonemes (Wilson, Saygin, Sereno, & Iacoboni, 2004). Finally, Watkins and Paus (2004) showed that during auditory speech perception, the increased size of the MEPs obtained by TMS over the face area of the primary motor cortex correlated with cerebral blood flow increase in Broca's area. This suggests that the activation of the MNS for facial gestures in the premotor cortex facilitates the primary motor cortex output to facial muscles, as evoked by TMS.

Not only speech perception, but also covert speech production activates the motor system. McGuigan and Dollins (1989) showed with EMG that tongue and lip muscles are activated in covert speech in the same way as during overt speech. An fMRI study by Wildgruber, Ackermann, Klose, Kardatzki, and Grodd (1996) showed primary motor cortex activation during covert speech. Finally, a recent study by Aziz-Zadeh, Cattaneo, Rochat, and Rizzolatti (2005) showed internal speech arrest after transient inactivation with repetitive TMS of the left primary motor cortex and left premotor BA 44.

The presence in Broca's region of both hand and mouth motor representations not only can cast some light on the evolution of language (Fadiga & Gallese, 1997; Rizzolatti & Arbib, 1998; Corballis, 2002, 2004; Arbib, 2005; Gentilucci & Corballis, 2006), but also on its ontogeny in humans. A tight relationship between the development of manual and oral motor skills has been repeatedly documented in children. Goldin-Meadow (1999) proposed that speech production and speech-related hand gestures could be considered as outputs of the same process. Canonical babbling in 6–8 months aged children is accompanied by rhythmic hand movements (Masataka, 2001). Hearing babies born to deaf parents display hand actions with a babbling-like rhythm. Manual gestures anticipate early development of speech in children and reportedly predict later success up to the two-word level (Iverson & Goldin-Meadow, 2005).

The same relationship between manual and oral language-related gestures persists in adulthood. Several works by Gentilucci and colleagues (Gentilucci, 2003; Gentilucci, Benuzzi, Gangitano, & Grimaldi, 2001; Gentilucci, Santunione, Roy, & Stefanini, 2004a; Gentilucci, Stefanini, Roy, & Santunione, 2004b) have shown a close relationship between speech production and the execution/observation of arm and hand gestures. This suggests that the system involved in speech production shares (and may derive from) the neural premotor circuit involved in the control of hand/arm actions.

In another related study, Gentilucci et al. (2004b) showed that different observed actions influence lip shaping kinematics and voice formants of the observer. The observation of grasping influences the first formant, which is related to mouth opening, while the observation of bringing to the mouth influences the second formant of the voice spectrum, related to tongue position. All of these effects are greater in children than in adults. As proposed by Gentilucci et al. (2004b), this mechanism may have facilitated the evolutionary shift from a primitive arm gesture communication system to speech. The phono-articulatory aspects of speech production, in principle as remote as possible from meaning, show unexpected connections with the execution/observation of socially meaningful arm motor acts.

In a very recent paper, Bernardis and Gentilucci (2006) showed that word and corresponding-in-meaning communicative arm gesture influence each other when they are simultaneously emitted. In sum, spoken words and symbolic communicative gestures are coded as a single signal by a unique communication system within the premotor cortex.

The involvement of premotor Broca's area in translating the representations of communicative arm gestures into mouth articulation gestures was recently confirmed by transient inactivation of BA 44 with repetitive TMS (Gentilucci, Bernardis, Crisi, & Volta, 2006). Since BA 44 is part of the MNS, it is likely to posit that through embodied simulation the communicative meaning of gestures is fused with the articulation of sounds required to express it in words. It appears that within premotor BA 44, 'vehicle' and 'content' of social communication are tightly interconnected. This is consistent with some tenets of 'constructionist' approaches to language, according to which *all* levels of linguistic descriptions involve pairing of forms with semantic/discourse functions (Goldberg, 2003).

9 Embodied Simulation and Language: Simulation at the Content Level

The meaning of a sentence, regardless of its content, has been classically considered to be understood by relying on symbolic, amodal mental representations (Pylyshyn, 1984; Fodor, 1998). An alternative hypothesis assumes that the understanding of language relies on 'embodiment' (Lakoff & Johnson,

1980, 1999; Lakoff, 1987; Glenberg, 1997; Rizzolatti & Gallese, 1997; Barsalou, 1999; Pulvermüller, 1999, 2002; Glenberg & Robertson, 2000; Gallese, 2003c; Gallese & Lakoff, 2005).

According to embodiment theory, the neural structures presiding over action execution should also play a role in understanding the semantic content of the same actions when verbally described. Empirical evidence shows this to be the case. Glenberg and Kaschak (2002) asked participants to judge if a read sentence was sensible or nonsense by moving their hand to a button requiring movement away from the body (in one condition) or toward the body (in the other condition). Readers responded faster to sentences describing actions whose direction was congruent with the required response movement. This clearly shows that action contributes to sentence comprehension.

The most surprising result of this study, though, was that the same interaction between sentence movement direction and response direction was also found with abstract sentences describing transfer of information from one person to another such as, "Liz told you the story" vs. "You told Liz the story." These latter results extend the role of action simulation to the understanding of sentences describing abstract situations (for similar results, see also Borghi, Glenberg, & Kaschak, 2004; Matlock, 2004).

A prediction of the embodiment theory of language understanding is that when individuals listen to action-related sentences, their MNS should be modulated. The effect of this modulation should influence the excitability of the primary motor cortex, henceforth the production of the movements it controls. To test this hypothesis we carried out two experiments (Buccino et al., 2005). In the first experiment, by means of single pulse TMS, either the hand or the foot/leg motor areas in the left hemisphere were stimulated in distinct experimental sessions, while participants were listening to sentences expressing hand and foot actions. Listening to abstract content sentences served as a control. Motor evoked potentials (MEPs) were recorded from hand and foot muscles. Results showed that MEPs recorded from hand muscles were specifically modulated by listening to hand action-related sentences, as were MEPs recorded from foot muscles by listening to foot action-related sentences.

In the second behavioral experiment, participants had to respond with the hand or the foot while listening to sentences expressing hand and foot actions, as compared to abstract sentences. Coherently with the results obtained with TMS, reaction times of the two effectors were specifically modulated by the effector-congruent heard sentences. These data show that processing sentences describing actions activates different sectors of the motor system, depending on the effector used in the listened action.[4]

---

[4] A discussion of the facilitatory or inhibitory nature of the specific modulation of the motor system during language processing is beyond the scope of this article, and therefore will not be dealt with here.

Several brain imaging studies have shown that processing action-related linguistic material in order to retrieve its meaning activates regions of the motor system congruent with the processed semantic content. Hauk, Johnsrude, and Pulvermüller, (2004) showed in an event-related fMRI study that silent reading of *words* referring to face, arm, or leg actions (e.g., lick, pick, kick) led to the activation of different sectors of the premotor-motor areas controlling motor acts of the body congruent with the referential meaning of the read action words. Tettamanti et al. (2005) showed that listening to *sentences* expressing actions performed with the mouth, the hand, and the foot, produces activation of different sectors of the premotor cortex, depending on the effector used in the action-related sentence listened to by participants. These activated premotor sectors correspond, albeit only coarsely, with those active during the observation of hand, mouth, and foot actions (Buccino et al., 2001). These results have been recently replicated and expanded by Aziz-Zadeh et al. (2006), who showed with a detailed voxel-based analysis that the same cortical regions activated by action observation were also activated by the understanding of action-related sentences.

The MNS is involved not only in understanding visually presented actions, but also in mapping acoustically or visually presented action-related sentences. It must be added, though, that the precise functional relevance of the MNS and embodied simulation in the process of language understanding remain unclear. One might argue that their involvement simply reflects motor imagery induced by the understanding process, which, in turn, might occur somewhere upstream, within the supposedly 'language-specific part of the brain', whatever it might be. The study of the spatio-temporal dynamic of language processing becomes crucial in settling this issue.

Evoked-Readiness-Potentials (ERP) experiments on silent reading of face-, arm- and leg-related words showed somatotopically-specific differential activations ~200 ms after word onset, with a strongest inferior frontal source for face-related words, and a maximal superior central source for leg-related words (Pulvermüller, Härle, & Hummel, 2000). This early differential activation can be hardly reconciled with the 'late motor imagery' hypothesis, while is more consistent with the embodied simulation account of language understanding.

This dissociation in brain activity patterns supports the idea of stimulus-triggered early lexico–semantic processes taking place within the premotor cortex. Pulvermüller, Shtyrov, and Ilmoniemi, (2003) using Magnetoencephalografy (MEG) showed that auditory areas of the left superior-temporal lobe became active 136 ms after the information in the acoustic input was sufficient for identifying the word, and activation of the left inferior-frontal cortex followed after an additional delay of 22 ms.

In sum, although these results are not conclusive on the relevance of embodied simulation of action for language understanding, they show that simulation is specific, automatic, and has a temporal dynamic compatible with such a function.[5] It must be added that several neuropsychological studies testify that frontal lesions including the premotor cortex produce deficits in the comprehension of action verbs (Bak, O'Donovan, Xuereb, Boniface, & Hodges, 2001; Bak et al., 2006; Bak & Hodges, 2003; Kemmerer & Tranel, 2000, 2003). We certainly need more research. More inactivation studies on healthy subjects and the careful neuropsychological study of patients with focal brain lesions will tell us more about the validity of this hypothesis.

10 The 'Neural Exploitation Hypothesis'

Let us finally address the wider implications of the MNS and embodied simulation for social cognition, by formulating the 'neural exploitation hypothesis.' My main claim is that the key aspects of human social cognition are produced by neural exploitation, that is, by the exaptation of neural mechanisms originally evolved for sensory-motor integration, later on also employed to contribute to the neurofunctional architecture of thought and language, while retaining their original functions as well (Gallese & Lakoff, 2005; see also Gallese, 2003c, 2007).

The execution of any complex coordinated action makes use of at least two cortical sectors – the premotor cortex and the motor cortex, linked by reciprocal neural connections. The motor cortex to a large extent controls individual synergies – relatively simple movements like extending and flexing the fingers, turning the wrist, flexing and extending the elbow, etc. The role of the premotor cortex is more complex: structuring simple motor behaviors into coordinated motor acts. The premotor cortex must thus provide a 'phase structure' to actions and specify the right parameter values in the right phases, for example, by activating the appropriate clusters of corticospinal neurons in the appropriate temporal order. This information is conveyed through neural connections by the premotor cortex to specific regions of the primary motor cortex. Similarly, as exemplified by the MNS, the same premotor circuitry controlling action execution instantiates the embodied simulation of the observed actions of others.

There is therefore 'structuring' neurofunctional architecture within the premotor system that can function according to two modes of operation. In the first operation mode, documented by some of the empirical evidence reviewed here, the circuit structures action execution and action perception, imitation, and imagination, with neural connections to motor effectors and/or other sensory cortical areas. When the action is executed or imitated, the corticospinal pathway is activated, leading to the excitation of muscles and the ensuing movements. When the action is observed or imagined, its actual execution is

[5] In the present paper I exclusively focus on action. Other studies, though, also show the involvement of the sensory-motor system in the mapping of other abstract domains, like the case of time mapped onto spatial metaphors (see Boroditsky, 2000; Boroditsky & Ramscar, 2002).

inhibited. The cortical motor network is activated (though, not in all of its components and, likely, not with the same intensity[6]), but action is not produced, it is only simulated.

In the second mode of operation, the same system is decupled from its action execution/perception functions and can offer its structuring output to non sensory-motor parts of the brain (Lakoff & Johnson, 1999; Gallese & Lakoff, 2005), among which the dorsal prefrontal cortex most likely plays a pivotal role. When engaged in the second mode of operation, the neurofunctional architecture of the premotor system might contribute to the mastering of the hierarchical structure of language and thought.[7] According to the neural exploitation hypothesis, the neural mapping of different hand/mouth goal-related motor acts, the 'words' of the premotor vocabulary (Rizzolatti et al., 1988) are not only assembled and chained to form intentional 'action sentences' (see the discussion of the MNS and action intentions); they can also be assembled and chained to structure language sentences and thoughts.

At present this is pure speculation. It is certainly possible that Broca's region and the ventral premotor cortex are multifunctional, and that the functional overlap testified by the activation of the same premotor cortical sectors during both language-related and action-related non-linguistic tasks is only apparent because of the poor spatial resolution of the currently available brain-imaging technology. However, it must be stressed that the neural exploitation hypothesis has the merit of offering elements for the neurofunctional grounding of the systematic relation observed between language and the activation of premotor sectors of the frontal cortex. We certainly cannot be satisfied of merely stating, for example, that syntax resides in Broca's region without explaining why, that is, without a clear understanding of what makes of Broca's region a 'syntax-committed' cortical area.

11 Conclusions

The social cognitive endowments of our species are likely the evolutionary outcome of the natural selection of mechanisms that were not mind reading specific. The neural exploitation hypothesis is parsimonious because it

---

[6] On average, the response of mirror neurons in monkeys is stronger during action execution than during action observation.

[7] Establishing a relation between the motor system and the structure of language is by no means a new idea. Lashley (1951) and Marsden (1984), for example, proposed a link between syntax and the action sequencing function of the basal ganglia. A discussion of the role played in syntax by sub-cortical motor centers and the premotor cortex is beyond the scope of this paper. However, it is worth noting that the present hypothesis is – at least partly – compatible with the procedural hypothesis of grammar proposed by Ullman (2001) according to which, aspects of grammar are subserved by a frontal/basal-ganglia procedural memory system that also underlies cognitive and motor skills.

postulates that the quantitative upgrading of pre-existing neurofunctional architecture can produce a qualitative evolution of different cognitive social skills, language included.

The MNS has been invoked to explain many different aspects of social cognition, like imitation (see Rizzolatti et al., 2001), action and intention understanding (see Rizzolatti, Fadiga, Gallese, & Fogassi, 2006), mind reading (see Gallese & Goldman, 1998; Gallese, 2007), empathy (see Gallese, 2001, 2003a, 2003b; de Vignemont & Singer, 2006; Sommerville & Decety, 2006) and its relatedness to aesthetic experience (see Freedberg & Gallese, 2007), and language (see Fadiga & Gallese, 1997; Rizzolatti & Arbib, 1998; Gallese & Lakoff, 2005; Arbib, 2005). The posited importance of mirror neurons for a better understanding of social cognition, together with a sort of mediatic over-exposure and trivialization, have stirred in some quarters of the cognitive sciences resistance, criticism, and even a sense of irritation.

I think a clarification is in order. The relevance of the MNS in so many different aspects of social cognition does not stem from a specific endowment of these neural cells, as if mirror neurons were 'magical neurons', so to speak. Mirror neurons derive their functional properties from the specific input-output connections they entertain with other populations of neurons in the brain.

The MNS and the functional mechanism describing their activity, embodied simulation, are involved in so many aspects of social cognition because the activation of the multiple and parallel cortico-cortical circuits instantiating mirror properties underpins a fundamental aspect of social cognition, that is, *the multi-level connectedness and reciprocity among individuals within a social group*. Such connectedness finds its phylogenetic and ontogenetic roots in the social sharing of situated experiences of action and affect. The MNS provides the neural basis of such sharing. Embodied simulation and the MNS certainly cannot provide a full and thorough account of our sophisticated social cognitive skills. However, I believe that the evidence presented here indicates that embodied mechanisms involving the activation of the motor system, of which the MNS is part, do play a major role in social cognition, language included. A second merit of this hypothesis is that it enables the grounding of social cognition into the experiential domain of existence, so heavily dependent on action (Gallese, 2007).

To imbue words with meaning requires a fusion between the articulated sound of words and the shared meaning of the experience of action. Embodied simulation does exactly that. Furthermore, and most importantly, the neural exploitation hypothesis holds that embodied simulation and the MNS provide the means to share communicative intentions and meaning, thus granting the parity requirements of social communication. This can provide neuroscientific grounding to aspects of language investigated by psycholinguistics like, among others, situation models and collaborative and interactive accounts of conversation, according to which the perception of shared environment and behaviors helps in maintaining alignment between conversational partners (see Clark & Wilkes-Gibbs, 1986; Pickering & Garrod, 2004).

By attributing to action the crucial role it plays in experientially grounding the meanings we share with others, the neural exploitation hypothesis stresses that the multi-level comparative study of the premotor system of primate brains is a necessary starting point for a better understanding of social cognition, and, more generally, for a better understanding of the human condition.

Acknowledgments This work was supported by MIUR (Ministero Italiano dell'Università e della Ricerca) and by the EU grants NESTCOM and DISCOS.

References

Anderson, M. L. (2003). Embodied cognition: A field guide. *Artificial Intelligence, 149*, 91–130.

Arbib, M. A. (2005). From monkey-like action recognition to human language: An evolutionary framework for neurolinguistics. *Behavioral and Brain Sciences, 28*, 105–168.

Aziz-Zadeh, L., Cattaneo, L., Rochat, M., & Rizzolatti, G. (2005). Covert speech arrest induced by rTMS over both motor and nonmotor left hemisphere frontal sites. *Journal of Cognitive Neuroscience, 17*, 928–938.

Aziz-Zadeh, L., Wilson, S. M., Rizzolatti, G., & Iacoboni, M. (2006). Congruent embodied representations for visually presented actions and linguistic phrases describing actions. *Current Biology, 16*, 1818–1823.

Bak, T.H., & Hodges, J.R. (2003). "Kissing and dancing" – a test to distinguish the lexical and conceptual contributions to noun/verb and object/action dissociations: Preliminary results in patients with frontotemporal dementia. *Journal of Neurolinguistics, 16*, 169–181.

Bak, T.H., O'Donovan, D.G., Xuereb, J.H., Boniface, S., & Hodges, J.R. (2001). Selective impairment of verb processing associated with pathological changes in Brodmann areas 44 and 45 in the motor neurone disease–dementia–aplasia syndrome. *Brain, 124*, 103–130.

Bak, T. H., Yancopoulou, D., Nestor, P. J., Xuereb, J. H., Spillantini, M. G., Pulvermuller, F., et al. (2006). Clinical, imaging and pathological correlates of a hereditary deficit in verb and action processing. *Brain, 129*, 321–332.

Barrett, L., & Henzi, P. (2005). The social nature of primate cognition. *Proceeding of the Royal Society. Series B. Biological Sciences. Biology, 272*, 1865–1875.

Barrett, L., Henzi, P., & Rendall, D. (2007) Social brains, simple minds: does social complexity really require cognitive complexity? *Philosophical Transactions of the Royal Society of London. Series B, Biological Sciences, 362*(1480), 561–575.

Barsalou, L. W. (1999). Perceptual symbol systems. *Behavioural and Brain Science, 22*, 577–609.

Bennett, M.R., & Hacker, P.M.S. (2003). *Philosophical foundations of neuroscience*. Blackwell Publishing.

Bernardis, P., & Gentilucci, M. (2006). Speech and gesture share the same communication system. *Neuropsychologia, 44*, 178–190.

Bird CM, Castelli F, Malik O, Frith U, Husain M. The impact of extensive medial frontal lobe damage on 'Theory of Mind' and cognition. Brain. **2004** Apr;127(Pt 4):914–28.

Blakemore, S.-J., Bristow, D., Bird, G. , Frith, C. and Ward J. (2005) Somatosensory activations during the observation of touch and a case of vision–touch synaesthesia. Brain 128: 1571–1583.

Bookheimer, S. (2002). Functional MRI of language: New approaches to understanding the cortical organization of semantic processing. *Annual Review of Neuroscience, 25*, 151–188.

Borghi, A. M., Glenberg, A. M., & Kaschak, M. P. (2004). Putting words in perspective. *Memory & Cognition, 32*, 863–873.

Boroditsky, L. (2000). Metaphoric structuring: understanding time through spatial metaphors. *Cognition, 75*, 1–28.
Boroditsky, L., & Ramscar, M. (2002). The roles of body and mind in abstract thought. *Psychological Science, 13*, 185–188.
Botvinick, M., Jha, A. P., Bylsma, L. M., Fabian, S.A., Solomon, P. E., and Prkachin, K. M. (2005) Viewing facial expressions of pain engages cortical areas involved in the direct experience of pain. *Neuroimage, 25*, 315–319.
Braten, S. (1988). Dialogic mind: The infant and the adult in protoconversation. In M. Carvallo (Ed.), Nature, *cognition and system* (Vol. I, pp. 187–205). Dordrecht: Kluwer Academic Publishers.
Braten, S. (1992). The virtual other in infants' minds and social feelings. In H. Wold (Ed.), *The dialogical alternative* (pp. 77–97). Oslo: Scandinavian University Press.
Braten, S. (2007). *On being moved: From mirror neurons to empathy* (p. 333). John Benjamins Publishing Company.
Bruner, J. (1986). *Actual minds, possible worlds*. Cambridge, MA: Harvard University Press.
Bruner, J. (1990). *Acts of meaning*. Cambridge, MA: Harvard University Press.
Buccino, G., Binkofski, F., Fink, G. R., Fadiga, L., Fogassi, L., Gallese, V., Seitz, R. J., Zilles, K., Rizzolatti, G., & Freund, H.-J. (2001). Action observation activates premotor and parietal areas in a somatotopic manner: An fMRI study. *European Journal of Neuroscience, 13*, 400–404.
Buccino, G., Lui, F., Canessa, N., Patteri, I., Lagravinese, G., Benuzzi, F., Porro, C. A., & Rizzolatti, G. (2004a). Neural circuits involved in the recognition of actions performed by nonconspecifics: An fMRI study. *Journal of Cognitive Neuroscience, 16*, 114–126.
Buccino, G., Vogt, S., Ritzl, A., Fink, G. R., Zilles, K., Freund, H.-J., & Rizzolatti, G. (2004b). Neural circuits underlying imitation learning of hand actions: An event-related fMRI study. *Neuron, 42*, 323–334.
Buccino, G., Riggio, L., Melli, G., Binkofski, F., Gallese, V., & Rizzolatti, G. (2005). Listening to action-related sentences modulates the activity of the motor system: A combined TMS and behavioral study. *Coginitive Brain Research, 24*, 355–363.
Buxbaum, L. J., Kyle, K. M. & Menon, R. (2005), On beyond mirror neurons: internal representations subserving imitation and recognition of skilled object-related actions in humans. *Cognitive Brain Research, 25*, 226–239.
Clark, A. (1997). *Being there: Bringing brain, body, and world together again*. Cambridge, MA: MIT Press.
Clark, H. H., & Wilkes-Gibbs, D. (1986). Referring as a collaborative process. *Cognition, 22*, 1–39.
Corballis, M. C. (2002). *From hand to mouth: The origins of language*. Princeton, NJ: Princeton University Press.
Corballis, M. C. (2004). FOXP2 and the mirror system. *Trends in Cognitive Sciences, 8*, 95–96.
de Saussure, F. (1973/1974). Course in general linguistics. In W. Baskins (Trans.), London: Fontana/Collins.
de Vignemont, F., & Singer, T. (2006). The emphatic brain: how, when, and why? *Trends in Cognitive Sciences, 10*, 435–441.
Fadiga, L., Craighero, L., Buccino, G., & Rizzolatti, G. (2002). Speech listening specifically modulates the excitability of tongue muscles: A TMS study. *The European Journal of Neuroscience, 15*, 399–402.
Fadiga, L., & Gallese, V. (1997). Action representation and language in the brain. *Theoretical Linguistics, 23*, 267–280.
Falck-Ytter, T., Gredeback. G., & von Hofsten, C. (2006). Infant predict other people's action goals. *Nature Neureuscience, 9*(7), 878–879.
Ferrari, P. F., Gallese, V., Rizzolatti, G., & Fogassi, L. (2003). Mirror neurons responding to the observation of ingestive and communicative mouth actions in the monkey ventral premotor cortex. *European Journal of Neuroscience, 17*, 1703–1714.

Ferrari, P. F., Maiolini, C., Addessi, E., Fogassi, L., & Visalberghi, E. (2005). The observation and hearing of eating actions activates motor programs related to eating in macaque monkeys. *Behavioural Brain Research, 161*, 95–101.
Flombaum, J. L., & Santos, L. R. (2006). Rhesus monkeys attribute perceptions to others. *Current Biology, 15*, 447–452.
Fodor, J. (1998). *Concepts*. Oxford: Oxford University Press.
Fogassi, L., Ferrari, P. F., Gesierich, B., Rozzi, S., Chersi, F., & Rizzolatti, G. (2005). Parietal lobe: From action organization to intention understanding. *Science, 302*, 662–667.
Freedberg, D., & Gallese, V. (2007). Motion, emotion and empathy in aesthetic experience. *Trends in Cognitive Sciences, 11*, 197–203.
Gallese, V. (2000). The inner sense of action: agency and motor representations. *Journal of Consciousness Studies, 7*, 23–40.
Gallese, V. (2001). The "shared manifold" hypothesis: from mirror neurons to empathy. *Journal of Consciousness Studies, 8*(5–7), 33–50.
Gallese, V. (2003a). The manifold nature of interpersonal relations: The quest for a common mechanism. *Philosophical Transactions of the Royal Society of London. Series B, Biological Sciences, 358*, 517–528.
Gallese, V. (2003b). The roots of empathy: The shared manifold hypothesis and the neural basis of intersubjectivity. *Psychopatology, 36*(4), 171–180.
Gallese, V. (2003c). A neuroscientific grasp of concepts: From control to representation. *Philosophical Transactions of the Royal Society of London. Series B, Biological Sciences, 358*, 1231–1240.
Gallese, V. (2005a). Embodied simulation: From neurons to phenomenal experience. *Phenomenology and the Cognitive Sciences, 4*, 23–48.
Gallese, V. (2005b). "Being like me": Self-other identity, mirror neurons and empathy. In S. Hurley & N. Chater (Eds.), *Perspectives on imitation: from cognitive neuroscience to social science* (Vol. 1, pp. 101–118). Cambridge, MA: MIT Press.
Gallese, V. (2006). Intentional attunement: A neurophysiological perspective on social cognition and its disruption in autism. *Cognitive Brain Research, 1079*, 15–24.
Gallese, V. (2007). Before and below theory of mind: Embodied simulation and the neural correlates of social cognition. *Philosophical Transactions of the Royal Society of London. Series B, Biological sciences. 362*, 659–669.
Gallese, V., Fadiga, L., Fogassi, L., & Rizzolatti, G. (1996). Action recognition in the premotor cortex. *Brain, 119*, 593–609.
Gallese, V., Fogassi, L., Fadiga, L., & Rizzolatti, G. (2002). Action representation and the inferior parietal lobule. In W. Prinz & B. Hommel (Eds.), *Attention and performance XIX* (pp. 247–266). Oxford University Press: Oxford.
Gallagher, H.L., & Frith, C.D. (2003). Functional imaging of 'theory of mind'. *Trends in Cognitive, 7(2)*, 77–83.
Gallese, V., & Goldman, A. (1998). Mirror neurons and the simulation theory of mind-reading. *Trends in Cognitive Sciences, 12*, 493–501.
Gallese, V., Keysers, C., & Rizzolatti, G. (2004). A unifying view of the basis of social cognition. *Trends in Cognitive Sciences, 8*, 396–403.
Gallese, V., & Lakoff, G. (2005). The brain's concepts: The role of the sensory-motor system in eason and language. *Cognitive Neuropsychology, 22*, 455–479.
Gentilucci, M. (2003). Grasp observation influences speech production. *The European Journal of neuroscience, 17*, 179–184.
Gentilucci, M., Benuzzi, F., Gangitano, M., & Grimaldi, S. (2001). Grasp with hand and mouth: a kinematic study on healthy subjects. *Journal of Neurophysiology, 86*, 1685–1699.
Gentilucci, M., Bernardis, P., Crisi, G., & Volta, R. D. (2006). Repetitive transcranial magnetic stimulation of Broca's area affects verbal responses to gesture observation. *Journal of Cognitive Neuroscience, 18*, 1059–1074.

Gentilucci M, Corballis MC. (2006) From manual gesture to speech: a gradual transition. Neuroscience and Biobehavioral Review 30: 949–60.

Gentilucci, M., Santunione, P., Roy, A. C., & Stefanini, S. (2004a). Execution and observation of bringing a fruit to the mouth affect syllable pronunciation. *European Journal of Neuroscience, 19*, 190–202.

Gentilucci, M., Stefanini, S., Roy, A. C., & Santunione, P. (2004b). Action observation and speech production: study on children and adults. *Neuropsychologia, 42*, 1554–1567.

Gibson, J. (1979). *The ecological approach to visual perception.* Hillsdale, NJ: Lawrence Erlbaum Associates.

Glenberg, A. M. (1997) What memory is for. *Behavioral and Brain Sciences, 20*, 1–55.

Glenberg, A. M., & Kaschak, M. P. (2002). Grounding language in action. *Psychonomic Bulletin & Review, 9*, 558–565.

Glenberg, A. M., & Robertson, D. A. (2000). Symbol grounding and meaning: A comparison of high-dimensional and embodied theories of meaning. *Journal of Memory and Language, 43*, 379–401.

Goldberg, A. E. (2003). Constructions: A new theoretical approach to language. *Trends in Cognitive Sciences, 7*, 219–224.

Goldin-Meadow, S. (1999). The role of gesture in communication and thinking. *Trends in Cognitive Science, 3*, 419–429.

Goldman, A. (2006). *Simulating Minds.* Oxford, UK: Oxford University Press.

Hare, B., Call, J., Agnetta, B., & Tomasello, M. (2000). Chimpanzees know what conspecifics do and do not see. *Animal Behavior*, 59, 771–785.

Hauk, O., Johnsrude, I., & Pulvermüller, F. (2004). Somatotopic representation of action words in human motor and premotor cortex. *Neuron 41*(2), 301–307.

Heidegger, M. (1925/1985). *History of the concept of time* (T. Kisiel, Trans.), Bloomington: Indiana University Press.

Heidegger, M. (1927/1962). *Being and time* (J. Macquarrie & E. Robinson, Trans.), New York: Harper & Row.

Heidegger, M. (1929/1995). *The fundamental concepts of metaphysics. World, finitude, solitude* (W. McNeill & N. Walker, Trans.). Bloomington: Indiana University Press.

Hjelmslev, J. (1959). *Essais linguistiques.* Copenhague: Circle linguistique de Copenhague.

Hutchison, W.D., Davis, K.D., Lozano, A.M., Tasker, R.R., & Dostrovsky, J.O. (1999). Pain related neurons in the human cingulate cortex. *Nature Neuroscience, 2*, 403–405.

Hutto, D. H. (2004). The limits of spectatorial folk psychology. *Mind and Language, 19*, 548–573.

Iacoboni, M., Molnar-Szakacs, I., Gallese, V., Buccino, G., Mazziotta, J., & Rizzolatti, G. (2005). Grasping the intentions of others with one's owns mirror neuron system. *PLOS Biology, 3*, 529–535.

Iacoboni, M., Woods, R. P., Brass, M., Bekkering, H., Mazziotta, J. C., & Rizzolatti, G. (1999). Cortical mechanisms of human imitation. *Science, 286*, 2526–2528.

Iverson, J. M., & Goldin-Meadow, S. (2005). Gesture paves the way for language development. *Psychological Science, 16*, 367–371.

Jackson, P. L., Meltzoff, A. N., & Decety, J. (2005). How do we perceive the pain of others: A window into the neural processes involved in empathy. *NeuroImage, 24*, 771–779.

Jacob, P., & Jeannerod, M., (2004). The motor theory of social cognition: A critique. Trends in Cognitive Neuroscience *9*, 21–25.

Kemmerer, D., & Tranel, D. (2000). A double dissociation between linguistic and perceptual representations of spatial relationships. *Cognitive Neuropsychology, 17*, 393–414.

Kemmerer, D., & Tranel, D. (2003). A double dissociation between the meanings of action verbs and locative prepositions. *Neurocase, 9*, 421–435.

Keysers C., & Gazzola, V. (2007). Integrating simulation and theory of mind: From self to social cognition. *Trends in Cognitive Science, 11(5)*, 194–196.

Keysers, C., Wickers, B., Gazzola, V., Anton, J-L., Fogassi, L., & Gallese, V. (2004). A touching sight: SII/PV Activation during the observation and experience of touch. *Neuron, 42*, 1–20 (April 22).
Knoblich, G., & Jordan, J. S. (2002). The mirror system and joint action. In M. I. Stamenov & V. Gallese (Eds.), Mirror neurons and the evolution of brain and language (pp. 115–124). Amsterdam: John Benjamins.
Kohler, E., Keysers, C., Umiltà, M. A., Fogassi, L., Gallese, V., & Rizzolatti, G. (2002). Hearing sounds, understanding actions: Action representation in mirror neurons. *Science, 297*, 846–848.
Lakoff, G. (1987). *Women, fire, and dangerous things: What categories reveal about the mind.* Chicago and London: University of Chicago Press.
Lakoff, G., & Johnson, M. (1980). *Metaphors we live by.* Chicago and London: University of Chicago Press.
Lakoff, G., & Johnson, M. (1999). *Philosophy in the flesh.* New York: Basic Books.
Lashley, K. S. (1951). The problem of serial order in behavior. In L. A. Jeffress (Ed.), *Cerebral mechanisms in behavior* (pp. 112–146). NY: Wiley.
Marsden, C. D. (1984). Which motor disorder in Parkinson's disease indicates the true motor function of the basal ganglia? In *Functions of the Basal Ganglia* (pp. 225–241). Ciba Foundation Symposium 108:. London: Pittman.
Masataka, N. (2001). Why early linguistic milestones are delayed in children with Williams syndrome: late onset of hand banging as a possible rate-limiting constraint on the emergence of canonical babbling. *Developmental Science, 4*: 158–164.
Matlock, T. (2004). Fictive motion as cognitive simulation. *Memory & Cognition, 32*, 1389–1400.
McGuigan, F. J., & Dollins, A. B. (1989). Patterns of covert speech behavior and phonetic coding. *The Pavlovian Journal of Biological Science, 24*, 19–26.
Meltzoff, A. N. (2007a) "Like me": A foundation for social cognition. *Developmental Science 10*, 126–134.
Meltzoff, A. N. (2007b). The "like me" framework for recognizing and becoming an intentional agent. *Acta Psychology, 12*, 26–43.
Meltzoff, A. N., & Brooks, R. (2001). "Like me" as a building block for understanding other minds: Bodily acts, attention, and intention. In B. F. Malle, L. J. Moses, & D. A. Baldwin (Eds.), *Intentions and intentionality: Foundations of social cognition* (pp. 171–191). Cambridge, MA: MIT Press.
Meltzoff, A.N., & Moore, M.K. (1977). Imitation of facial and manual gestures by human neonates. *Science, 198* (4312), 74–78.
Meltzoff, A. N., & Moore, M. K. (1998). Infant inter-subjectivity: Broadening the dialogue to include imitation, identity and intention. In S. Braten (Ed.), *Intersubjective communication and emotion in early ontogeny* (pp. 47–62). Paris: Cambridge University Press.
Merleau-Ponty, M. (1960/1964). *Signs* (R. C. McClearly, trans.). Evanston, IL: Northwestern University Press.
Nelissen, K., Luppino, G., Vanduffel, W., Rizzolatti, G., & Orban, G. A. (2005). Observing others: Multiple action representation in the frontal lobe. *Science, 310*, 332–336.
Niedenthal, P. M., Barsalou, L. W., Winkielman, P., Krauth-Gruber, S., & Ric, F. (2005). Embodiment in attitudes, social perception, and emotion. *Personality and Social Psychology Review, 9*, 184–211.
Nishitani, N., Schurmann, M., Amunts, K., & Hari, R. (2005). Broca's region: From action to language. *Physiology 20*, 60–69.
Oberman, L.M., & Ramachandran, V.S. (2007). The simulating social mind: Mirror neuron system and simulation in the social and communicative deficits of Autism Spectrum Disorder. *Psychological Bulletin 133*, 310–327.
Onishi, K. H., & Baillargeon, R. (2005). Do 15 months-old understand false beliefs? *Science, 308*, 255–258.

Pacherie, E., & Dokic, J. (2006). From mirror neurons to joint action. *Cognitive Systems Research*, 7,101–112.

Paukner, A., Anderson, J. R., Borelli, E., Visalberghi, E., & Ferrari, P. F. (2005). Macaques (Macaca nemestrina) recognize when they are being imitated. *Biology Letters. 1*, 219–222.

Pickering, M. J., & Garrod, S. (2004). Toward a mechanistic psychology of dialogue. *Behavioral and Brain Science*, *27*, 169–226.

Povinelli, D. J., & Eddy, T. J. (1996). What young chimpanzees know about seeing. *Monographs Society Research Child Development*, *61*, 1–152.

Povinelli, D., Vonk, J. (2003). Chimpanzee minds: Suspiciously human? *Trends in Cognitive Science*, *7*, 157–160.

Pulvermüller, F. (1999). Word in the brain's language. *Behavioral Brain Sciences*, *22*, 253–336.

Pulvermüller, F. (2002). *The neuroscience of language*. Cambridge, UK: Cambridge University Press.

Pulvermüller, F., Härle, M., & Hummel, F. (2000). Neurophysiological distinction of verb categories. *Neuroreport*, *11*, 2789–2793.

Pulvermüller, F., Shtyrov, Y., & Ilmoniemi, R. J. (2003). Spatio-temporal patterns of neural language processing: An MEG study using Minimum-Norm Current Estimates. *Neuroimage*, *20*, 1020–1025.

Pylyshyn, Z. W. (1984). *Computation and cognition: Toward a foundation for cognitive science*. Cambridge, MA: MIT Press.

Ricoeur, P. (1986/1991). *From text to action. Essays in hermeneutics*, II. (K. Blamey & J. B. Tompson, trans.), Evanston, Illinois: Northwestern University Press.

Rizzolatti, G., & Arbib, M. A. (1998). Language within our grasp. *Trends in Neurosciences*, *21*, 188–194.

Rizzolatti, G., Camarda, R., Fogassi, M., Gentilucci, M., Luppino, G., & Matelli, M. (1988). Functional organization of inferior area 6 in the macaque monkey: II. Area F5 and the control of distal movements. *Experimental Brain Research*, *71*, 491–507.

Rizzolatti, G., & Craighero, L. (2004). The mirror neuron system. *Annual Review of Neuroscience*, *27*, 169–192.

Rizzolatti, G., Fadiga, L., Gallese, V., & Fogassi, L. (1996). Premotor cortex and the recognition of motor actions. *Coginitive Brain Research*, *3*, 131–141.

Rizzolatti, G., Fogassi, L., & Gallese, V. (2001). Neurophysiological mechanisms underlying the understanding and imitation of action. Nature Neuroscience Reviews, 2, 661–670.

Rizzolatti, G., Fogassi, L., & Gallese, V. (2006). Mirrors in the mind. *Scientific American Nov* *295*(5), 54–61.

Rizzolatti, G., & Gallese, V. (1997). From action to meaning. In: J.-L. Petit (Ed.). *Les Neurosciences et la Philosophie de l'Action* (pp. 217–229). Librairie Philosophique J. Vrin: Paris.

Santos, L. R., Nissen, A. G., & Ferrugia, J. A. (2006). Rhesus monkeys, Macaca mulatta, know what others can and cannot hear. *Animal Behavior*, *71*, 1175–1181.

Saxe, R. (2005). Against simulation: The argument from error. *Trends in Cognitive Sciences*, *4*, 174–179.

Singer, T., Seymour, B., O'Doherty, J., Kaube, H., Dolan, R.J., & Frith, C.F. (2004). Empathy for pain involves the affective but not the sensory components of pain. *Science* *303*, 1157–1162.

Sommerville, J. A., Decety, J. (2006). Weaving the fabric of social interaction: articulating developmental psychology and cognitive neuroscience in the domain of motor cognition. *Psychonomic Bulletin & Review*, *13*, 179–200.

Sommerville, J. A., Woodward, A. (2005). Pulling out the intentional structure of action: the relation between action processing and action production in infancy. *Cognition*, *95*(1), 1–30.

Sommerville, J. A., Woodward, A., & Needham, A. (2005). Action experience alters 3-month-old perception of other's actions. *Cognition*, *96*(1), 1–11.

Stern, D. N. (1985). *The interpersonal world of the infant*. London: Karnac Books.
Tettamanti, M., Buccino, G., Saccuman, M. C., Gallese, V., Danna, M., Scifo, P., Fazio, F., Rizzolatti, G., Cappa, S. F., & Perani, D. (2005). Listening to action-related sentences activates fronto-parietal motor circuits. *Journal of Cognitive Neuroscience, 17*, 273–281.
Tomasello, M., Carpenter, M., Call, J., Behne, T., & Moll, H. (2005). Understanding and sharing intentions: the origins of cultural cognition. *Behavioral of Brain Science, 28*, 675–691.
Trevarthen, C. (1979). Communication and cooperation in early infancy: A description of primary intersubjectivity. In: M. Bullowa (Ed.), *Before speech: The beginning of interpersonal communication* (pp. 321–347). New York: Cambridge University Press.
Trevarthen, C. (1993). The self born in intersubjectivity: An infant communicating. In U. Neisser (Ed.), *The perceived self* (pp. 121–173). New York: Cambridge University Press.
Tronick, E. (1989). Emotion and emotional communication in infants. *American Psychologist, 44*, 112–119.
Uddin, L. Q., Iacoboni, M., Lange, C., & Keenan, J. P. (2007). The self and social cognition: the role of cortical midline structures and mirror neurons. *Trends in Cognitive Sciences, 11*, 153–157.
Uddin, L. Q., Kaplan, J. T., Molnar-Szakacs, I., Zaidel, E., & Iacoboni, M. (2005). Self-face recognition activates a frontoparietal "mirror" network in the right hemisphere: an event-related fMRI study. *Neuroimage, 25*, 926–935.
Uddin, L. Q., Molnar-Szakacs, I., Zaidel, E., & Iacoboni, M. (2006). rTMS to the right inferior parietal lobule disrupts self-other discrimination. *Social Cognitive and Affective Neuroscience, 1*, 65–71.
Ullman, M. T. (2001). A neurocognitive perspective on language: The declarative/procedural model. *Nature Reviews Neuroscience, 2*, 717–726.
Umiltà, M. A., Kohler, E., Gallese, V., Fogassi, L., Fadiga, L., Keysers, C., & Rizzolatti, G. (2001). "I know what you are doing": A neurophysiological study. *Neuron, 32*, 91–101.
Watkins, K. E., & Paus, T. (2004). Modulation of motor excitability during speech perception: the role of Broca's area. *Journal of Coginitive Neurosciences, 16*, 978–987.
Watkins, K. E., Strafella, A. P., & Paus, T. (2003). Seeing and hearing speech excites the motor system involved in speech production. *Neuropsychologia, 41*(8), 989–994.
Wicker, B., Keysers, C., Plailly, J., Royet, J-P., Gallese, V., and Rizzolatti, G. (2003). Both of us disgusted in my insula: The common neural basis of seeing and feeling disgust. *Neuron, 40*, 655–664.
Wildgruber, D., Ackermann, H., Klose, U., Kardatzki, B., & Grodd, W. (1996). Functional lateralization of speech production at primary motor cortex: A fMRI study. *NeuroReport, 7*, 2791–2795.
Wilson, S. M., Saygin, A. P., Sereno, M. I., & Iacoboni, M. (2004). Listening to speech activates motor areas involved in speech production. *Nature Neuroscience 7*, 701–702.
Wittgenstein, L. (1953). *Philosophical investigations*. Oxford University Press: Oxford.
Woodward, A. L. (1998). Infants selectively encode the goal object of an actor's reach. *Cognition, 69*, 1–34.
Zoia, S., Blason, L., D'Ottavio G., Bulgheroni M., Pezzetta E., Scabar A., & Castiello U. (2007). Evidence of early development of action planning in the human foetus: A kinematic study. *Experimental Brain Research, 176*, 217–226.

From Imitation to Reciprocation and Mutual Recognition

Philippe Rochat and Claudia Passos-Ferreira

Abstract Imitation and mirroring processes are necessary but not sufficient conditions for children to develop human sociality. Human sociality entails more than the equivalence and connectedness of perceptual experiences. It corresponds to the sense of a shared world made of shared values. It originates from complex 'open' systems of reciprocation and negotiation, not just imitation and mirroring processes that are by definition 'closed' systems. From this premise, we argue that if imitation and mirror processes are important foundations for sociality, human inter-subjectivity develops primarily in reciprocation, not just imitation. Imitation provides a basic sense of social connectedness and mutual acknowledgment of existing with others that are 'like me.' However, it does not allow for the co-construction of meanings with others. For human sociality to develop, imitation and mirroring processes need to be supplemented by an open system of reciprocation. Developmental research shows that from the second month, mirroring, imitative, and other contagious emotional responses are by-passed. Imitation gives way to first signs of reciprocation (primary intersubjectivity), joint attention to objects (secondary intersubjectivity), the emergence of values that are jointly represented and negotiated with others (tertiary intersubjectivity), and eventually the development of an ethical stance accompanying theories of mind by 4 years of age. We review this development and propose that if mirroring processes enable individuals to bridge their subjective experiences, human inter-subjectivity proper develops from reciprocal social exchanges that lead to value negotiation and mutual recognition, both cardinal trademarks of human sociality.

Keywords Sociality · Reciprocation · Inter-subjectivity · Co-construction

---

P. Rochat
Department of Psychology, Emory University, Atlanta, GA, USA

J.A. Pineda (ed.), *Mirror Neuron Systems*, DOI: 10.1007/978-1-59745-479-7_9,

1 Introduction

Human sociality is inseparable from the elusive, yet powerful sense of a shared world made of shared values. This sense arises from the interaction with others via complex 'open' systems of reciprocation and negotiation. It cannot be reduced to imitation and mirroring processes that are, in a strict etymological sense, 'closed' systems, in themselves copying mechanisms like mirrors reflecting whatever is facing them.

In the strict, literal sense, the word imitation derives from the Latin word 'imitatus', the past participle of 'imitati' which means *to copy*. By imitation, we thus refer literally to a mechanism of copying, a system that does not account for any selective process as to *what* is copied, or *why* it is copied; in other words what new meanings might grow out of the copying process. Taken literally, imitation thus stands for a system of direct reflection of what is out there, impoverishing of the process by which we actually relate and understand each other, a process that is in essence, selective, and creative of new meanings (ideas, feelings, values) that arise from on-going social exchanges. Rather than mirroring, or imitation in the strict sense of copying, other metaphors are needed to account for the foundation of human sociality and social cognition.

Imitation and mirror processes are probably important foundations for sociality (i.e., the capacity to relate, interact, and possibly re-present or simulate, hence 'bridge' self with others' experience). But the sense of shared experience and of shared values develops primarily in a process of *reciprocation* that adds to the process of imitation and mirroring as copying.

The inclination to copy and simulate the behaviors of other might provide a basic sense of social connectedness and mutual acknowledgment of existing with others that are 'like me.' But without other mechanisms, the process of reproducing or copying the behaviors of others is essentially not creative, leading nowhere in itself. In a strict sense, imitation and mirroring are closed loop 'tit for tat' systems. More processing is needed to allow for the social construction of meanings that drive human transactions (e.g., shared ideas or values such as trust, guilt, the sense of what's right and what's wrong, who is to be admired and emulated, who is commendable and has prestige, who is to be avoided and despised).

The gist of the argument put forth in this chapter is that for human sociality to develop, imitation, and mirroring processes need to be supplemented by an open system of reciprocation. The reflection arising from mirroring processes needs to be broken down and somehow by-passed. As a case in point, young children, not yet showing any signs of self-recognition, when faced with their own mirror reflection, often try desperately to break away from the perfect contingency of the specular image which is limited in providing only absolute imitation and no conversation proper (Amsterdam, 1972; Rochat & Striano, 2002). We try to show that in early ontogeny, particularly starting the second month, mirroring, imitation, and other contagious emotional responses tend to become more subtly attuned to interactive others. This first social register of the neonate is by-passed in 'proto' conversation with others, in the context of first reciprocal exchanges that form open, as opposed to closed, loop systems.

In short, here we argue that if mirroring processes might enable individuals to bridge their subjective experiences via embodied simulation (Gallese, 2007), human inter-subjectivity properly develops from reciprocal social exchanges and the constant negotiation of values with others. The general 'developmental' message we would like to get across in the context of this book on the role of mirroring processes in social cognition is that infants and young children develop to become *Homo Negotiatus* (Rochat & Passos-Ferreira, 2008), not just to become *Homo Mimesis*.

Imitation and mirroring processes are necessary but not sufficient mechanisms for children to develop inter-subjectivity and sociality. We argue that human sociality (i.e., the inclination to associate with or be in the company of others[1]) entails more than the equivalence and connectedness of perceptual experiences. It entails a sense of reciprocity that is more than the 'like-me stance' or embodied simulation that researchers derive from early imitation (Meltzoff, 2007) or from the recent discovery of mirror neuron systems in the brain (Gallese, Fadiga, Fogassi, & Rizzolatti, 1996; Rizzolatti, Fadiga, Gallese, & Fogassi, 1996; Gallese, Fogassi, Fadiga, & Rizzolatti, 2002; Rizzolatti & Craighero, 2004; Fogassi et al., 2005; Goldman & Sripada, 2005).

From a developmental perspective, by 2 months infants already appear to transcend basic mirroring processes by manifesting first signs of reciprocation in face-to-face exchanges (primary intersubjectivity). They soon engage in triadic intentional communication with others about objects (secondary intersubjectivity starting approximately 9 months) and eventually begin to negotiate with others about the values of things, including the self as shared representations (tertiary intersubjectivity, starting approximately 20 months). This development culminates with the ethical stance that children begin to take around their fourth birthday when they begin to manifest explicit rationale about what is right and what is wrong, as well as 'theories' regarding the mind of others.

In what follows, we distinguish levels of 'inter-subjectivity' beyond the primary vs. secondary distinction introduced years ago by Trevarthen & Hubley (1978), Trevarthen (1979) and Bruner (1983). We review this development up to 5 years of age when children show explicit understanding of the mental states that drive others in their behaviors, beliefs, and decisions (i.e., 'theories of mind' in Wellman, 2002). This development leads the child from neonatal imitation to the development of reciprocation starting at 2 months of age, and ultimately toward an 'ethical stance' from 4 to 5 years on, according to our own recent research.

---

[1] This is the first definition of 'sociability' offered by the Unabridged Random House Dictionary (2nd Edition). 'Sociality' is defined as the state or quality of being sociable (third definition). This is the generic sense of these terms used here.

Table 1 Five levels of Social Connectedness in early development

	Type	Context	Behavioral index	Process	Age
I	Mirroring	Face-to-face engagement	Imitation	Automatic simulation	Birth
II	Primary Inter-subjectivity	Reciprocal dyadic exchanges	Proto-conversation, social expectations	Emotional co-regulation	2 months
III	Secondary Inter-subjectivity	Triadic exchanges about things	Joint attention; social referencing	Intentional communication and intentional co-experience	9 months
IV	Tertiary Inter-subjectivity	Triadic exchanges about the value of things	Self-recognition and embarrassment, use of possessives, claim of ownership, pro-social behaviors	Projection and identification of self onto others	20 months
V	Ethical stance	Decision regarding the value of things, what is right vs. wrong	Sense of property, sharing, distributive justice, theories of mind	Value negotiation with others, narration, meta-representation of reputation	From 4 years

Table 1 summarizes the various levels of social connectedness associated with this development in relation to context, behavioral index, putative underlying process, and chronological age. It is a proposed road map that would take the healthy child, starting the second month after birth, beyond the basic endowment of mirroring and imitative processes. It leads toward reciprocation, social negotiation, and ultimately the sense of mutual recognition and the explicit moral sense that is unique to our species (i.e., reasoned codes of conducts toward others and other juridical rationales).

2 Imitation as Source of Innovation

The idea that imitation plays a central role in human psychic and social life is perennial in the history of social sciences. Psychologists and sociologists of the 19th-early 20th centuries offer theories which state that the propensity of individuals to copy and echo each other is a cornerstone of what makes individuals understand and feel for each other and is also a major source of learning and novelty. Developmental and comparative theorists see imitation as the basic mechanism by which children develop empathy and the capacity to represent, think and speak. Imitation has also been considered for a long time as a mechanism by which children develop theories of mind, in addition to being the source of social connection and affiliation. It is also seen as the source of behavioral synchronization among individuals as well as a major social learning mechanism, a source of innovation in group living. Early on, theorists understood the importance of imitation, not only as a strict copying mechanism, but also as the source of innovation and developments that are unique to our species.

In his account of human evolution, Merlin Donald (1991) writes: “Human children routinely re-enact the events of the day and imitate the actions of their parents and siblings. They do this very often without any apparent reason other than to reflect on their representation of the event. This element is largely absent from the behavior of apes” (1991, p. 172). The idea that imitation or mimesis and the ability to simulate are at the core of what distinguish humans from other animals is a recurrent theoretical proposal in philosophical, psychological, and comparative theories.

Over a century ago, sociologist and social psychologist Gabriel De Tarde (1843–1904) emphasized the central role played by imitation in the dynamic and reorganization of group living. The basic propensity of the human individual to imitate reverberates and impacts on a more macro, ‘societal’ level via the formation of normative opinions, blind group beliefs, propaganda, and other crowd behaviors. More importantly for Tarde, imitation would be the mechanism by which group customs and traditions, but also novel ideas, propagate over time. A more modern version of this account is proposed by Sperber who draws an analogy with epidemiological models in biology to account for the ‘contagion’ of ideas in cultural evolution (Sperber, 1996).

It is worth noting that contrary to the assumptions of political economists of the 19th century like Marx and Engels who posited that human society was born with the first exchange of goods, Tarde in his book '"Les lois de l'imitation' (1890/1993) proposes as an alternative to these theories that society began from the moment one individual copied another. Tarde considered imitation as the matrix of all principles in sociology, a mechanism that reverberates from the individual to the group at large, and is the driving force behind cultural evolution and societal changes, thus innovation.

There is a fundamental paradox between imitation as an act of reproduction and imitation as the creation of changes, hence variation that is the essence of evolution. This is what Tarde was interested in, in his research, trying to reconcile repetition and reproduction with innovation. He placed imitation in an open loop transmission of customs, beliefs, and desires.

For Tarde, behaviors and ideas transmitted by imitation are not just copied as mirrors copy the world in their reflections. Imitation is active in the sense of being selective. It is intentional, not just a source of contamination by reproduction, the main source of *novelty and discovery* characterizing the rapid evolution of modern human societies (e.g., the making and use of tools, new knowledge, myths, skills, or customs).

In a dissertation defended in 1911, Icelandic sociologist Gudmundur Finnbogason (1873–1944) came out with a theory on 'sympathetic intelligence' that is, a remarkable intuition of all the current simulation and imitation theories in social cognition that now find neurobiological validation in the discovery of mirror neuron systems. Finnbogason laid down a theory that reduce imitative motor acts to perception, a theory that explicitly posits that performing a motor act or seeing it performed by a model can *de facto* be the same. This is basically what the discovery of mirror neurons tells us today. Finnbogason already had the intuition of the core ingredients of modern simulation theories (Goldman & Sripada, 2005; Gallese et al., 2002; Meltzoff, 1995, 2007; Harris, 1992).

In '*The Mind and the World Around Us'* (1912), Finnbogason writes:

> "The expression, way of acting and all apparent behavior of other people could be echoed in us if we observed them closely, and that from this echo, this involuntary miming or tendency to mime others, originated our comprehension of the mental life of our fellow men; we could sense their expression acquire a grasp on our own faces and simultaneously become aware of their personality entering us." Finnbogason goes on: "This also opens the possibility of interiorizing the individual characteristics of others, for if we manage to simulate their expressions, posture, motions and actions – in fact everything external about them – then we will have positioned ourselves in the same manner toward the outside world and can to some extent acquire the same perspective and same feelings about it" (Finnbogason, 1912, pp. 250 and 262; cited by Hauksson, J. 2000, Acta Sociologica, 43, 307–315).

In relation to cognitive development, Piaget (1962) places the ontogenetic origins of mental imagery, symbols, and representations in the act of reproducing events perceived by the child; hence, in imitation. These copying acts eventually become internalized to form representations and objects of thoughts. Piaget writes for example: "Mental imagery or the symbol as internalized copies of an object is a product of imitation" (1962, p. 71). In Piaget's view, imitation in the broad sense is at the core of what allows children to become symbolic and what enables them to eventually learn and communicate via the abstract sign systems that are human languages.

In all, for a long time theorists have seen in imitation a central mechanism, *the* mechanism driving children's development, the evolution of human societies, and those abilities that set us apart as a species (e.g., complex abstract languages, explicit ethics, empathic feelings, technological inventions, and their cultural transmission). What these theories all stress one way or another is that *imitation is not only a copying capacity, it is but also a source of novelty, a source of innovation.* It allows individuals to connect, builds intersubjectivity and ultimately feels what other individuals feel, as suggested by Finnbogason a century ago. It also allows people to transmit and to create new knowledge and new skills as suggested by Tarde a few years earlier. For Piaget (1962), imitation is a source of major progress in cognitive development, no less than the origin of mental imagery, pretend play, and symbolic functioning. It is much more than the ability to mirror the world. For Tarde as well as for Piaget, imitation is action and selection. It is intentional, not just automatic. If imitation is a source of novelty, then it is much more than mirroring or mimicking in the strict sense. The mirror metaphor should be replaced by the dynamic, open ended, and relational concept of *reciprocation.*

3 Reciprocation

In reproducing the behavior of others we create inter-subjectivity, bridging self, and others' experience as suggested by Finnbogason and current simulationist theories that find validation in the discovery of mirror neurons. If imitation in the strict sense is a source of vicarious experiences that give individuals the opportunity to get 'into the shoes of others' and possibly empathize with them, it is also a source of discovery and learning. New skills can be learned by imitation following periods of passive observation. In Japan, for example, it is said that the apprentice cook watches the Sushi Master cutting fish for months before he is handed over a knife and allowed to do it himself. In most Non Western small-scale traditional and rural communities from all over the world observational and imitative learning prevail. Children learn primarily by observing, via observational and imitative learning, rarely if not at all, via the explicit instruction that prevails in Western cultures (Odden & Rochat, 2004; Rogoff, 1995; Boggs, 1985; Lancy, 1996). What is important to note is that observational and imitative learning is *selective* and *intentional.* New skills are not just learned by accident, or rarely so, typically scaffold by more advanced individuals who transmit their skills and knowledge to the apprentice or novice learner (Lave, 1988; Rogoff, 1990), a process that contributes to cultural learning in general (Tomasello, Kruger, & Ratner, 1993).

For novelty to emerge and knowledge to be transmitted via observation and imitation, as in the case of the apprentice and his Sushi Master, entails more than passive 'random' and incidental learning. It entails *reciprocation* in the following basic sense. For learning to take place there is a mutual willingness on the part of the novice to observe the expert and on the part of the expert to be observed by the novice. Both protagonists meet in the reciprocal willingness to share attention toward each other, the novice observing the expert, and the expert modeling for the novice. When imitative learning cannot be considered as purely incidental and automatic like in instances of crowd behaviors, there is indeed a mutual, reciprocal willingness to either imitate or model on the part of the protagonists, each acting one side of the same process.

The reciprocal willingness to learn and to teach that is constitutive of imitative learning, when not purely incidental, makes the process break away from imitation in the strict sense of copying, mirroring or the direct '"shadowing' of the other. Mutual attention and intention are involved. This is expressed in the reciprocal sharing of attention, each protagonist aware of and monitoring the other.

In this context, imitation becomes a source of selective transmission and learning, not just a mechanism by which individuals can create an intersubjective bridge by simulating the subjective experience of others. Once again, when not accidental or linked to automatic contagion as in the case of irrepressible yawning after witnessing someone else's yawn, our tendency to open our own mouth while spoon-feeding someone else, or the echoing of individual behaviors to those of a crowd, imitation becomes more than an automatic mirroring process. It is a source of learning and novelty that is *co-created,* based on exchanges that are reciprocal. Imitation is transformed into reciprocation.

George Herbert Mead (1934) emphasized the mutual aspect of communication in which he saw the origins of how individuals construct an explicit sense of self. For Mead, self-identity (who one conceives as 'Me') is the product of what we see in others responding to us, where others are viewed as the social mirror in which the self can be objectified and eventually conceptualized. But the social mirror is a two-way mirror reflecting an image that is not on its surface but rather at the intersection or meeting point between others as mirror of the self and the self as mirror of others. It is an image that is *co-constructed,* reflecting back simultaneously to all the interacting individuals.

This process of co-construction that Mead applies to self-identity which can be generalized to all meanings arising from reciprocal communicative exchanges, whether these meanings correspond to ideas, gifts, instructions, requests, or insults. They are products of complex on-going mutual monitoring processes that entail much more than mirroring. It is an open loop, dynamic, and creative system. It is creative because new meanings are constantly ratcheting up, feeding on each other, and finding new equilibrium until some kind of agreement is reached.

Reciprocal exchanges consist canonically in an exchange of bids and counter-bids until some agreement is reached. Closure is reached when the protagonists recognize themselves in the agreement that, for example, the instruction is followed or understood, the gesture acknowledged, the gift received, and appreciated. In general, closure is reached when a meta-agreement (agreement on agreement) is expressed by all concerned, hands are shaken, papers are mutually signed, or hugs unfold in mutual recognition. Hand shakes considered as prototypical sign of agreement might be a mirroring gesture, in which each hand becomes simultaneously agent and patient of the shake, motor, and perceptual like mirror neurons. But the agreement they express in a mirroring way does not arise from straight mirroring. It always arises from an on-going, open process of reciprocation, and negotiation. Handshakes punctuate such process in which meanings are always put back on the table for further negotiation.

4 Mutual Recognition

Hands that are shaken as a mutual, mirroring gesture punctuating negotiation is nothing more than the explicit (public) expression of a shared understanding regarding the value of things. In such a public manifestation of agreement, reciprocal understanding is temporarily reached as to the relative value of two or more things, be there material or non-material things, such as ideas, beliefs, or feelings. In general, constant on-going negotiation characterizes most of our social exchanges, aimed primarily at co-constructing a shared sense of equivalence among things with others. In human social affairs, most time is spent adjusting and readjusting bids and counter-bids to reach an elusive sense of equity, the latter being the main motor of human transactions.

Early anthropologists like Mauss (1967) or Malinowski (1932), following the pioneer work of Franz Boas on native North American tribes, demonstrated that small scale traditional societies from all over the world tend to be organized around gift systems. In such systems, individuals acquire properties for the sole purpose of relinquishing it following particular rituals. By way of elaborate gifting rituals, individuals, and groups build social ties and reputation. Ritualized gift systems provide ways of establishing a sense of mutual trust and also a means to monitor this trust on the assumption that each gift will be *reciprocated,* their sole function being to circulate among individuals and groups of individuals. For example, in traditional cultures that still prevail in the South Pacific, there is a class of objects that are endowed with the sole function of being offered and received. In these cultures, many daily activities are dedicated to the time consuming confection of gift objects like the fine straw mats of Polynesia (Shore, 1982). These objects have essentially an affective rather than a monetary value.

In their circulation, individuals and groups can monitor and control their social situation, in particular how they are recognized and valued by others. Following the assumption of reciprocation, when one gives an item demonstrating a particular value in the amount that is given, it is expected that others will return the same amount or more. They are challenged to do the same when it comes to be their turn (e.g., the famous 'potlatch' ceremonies found in native North American tribes described by Franz Boas in the early 20th century). The tallying of such exchanges becomes an objective measure of social affiliation. It is also a way to measure the regards others have for the self and to what extent there is some equivalence between these regards, whether they are mutual and represent a comparable value; in other words, whether they tend to 'mirror each other' and express a two-way, *mutual recognition.*

Sociality or the quality of being sociable is inseparable from the elusive feeling of being included and having a causal role or impact on the life of others (Rochat, 2008/in press). It is about being 'connected', visible rather than invisible, and recognized rather than ignored or ostracized (see Honneth, 1995, for a philosophical elaboration of the idea). In this view, sociality rests on *mutual recognition*. The dramatic experience of trying to engage and interact with a person suffering from a lack of sociality gives clinical support to such account.

Kanner (1943), in his description of 'infantile autism', notes that these children appear to have "an innate inability to form the usual biologically provided affective contact with people, just as other children come into the world with innate physical and intellectual handicaps." Kanner goes on insisting on what he sees as the 'extreme autistic aloneness' of these children, their social isolation. Interestingly, for novice, yet well intended healthy adults who might try to engage a child diagnosed with autism, there is always a great deal of discomfort, frustration, and the sense of being 'thwarted,' of becoming unsettled and unsure of themselves (Sigman & Capps, 1997; Greenspan & Wieder, 2006). These children are difficult to figure out, removed, unpredictable, un-reachable. Looking through or beside you, they behave as if you were transparent, *invisible,* non existent, *non consequential,* an experience that is a typical source of great discomfort for the well intended parent or caretaker, and presumably a permanent discomfort for the autistic child withdrawn into his world.

The symptomatic trademark of autistic children is the depleted 'sociality' experience by anybody trying to engage them and connect with them. The social current and co-creation of meanings that normally arise among communicating individuals are either hindered or plainly absent. It takes a great deal of expertise and exercises from parents, educators, and therapists to contact these children, a difficult and courageous enterprise that requires sometimes infinite patience (e.g., Greenspan & Wieder, 2006).

What makes the raising of an autistic child so much more difficult and exhausting compared to raising a healthy, even hyperactive child is the fact that there is no room for mutual recognition, no room for reciprocal acknowledgment of each other. The love parents of autistic children might express, often inexhaustibly, remains unmatched in its return. In this context, parents have difficulties recognizing themselves in the impact they have on their child. Inversely, the child is impaired in recognizing himself in what he does to his parents. Autism causes *mutual* blind mindedness, mutual invisibility, and it is a source of great discomfort, obviously for the trying parents, but also for the disconnected child.

5 From Basic Mirroring to Reciprocation and Social Expectations

The sense of reciprocity is expressed very early in the life of the healthy child. By two months, infants start to engage in face-to-face proto-conversations, first manifesting signs of socially elicited smiles toward others (Wolff, 1987; Sroufe, 1996; Rochat, 2001). Such emotional co-regulation and affective attunement are more than the mirroring process underlying neonatal imitation and emotional contagion evident immediately after birth (Meltzoff & Moore, 1977; Simner, 1971; Sagi & Hoffman, 1976). From this point on, infants express a new sense of shared experience with others in the context of interactive, typically face-to-face plays, what Colwyn Trevarthen (1979) first coined as 'primary intersubjectivity.'

When infants start to engage in proto-conversation, they are quick to pick up cues regarding what to be expected next from the social partner. In general they are quick to expect that following an emotional bid on their part, be it via a smile, a gaze, or a frown, the other will respond in return. Interestingly, adult caretakers in their response are typically inclined to reproduce, even exaggerate the bid of the child. If the child smiles or frowns, we are inclined to smile or frown back at her with amplification and additional sound effects. There is some kind of irrepressible affective mirroring on the part of the adult (Gergely & Watson, 1999).

The complex mirror game underlying social cognition does manifest itself from approximately two months of age and from then on, infants develop expectations and representations as to what should happen next in this context. The still-face experimental paradigm that has been extensively used by infancy researchers for over 30 years provides good support for this assertion (see the original study by Tronick, Als, Adamson, Wise, & Brazelton, 1978). Infants are disturbed when the interactive partner suddenly freezes while staring at them (Rochat & Striano, 1999). They manifest unmistakable negative affects, frowning, suppressing bouts of smiling, looking away, and sometimes even starting to cry. In general, they become avoidant of the other person, presumably expecting them to behave in a different, more attuned way toward them.

This reliable phenomenon is not just due to the sudden stillness of the adult, as the infant's degree of negative responses varies depending on the kind of facial expression (i.e., happy, neutral, or fearful) adopted by the adult while suddenly still (Rochat, Striano, & Blatt, 2002). Also, it appears that beyond seven months old, infants become increasingly active, rather than avoidant, and unhappy, showing initiative in trying to re-engage the still-faced adult. Typically, they touch her, tap her, or clap hands to bring the still-faced adult back into the play, with an intense gaze toward her (Striano & Rochat, 1999).

Numerous studies based on this still-face paradigm and studies using the double video paradigm, in which the infants interact with his mother seen on a TV (either live or in replay) (Murray & Trevarthen, 1985; Nadel, Carchon, Kervella, Marcelli, & Réserbat-Plantey, 1999; Rochat, Neisser, & Marian, 1998), all show that early on, infants develop social expectations as to what should happen next or what should happen while interacting with others. The difficult question is what do these expectations actually mean psychologically for the child. What does it mean for a 2-month-old to understand that if he smiles toward an individual, this individual should 'normally' smile back at him? What does it mean that he picks up the fact that amplified and synchronized mirroring from the adult is an invitation for more bouts of interaction?

One could interpret these expectations as basic, possibly sub-personal and automatic. Accordingly, face-to-face interactions are information-rich events for which infants are innately wired to pick up information, attuned, and prepared from birth to attend to and eventually recognize familiar voices and faces (e.g., DeCasper & Fifer, 1980; Morton & Johnson, 1991). From birth, infants would be attuned to perceptual regularities and perceptual consequences of their own actions, wired to prefer faces, human voices, and contingent events as opposed to any other objects, any other noises, or any other random events. Accordingly, this would be enough for young infants to build social expectations and manifest apparent eagerness to be socially connected as shown by studies using the still-face experimental paradigm or the double video system. But there is more than what meets the eyes of an 'engineering look' at the phenomenon (Rochat, 2008/in press). It is more than just mechanical and requires another, richer look to capture its full psychological meaning.

This proposal is based on evidence of major developmental changes in the ways that children appear to connect with others and reciprocate. Infants rapidly go beyond mirroring and imitation to reciprocate with others in increasingly complex ways, adding the explicit social negotiation of *values* to the process. This development corresponds to the unfolding of primary and secondary (i.e., triadic exchanges of the infant with people in reference to objects in the environment by 7–9 months), and also a *tertiary* level of inter-subjectivity from at least 3 years of age.

Next, we focus on this latter level that we introduce as a major extension of the first two, both well accounted for in the literature (Bruner, 1983; Trevarthen & Hubley, 1978; Trevarthen, 1979; Tomasello, 1995; see Table 1). At the tertiary level of inter-subjectivity, objects and situations in the environment are not just jointly attended to (secondary inter-subjectivity), they become also *jointly evaluated* via negotiation, until eventually some kind of a mutual agreement is reached.

6 From Secondary to Tertiary Inter-Subjectivity

With the intentional communication about objects that emerges by 9 months via social initiatives and explicit bouts of joint attention (secondary inter-subjectivity), infants break away from the primary context of face-to-face exchanges. They become referential beyond the dyadic exchanges to include objects that surround the relationship. Social exchanges also include conversations about things outside of the relationship, becoming triadic in addition to being dyadic. Exchanges become object oriented or objectified, in addition to being the expression of a process of emotional co-regulation. Infants now willfully try to capture and control the attention of others in relation to themselves via objects in the environment. At this point, however, the name of the game is limited to the sharing of attention just for the sake of it. Children measure the extent to which others are paying attention to them and what they are doing. They begin to check back and forth between the person and the object they are playing with (Tomasello, 1995); or they begin to bring an event to the attention of others by pointing or calling for attention to share the experience with them. However, such initiative ends there, and is typically not followed through in further conversation or co-regulation. For infants, secondary inter-subjectivity in triadic exchanges is a new means to control their social environment, in particular the proximity of others as they gain new degrees of freedom in roaming about the environment (Rochat, 2001). By becoming referential, infants also open the gate of symbolic development. They develop a capacity for dual representation whereby communicative gestures stand for and become the sign of something else (e.g., a pointing gesture as standing for a thing out there to be shared with others). Communication becomes intentional, transcending the process of emotional co-regulation and affective attunement that characterizes early face-to-face, proto-conversational exchanges (i.e., primary inter-subjectivity). Yet, it remains restricted to the monitoring of whether others are, or are not, co-experiencing with the child.

Nevertheless, with the emergence of intentional communication and the drive to co-experience events and things in the environment, infants learn and begin to develop shared meanings about things. To some extent, they also begin to develop shared values about what they experience of the world, but this development remains limited. For example, when facing dangers or encountering new situations in the environment, they are now inclined to refer to the facial expressions of others that are paying attention to the same events (Campos & Sternberg, 1981; Striano & Rochat, 2000). The meaning of a perceived event (e.g., whether something is dangerous or threatening) is now referred to others' emotional responses, to some extent evaluated in relation to others, but it ends there. The process does not yet entail any kind of negotiation regarding the value of what is experienced. The world is essentially divided into either good (approach) or bad (avoidance) things and events. Such basic social referencing emerges at around 9 months, in parallel to the propensity of infants to share attention with others and to communicate with them intentionally (Tomasello, 1999; Rochat & Striano, 1999).

By the middle of the second year, triadic exchanges develop beyond basic social referencing and the sense of co-experience with others that is the trademark of secondary inter-subjectivity. The child now begins to engage in active negotiation regarding the values of things co-experienced with others. They manifest *tertiary inter-subjectivity,* a sense of shared experience that rests on complex on-going exchanges unfolding over time: things that happened in the past are manifest in the present and are projected by the child into the future. The prototypical expression of this new level of inter-subjectivity is the expression of secondary emotions such as embarrassment or guilt.

In relation to the self, by 20 months, children begin to represent what others perceive of themselves and gauge this representation in relation to values that are negotiated. If they see themselves in a mirror and notice a mark surreptitiously put on their face, they will be quick to remove it and often display coy behaviors or acting out (Amsterdam, 1972; Rochat, 2003). They begin to pretend and mask their emotions (Lewis, 1992). In general, they become self-conscious, negotiating, and actively manipulating what others might perceive and evaluate of themselves (Lewis, 1992; Rochat, 2008/in press). From this point on (18–20 months), children project and manipulate a public self-image, the image they now identify, and recognize in the mirror. It is an image that is objectified and shared with others, a represented 'public' self-image that from now on will be constantly updated and negotiated in relation to others. Interestingly, by 20 months, children's linguistic expressions begin also to include the systematic use of possessives, children starting to claim ownership over things with imperative expressions such as "mine!" (Bates, 1990; Tomasello, 1998). Such expressions demarcate the value of things that are jointly attended in terms of what belongs to the self and what belongs to others. This value begins to be negotiated in the context of potential exchanges, bartering, or donations. With the explicit claim and demarcation of property, the child develops a new sense of reciprocity in the context of negotiated exchanges of property, whether objects, feelings, or ideas. At around the same age, children also begin to demonstrate pro-social behaviors, engaging in acts of giving and apparent benevolence by providing help or spontaneously consoling distressed others (Zahn-Waxler, Radke-Yarrow, Wagner, & Chapman, 1992). Self-concept, ownership claim, and a new concern for others bring the child to the threshold of moral development and the progressive construction of an explicit sense of justice (Damon, 1994). What follows in development is a new level of social reciprocity that is increasingly organized around an ethical stance taken by the child. But this ethical level of reciprocity develops between 3 and 5 years of age, and beyond as shown by our recent investigation of young children's sense of fairness in sharing across cultures.

7 Emergence of an Ethical Stance

There is a developmental trend from a reluctance to share, to subtle, more reciprocal exchanges. For example, children between 3 and 5 years of age become significantly more flexible and systematic in adjusting their successive bids while engaged in bartering exchanges of stickers or toys. They are increasingly inclined to up their bids until an agreement is reached. This developmental trend is also associated with an increased understanding by the child of others' mental states, a trend that appears to cut across cultures (Callaghan et al., 2005).

As children start to claim explicit ownership and invest affects into objects of devotion, they do so by first manifesting unmistakable exclusivity in their possession, a blunt reluctance to exchange. They show overwhelming egocentrism. When the child begins to say "mine!" she does not only imply that "it is not yours." It is also an explicit statement of defensive exclusivity, a reluctance to even contemplate sharing, and an unmistakable claim that they want to keep it for themselves.

In recent cross-cultural observations (Rochat et al., 2008, in press), we confirmed that this 'egocentric' trend is a universal trend. We found it in 3 year-olds and to a lesser extent in 5 year-old children from all over the world, growing up in highly contrasted physical, social, economical, and cultural environments. It happens in children living in rich or poor neighborhoods and in cultures fostering radically different values regarding private property. We observed this trend in children from small, highly collectivist villages of rural Peru, or from small isolated fishing communities in Fiji. This same trend occurs in children growing up in violent and lawless as well as affluent neighborhoods of Rio de Janeiro; unschooled kids begging and living on the streets of Recife in Brazil, as well as young children attending a Communist Party controlled preschool in Shanghai, China, or in middle class North American children of Atlanta.

In general, we found that across cultures, between 3 and 5 years, there is a robust developmental trend toward more equity in sharing. In conditions, where the child was one of the two recipients, 3 year-olds tended to distribute overwhelmingly more candies to themselves, whether equity was possible or not. By 5 years, however, this trend was still evident but significantly tamed. Children continued to favor themselves and are selfish but markedly less. Interestingly this trend was the same in children from all cultures, but reduced in children growing up in small rural and collective communities (i.e., Peru and Fiji in our sample). In development, there is thus a universal drift in active sharing from massive to reduced selfishness between 3 and 5 years of age, a trend that is moderated by the cultural environment of the child. Despite the significance of cultural factors, the trend toward increased altruism/pro-sociality in sharing is remarkably robust from the time children begin to be explicit in claiming ownership.

Culture appears to play a role in the developmental pace at which the child becomes inclined to share with greater equity, but the general trend is there regardless of marked variations. In China, children were tested in a preschool that emphasized primarily group activities and sharing. Children always play, sing, and learn as members of a group, rarely as individuals isolated from the group. Such attempts are much less frequent in middle class North American preschools, such as those of the children we tested in Atlanta. In Fiji, or in Peru, the tested children lived in small, close knit communities where public and shared properties dominate over ostentatious private ownership. When they exist, preschools in these regions are known to emphasize synchronized group activities in children.

The stability of this developmental trend is particularly striking when considering the three groups of Brazilian children. Each group grows in highly contrasted economical and social circumstances within the same national and cultural borders. A group of children lived in the poor and insecure environment of a favela in Rio de Janeiro, an environment dominated by young drug lords that terrorize and dictate law and order. Another group was composed of privileged children, of the same age, from an affluent private preschool situated just a few miles away from the favela. The third group of Brazilian children was composed of 3 and 5 year-old unschooled street kids from the city of Recife, a few hundred miles North-East of Rio. These children spent their days unsupervised by adults, begging on the street, collecting refuse, and typically spending the night with an extended family living in precarious, unsanitary slums close to public dumps.

One could easily presume that the drive to own, and not to share, in the young children of the favela, and particularly the street kids of Recife, might be different compared to the privileged children of Rio. Our research shows that it is not the case. All of these children demonstrate the same developmental trend toward a significant decrease in selfishness and increase in more equitable sharing between 3 and 5 years. But why is that?

Young children develop to become more equitable in their sharing, regardless of their economical and cultural circumstances because they enter the culture of their species (*Homo Negotiatus*), a culture that is fundamentally based on reciprocal exchanges. Hoarding and coercion are antithetical to this culture. If it exists, it is an anomaly, due to particularly stressful circumstances (war, disaster, rebellion, madness). It is not cardinal to the culture of *Homo Negotiatus*, unlike any other animal species that are not designed to have others in mind in their social exchanges and their sharing of resources.

We construct equity as well as agree on values by an active process of approximation and mutual monitoring. This process takes form within reciprocal exchanges. We do so by negotiation and ultimately by caring about reputation, namely our relative proximity with others. What happens between

3 and 5 years, is a marked progress in this process that channels children away from greed and immediate gratification. The product of this development is the emergence of a moral space in which children begin to care about reputation.

Children between 3 and 5 years develop an understanding that they are potentially liable and that they build a history of transactions with others. Needless to say that parents and educators foster this development in all cultures, but this fostering is essentially the enforcement of the basic rules of reciprocity, the constitutive elements of human exchanges. Children are channeled to adapt to these rules they depend on to maintain proximity with others. From this, they begin to build a moral space in relation to others, a moral space that is essentially based on the basic rules of reciprocity (Rochat, 2008/in press). It is a moral space that is constantly in the making, constantly revised, and in which equity is endlessly approximated by way of negotiation.

8 Conclusions: Human Sociality Buds in Imitation But Blossoms in Reciprocation

In this chapter, our intention was to show that basic mirroring processes expressed in neo-natal imitation and emotional contagion at birth are necessary, but not sufficient, to account for the early development of reciprocal exchanges that takes place from the second month. Imitation and emotional contagion, taken literally as close-loop automatic mirror systems, are soon transformed into dynamic, ultimately creative exchanges that take the form of open-ended proto-conversations ruled by principles of *reciprocation*.

The basic mirror processes expressed at birth probably correspond to innate social binding mechanisms. They are basic resonance processes (Gallese, 2003) that allow the child, from the outset, to match self and others' experience. These mechanisms allow for a necessary starting state of implicit inter-subjective equivalence. Endowed with, and capable of such processes, infants from birth would automatically perceive others as 'like them.' This basic, obligatory perception would be mediated by sub-personal innate mirror mechanisms (i.e., neural mirror systems). However, we tried to show that the way infants and young children connect to the social world develops dramatically with the emergence of active, creative, and increasingly complex reciprocal exchanges.

We argued that from approximately 2 months following birth, there is a major qualitative shift that can be equated to a functional 'transcendence' of the mirroring processes expressed at birth. These basic processes soon become integrated into complex, open-ended systems of *reciprocation,* first in dyadic exchanges from the second month, and eventually in triadic exchanges that include objects from 9 months. Active emotional co-regulation, as opposed to strict mirroring, underlies the first open-ended reciprocal exchanges that emerge by 2 months in the healthy child. Intentional communication and the drive to include others in the experience of the physical world underlie the triadic reciprocal exchanges emerging by 9 months. It is in the development of reciprocal social exchanges that infants learn and eventually find their way in a world not only made of objects and people, but more importantly, a world made of 'shared values.'

Beyond 9 months, and in particular by the end of the second year, children become increasingly conceptual in their reciprocation. They recognize themselves in mirrors and become explicit about what is theirs as opposed to others, starting to use possessives and manifesting ostentatious acts of appropriation. They also start to show concerns, embarrassment, as well as signs of guilt. In short, from this point on, not only do they interact with others in reciprocation for the sake of co-experiencing things, but they begin also to represent how others perceive and evaluate them in the process. They become newly self-conscious and *co-evaluative* in their social exchanges.

From the time children begin to show concerns regarding the extent to which they are recognized for what they do, feel, and what they achieve, they enter a world of values that are constantly negotiated in interaction with others. It is in this context that children develop an ability to construe what is happening in the mind of others. Theories of mind emerging by 4–5 years are probably by-products or spin offs of the tertiary inter-subjectivity developing by the second year (see Rochat, 2006a, for further discussion).

Following the roadmap proposed in the introduction (Table 1), by the end of the second year, children show signs of active projection and identification with others. For example, they begin to display active empathic feelings (e.g., Zahn-Waxler et al., 1992) and to detect when others are intentionally mimicking them (e.g., Agnetta & Rochat, 2004). They perceive others as disposed in certain ways toward them, but also whether they deserve contempt, help, or comfort. From this time on, the child's social binding becomes deeply evaluative, beyond the mere drive to interact harmoniously with others, or to share and synchronize attention toward things. In becoming evaluative, children develop the need to agree with others on the values of things via open-ended negotiation, a process that from now on dominates reciprocal exchanges and is arguably the trademark of all human cultures (Rochat, 2006a,b).

By 4–5 years, universally, children begin to predict the behavior of others based on their construal of what's on their mind: what they might feel, think, or believe (e.g., Theories of mind, Callaghan et al., 2005). More importantly, they also begin to construe others in their vulnerability to be unjustly treated and feel hurt, anticipating potential long-term reprisal over un-equitable treatment.

Our research shows that between 3 and 5 years children develop to inhibit their inclination to maximize their own gain when asked to share desirable goods such as candies. This appears to be a universal trend across highly contrasted cultures (Rochat et al., submitted). Compared to 3 year-olds, children at 5 manifest more explicit fairness in distributive justice. They develop an ethical stance toward others and are increasingly concerned with what is right and what is wrong within the particular context of their culture. Reciprocal exchanges now take place within a moral space in which children develop their own situation, constantly negotiating with others the value of things, in search of mutual agreements that are endlessly re-examined and revised. From then on, this process will dominate social exchanges throughout the lifespan.

To conclude, if mirroring processes form a necessary basis for social binding, they are in themselves not sufficient to account for the rapid development of open-ended and creative levels of reciprocation that take place starting the second month after birth. Mutual recognition in a moral space is arguably the measure of human social affiliation. Such recognition might find its roots in imitation and mirroring processes. However, we argue that these basic processes are only a seed that can only grow in the context of reciprocal exchanges with more advanced and cultured others.

Acknowledgments While writing this chapter, the first author was supported by a 2007 John Simon Guggenheim fellowship. Much appreciation is expressed to Britt Berg for editorial help and comments on an earlier draft.

References

Agnetta, B., & Rochat, P. (2004). Imitative games by 9-, 14-, and 18-month-old infants. *Infancy, 6*(1), 1–36.

Amsterdam, B. (1972). Mirror self-image reactions before age two. *Developmental Psychobiology*, 5, 297–305.

Bates, E. (1990). Language about me and you: Pronominal reference and the emerging concept of self. In D. Cicchetti & M. Beeghly (Eds.), *The self in transition: infancy to childhood* (pp. 165–182). Chicago: University of Chicago Press.

Boggs, S. T. (1985). *Speaking, relating, and learning: A study of Hawaiian children at home and at school.* Norwood, NJ: Ablex Publishing Co.

Bruner, J. S. (1983). *Child's talk.* New York: Norton.

Callaghan, T., Rochat, P., Lillard, A., Claux, M. L., Odden, H., Itakura, S., Tapanya, S., & Singh, S. (2005). Synchrony in the onset of mental-state reasoning: Evidence from five cultures. *Psychological Science, 6*(5), 378–384.

Campos, J., & Sternberg, C. (1981). Perception, appraisal, and emotions: The onset of social referencing. In M. Lamb & L. Shjerrod (Eds.). *Infant social cognition: Empirical and theoretical considerations* (pp. 273–314). Hillsdale, NJ: Lawrence Erlbaum Publishers.

Damon, W. (1994). Fair distribution and sharing: The development of positive justice. In B. Puka (Ed.), *Fundamental research in moral development* (pp. 189–254). N.Y.: Garland Publishing.

DeCasper, A. J., & Fifer, W. P. (1980). Of human bonding: Newborns prefer their mother's voices. *Science, 208,* 1174–1176.

Donald, M. (1991). *Origins of the modern mind: three stages in the evolution of culture and cogntion.* Cambridge, Mass: Harvard University Press.

Finnbogason, G. (1912). *The mind and the world around us,* (pp. 250 and 262, cited by Hauksson, J. 2000). *Acta Sociologica, 43,* 307–315 .

Fogassi, L., Ferrari, P. F., Gesierich, B., Rozzi, S., Chersi, F., & Rizzolatti, G. (2005). Parietal lobe: from action organization to intention understanding. *Science, 308*(5722), 662–667.

Gallagher, S., & Hutto, D. (2007). Understanding others through primary interaction and narrative practice. In: J. Zlatev, T. P. Racine, C. Sinha & E. Itkonen (Eds.), *The shared Mind: Perspectives on intersubjectivity* (pp. 17–38). Amsterdam: John Benjamins.

Gallese, V., Fadiga, L., Fogassi, L., & Rizzolatti, G. (1996). Action recognition in the premotor cortex. *Brain, 119*, 593–609.

Gallese, S., Fogassi, L., Fadiga, L., & Rizzolatti, G. (2002). Action representation in the inferior parietal lobule. In W. Prinz & B. Hommel (Eds.), *Attention and performance* (Vol. 19, pp. 247–266). New York: Oxford University Press.

Gallese, V. (2003). The Roots of empathy: The shared manifold hypothesis and the neural basis of intersubjectivity. *Psychopathology, 36*, 171–180.

Gallese, V. (2007) Before and below "theory of mind": embodied simulation and Philosophical Transaction of the Royal Society *B, 362*, 659–669.

Gergely, G., and Watson, J. S. (1999). Early social-emotional development: Contingency perception and the social-biofeedback model. In P. Rochat (Ed.), *Early social cognition* (pp. 101–136). Hillsdale: Erlbaum Publishers.

Goldman, A., & Sripada, C. S. (2005). Simulationist models of face-based emotion recognition. Cognition, *94*, 193–213.

Greenspan, S., & Wieder, S. (2006). *Engaging autism*. Cambridge: Da Capo Press.

Harris, P. L. (1992). From simulation to folk psychology: The case for development. *Mind and language, 7*, 120–144.

Honneth, A. (1995). *The struggle for recognition – The moral grammar of social conflicts*. Cambridge (Mass): M.I.T. Press.

Kanner, L. (1943). Autistic disturbances of affective contact. *The nervous child, 2*, 217–250.

Lancy, D. (1996). *Playing on the mother-ground: Cultural routines for children's development*. New York: Guilford Press.

Lave, J. (1988). *Cognition in practice*. Cambridge: Cambridge University Press.

Lewis, M. (1992). *Shame: the exposed self*. New York: Free Press.

Malinowski, B. (1932). *Argonauts of the Western Pacific: An account of native enterprise and adventure in the archipelagoes of Melanesian New Guinea*. London: Routledge & Sons.

Mauss, M. (1967). *The Gift: Forms and functions of exchange in archaic societies*. New York: Norton.

Mead, G. H. (1934). *Mind, self and society*. Chicago: University of Chicago Press.

Meltzoff, A. N., & Moore, M. K. (1977). Imitation of facial and manual gestures by human neonates. *Science, 198*, 75–78.

Meltzoff, A.N. (1995). Understanding the intentions of others: Re-enactment of intended acts by eighteen-month-old children. *Developmental Psychology 31*(5), 838–850.

Meltzoff, A. N. (2007). The "like me" framework for recognizing and becoming an intentional agent, *Acta Psychologica, 124*(1), 26–43.

Morton, J., & Johnson, M. H. (1991). CONSPEC and CONLERN: A two-process theory of infant face recognition. *Psychological Review*, 98(2), 164–181.

Murray, L., & Trevarthen, C. (1985). Emotional regulation of interactions between two-months-old and their mothers. In T. M. Field & N. A. Fox (Eds.), *Social perception in infants* (pp. 177–197). Norwood, NJ: Ablex.

Nadel, J. Carchon, I. Kervella, C., Marcelli, D., Réserbat-Plantey, D. (1999). Expectancies for social contingency in 2-month-olds. *Developmental Science, 2*(2), 164–173(10).

Odden, H., & Rochat, P. (2004). Observational learning and enculturation. *Educational and Child Development, 21*(2), 39–50.

Piaget, J. (1962). *Play, dreams, and imitation in childhood*. N.Y.: Norton.

Rizzolatti, G., Fadiga, L., Gallese, V., Fogassi, L. (1996). Premotor cortex and the recognition of motor actions. *Brain Research. Cognitive Brain Research, 3*(2), 131–141.

Rochat, P. (2008/in press). *Others in Mind–Social Origins of Self-Consciousness*. New York: Cambridge University Press.

Rochat, P., Neisser, U., & Marian, V. (1998). Are young infants sensitive to interpersonal contingency? *Infant behavior and development, 21*(2), 355–366.

Rochat, P., & Striano, T. (1999). Social cognitive development in the first year. In P. Rochat (Ed.), *Early social cognition* (pp. 3–34). Lawrence Erlbaum Associates.

Rochat, P., & Striano, T. (2002). Who is in the mirror: Self-other discrimination in specular images by 4- and 9-month-old in fants. *Child Development, 73*, 35–46.
Rochat, P., Striano, T., & Blatt, L. (2001). Differential effects of happy, neutral, and sad still-faces on 2-, 4-, and 6-month-old infants. *Infant and Child Development, 11*(4), 289–303.
Rochat, P. (2001) *The infant's world*. The Developing Child Series. Harvard University Press.
Rochat, P. (2003). Five levels of self-awareness as they unfold early in life. *Consciousness and Cognition, 12*(4), 717–731.
Rogoff, B. (1990). *Apprenticeship in Thinking: Cognitive Development in Social Context*. New York: Oxford University Press.
Rizzolatti, G., & Craighero, L. (2004). The mirror-neuron system. *Annual Review of Neuroscience 27*, 169–192.
Rochat, P. (2006a). Humans evolved to become *Homo Negotiatus* ... the rest followed. *Behavioral and Brain Sciences, 28*, 714–715.
Rochat, P. (2006b). What does it mean to be human? *Anthropological Psychology, 17*, 100–108.
Rochat, P. (2008/in press). Mutual recognition as foundation of sociality and social comfort. In Striano, T., & Reid, V. (Eds.), *Social cognition: Development, neuroscience and autism*. Oxford: Blackwell Publishing.
Rochat, P. (2008/in press) Others in mind – Fear of rejection and the social origins of self-consciousness.
Rochat, P., Dias, M. D. G., Guo, L., Broesch, T., Passos-Ferreira, C., & Winning, A. (2008/in press). Fairness in distributive justice by 3- and 5-year-olds across 7 cultures. Journal of Cross-Cultural Psychology.
Rochat, P., & Passos-Ferreira, C. (2008). Homo Negotiatus. Ontogeny of the unique ways humans own, share and deal with each other. In Itakura, S., & Fujita, F. *Origins of the social mind: Evolutionary and developmental views*. New York: Springer-Verlag.
Rogoff, B. (1995). Observing sociocultural activity on three planes: participatory appropriation, guided participation, and apprenticeship. In J. Wertsch, P. del Rio, & A. Alvarez (Eds.), *Sociocultural studies of mind*, (pp. 139–164). Cambridge: Cambridge University Press.
Sagi, A., & Hoffman, M. L. (1976). Empathic distress in the newborn. *Developmental Psychology, 12*, 175–176.
Shore, B. (1982). *Salailua: A samoan mystery*. New York: Columbia University Press.
Sigman, M., & Capps, L. (1997). *Children with autism – A developmental perspective*. Cambridge: Harvard University Press.
Simner, M. L. (1971). Newborn's response to the cry of another infant. *Developmental Psychology*, 5, 136–150.
Sperber, D. (1996). *La Contagion des Idées. Theorie naturaliste de la culture*. Paris: Odile Jacob.
Sroufe, L. A. (1996). *Emotional development: The organization of emotional life in the early years*. New York: Cambridge University Press.
Striano, T., & Rochat, P. (1999). Developmental link between dyadic and triadic social competence in infancy. *British Journal of Developmental Psychology*, 17, 551–562.
Striano, T., & Rochat, P. (2000). Emergence of selective social referencing in infancy. *Infancy, 2*, 253–264.
Tarde, G. (1890/1993). *Les lois de l'imitation*. Paris: Editions Kimé.
Tomasello, M., Kruger, A. C., & Ratner, H. H., (1993). Cultural learning. *Behavioral and brain sciences, 16*(3), 495–552.
Tomasello, M. (1995). Joint attention as social cognition. In C. Moore & P. Dunham (Eds.), *Join attention: Its origins and role in development* (pp. 103–130). Hillsdale, NJ: Erlbaum.
Tomasello, M. (1998). One child early talk about possession. In J. Newman, (Ed.). *The linguistic of giving*. Amsterdam: John Benjamins.
Tomasello, M. (1999). *The cultural origins of human cognition*. Cambridge, MA: Harvard University Press.

Trevarthen, C., & Hubley, P. (1978). Secondary intersubjectivity: confidence, confiding and acts of meaning in the first year. In A. Lock (Ed.), *Action, gesture and symbol: The emergence of language* (pp. 183–229). London: Academic Press.

Trevarthen, C. (1979). Communication and cooperation in early infancy: A description of primary intersubjectivity. In M. M. Bullowa (Ed.), *Before speech: The beginning of interpersonal communication* (pp. 321–347). New York: Cambridge University Press.

Tronick, E. Z., Als, H., Adamson, L., Wise, S., & Brazelton, T. B. (1978). The infant's response to entrapment between contradictory messages in face-to-face interaction. *Journal of the American Academy of Child Psychiatry,* 17, 1–13.

Wellman, H. M. (2002). Understanding the psychological world: Developing a theory of mind. In Goswami, U. (Ed.), *Blackwell handbook of child cognitive development* (pp. 167–187). Blackwell Publishing.

Wolff, P. (1987). *The development of behavioral states and the expression of emotions in early infancy*. Chicago: The University of Chicago Press.

Zahn-Waxler, C., Radke-Yarrow, M., Wagner, E., & Chapman, M. (1992). Development of concern for others. *Developmental Psychology,* 28, 126–136.

Automatic and Controlled Processing within the Mirror Neuron System

Trevor T.-J. Chong and Jason B. Mattingley

Abstract The human ability to recognize the actions and gestures of others is fundamental to communication and social perception. Evidence suggests that this ability is supported by the mirror neuron system, the primary function of which is to mentally simulate a perceived action in the observer's own motor system. Traditionally, the processing that occurs within this network is considered to be automatic and stimulus-driven, but neurophysiological data from macaques suggest that even the activity of single mirror neuron units maybe modulated by attention and context. Similarly, in humans, there is a growing body of evidence to indicate that the mirror system is also vulnerable to top-down processes such as cognitive strategy, learned associations and selective attention. In this chapter, we review the evidence that indicates observed actions are processed automatically, and contrast these data with those that indicate a susceptibility of action processing to top-down factors. We suggest that the assumption that observed actions are processed involuntarily arose largely because most studies have not explicitly challenged the automaticity of the visuomotor transformation process. The frontoparietal mirror system should therefore be viewed in the context of a larger network of areas involved in action observation and social cognition, whose activity may mutually inform and be informed by the mirror system itself. Such reciprocal connections maybe critical in guiding ongoing behavior by allowing the mirror system to adapt to concurrent task demands and inhibit the processing of task-irrelevant gestures.

Keywords Automaticity · Selective attention · Cognitive strategy · Prior exposure

---

T.T.-J. Chong
Macquarie Centre for Cognitive Science, Macquarie University, Sydney, New South Wales 2109, Australia
e-mail: trevor.chong@maccs.mq.edu.au

J.A. Pineda (ed.), *Mirror Neuron Systems*, DOI: 10.1007/978-1-59745-479-7_10,

1 Introduction

> We may lay it down for certain that every mental representation of a movement awakens to some degree the actual movement which is its object; and awakens it in a maximum degree whenever it is not kept from so doing by an antagonistic representation present simultaneously to the mind
>
> – William James (1980). *Principles of Psychology*, p. 526, Vol. II

After more than a century of scientific investigation, we now have a substantial body of evidence in broad agreement with James' ideomotor principle of voluntary action – that the perception of an action results in co-activation of visual and motor circuits within the observer (James, 1890). This principle has had a lasting influence on more modern theories of motor cognition, most of which can be considered direct descendents of James' initial ideomotor principle (Brass, Bekkering, Wohlschläger, & Prinz, 2000; Greenwald, 1970, 1972; Hommel, Müsseler, Aschersleben, & Prinz, 2001; Jeannerod, 2001; Kornblum & Lee, 1995; Prinz, 1990, 1997, 2002; Wohlschläger, Gattis, & Bekkering, 2003). The discovery of the primate mirror neuron system represented a significant advance in the field of motor cognition, by providing many of these theories with a plausible and parsimonious neuroanatomical substrate.

As has been discussed in the preceding chapters, the mirror system is a network of parieto-premotor areas which responds to both the observation and the execution of a movement. The core of this network comprises the posterior part of the inferior frontal gyrus (IFG, corresponding to Brodmann's areas 44/45) and the rostral part of the inferior parietal lobe (IPL, BA 40), which are believed to be the human homologues of macaque mirror areas F5 and PF/PFG, respectively. Together, these areas are thought to underpin such complex sociocognitive phenomena as observational learning (Arbib, 2002; Buccino, Vogt et al., 2004), empathy (Carr, Iacoboni, Dubeau, Mazziotta, & Lenzi, 2003; Gallese, 2003b; Gallese, Rizzolatti & Keysers, 2004), theory of mind (Gallese & Goldman, 1998), socialization (Gallese & Goldman, 1998), and the evolution of human language (Arbib, 2005; Arbib & Rizzolatti, 1999).

At the most fundamental level, all of these postulated functions of the mirror system presuppose that it provides a means by which an observed action can be directly matched with its corresponding representation in the observer's own motor repertoire. Thus, the observation of an action is said to cause the motor system of the observer to 'resonate' (Rizzolatti, Fadiga, Fogassi, & Gallese, 1999) and, through this process of 'Direct Matching,' allow an observed action to be recognized and/or overtly imitated by the individual (Kilner, Paulignan, & Blakemore, 2003; Rizzolatti & Craighero, 2004). This Direct Matching Hypothesis is consistent with patient data, which show that patients with a congenital paralysis of their facial musculature (Moebius syndrome) are also impaired in recognizing the emotive facial expressions of others (Cole, 1999, 2001).

Given the biological and human significance of action recognition and imitation, it is critical for observed gestures to be processed efficiently. This has led to the prevalent assumption in the literature that the activation of motor representations within the observer is direct and highly automatic (Buccino, Binkofski, & Riggio, 2004; Coricelli, 2005; Gallese, 2003a; Gallese, Fadiga, Fogassi, & Rizzolatti, 1996; Gallese & Metzinger, 2003; Rizzolatti & Craighero, 2004; Rizzolatti et al., 1996; Wilson & Knoblich, 2005). In fact, this assumption echoes James' implicit assertion in his ideomotor principle that, '*every* mental representation of a movement awakens *to some degree* the actual movement which is its object' (James, 1890, p. 526, Vol. II; emphasis ours). A more recent formulation of the way in which mirror neurons mediate action understanding has been synopsized as follows: 'Each time an individual sees an action done by another individual, neurons that represent that action are activated in the motor cortex. This *automatically induced*, motor representation of the observed action corresponds to what is spontaneously generated during active action and whose outcome is known to the acting individual. Thus, the mirror system transforms action into knowledge' (Rizzolatti & Craighero, 2004, p. 172; emphasis ours). Some authors have gone further in their claims by suggesting that this motor simulation proceeds, not only automatically, but also implicitly and unconsciously (e.g., Gallese, 2001).

What does it mean for mirror neuron activity to be 'automatic'? Traditional views of human cognition have distinguished automatic cognitive processes from those that are controlled. Typically, controlled processes are voluntary, require attention, and are relatively slow (Cohen, Dunbar, & McClelland, 1990). By contrast, automatic processes usually arise without conscious effort and may occur outside of awareness, are triggered involuntarily, do not require attention for their execution, and are relatively fast (Bargh, 1992; Cohen et al., 1990; Hasher & Zacks, 1979; Posner, 1978). A corollary of this definition is that automatic processes do not draw on general cognitive resources. Thus, they do not interfere with, nor are they subject to, interference from other concurrent perceptual or cognitive demands. As a result, several such processes can operate in parallel without capacity limits (Pashler, 1998). Early research on cognition considered automaticity to be an all-or-nothing phenomenon that proceeds independently of controlled processes. However, a large body of empirical data suggests that this dichotomy is an oversimplification (Kahneman & Henik, 1981; MacLeod & Dunbar, 1988; Shiffrin & Schneider, 1977). Kahneman and Treisman (1984), for example, distinguish between three types of automatic processes – those that are 'strongly,' 'partially,' or 'occasionally' automatic – depending on the amount of attention required for those processes to be completed.

Is the motoric simulation of an observed action a strongly automatic process, or must attention be directed toward the observed actions for it to occur? Surprisingly, this important issue in ideomotor theory, and in the theory of action perception in general, has not been extensively explored or discussed. Because the original work on the mirror system emphasized its capacity to automatically process perceived actions, the potential role of conscious, controlled, or executive processing on the operation of this network has been largely overlooked. However, a complete account of the mirror system and its role in action understanding and imitation clearly requires an examination of how top-down factors influence the activity of this system. Our goal in this chapter is to first consider the extent to which the mirror system operates involuntarily, prior to discussing the way in which voluntary and controlled processes, such as strategy, learning, and selective attention, may also modulate its activity.

2 Automatic Processing Within the Human Mirror System

2.1 Behavioral Studies

The activity of single mirror neuron units is often considered to arise 'automatically,' in the sense that they respond to an observed action even when the animal has no intention of acting upon the target object (Gallese et al., 1996). Indeed, the activity of individual mirror neurons would *prima facie* seem to be a straightforward matter of 'monkey see, monkey do' (Carey, 1996). The majority of human studies on action observation tend to be consistent with the proposal that the visuomotor matching process is relatively stimulus-driven and unaffected by strategic processes. Behavioral evidence for this proposition comes from several experimental paradigms that reveal involuntary response priming during action observation (Heyes, Bird, Johnson, & Haggard, 2005; Kilner et al., 2003; Press, Bird, Flach, & Heyes, 2005; Stürmer, Aschersleben, & Prinz, 2000).

If a perceived action automatically activates its equivalent motor representation via the mirror system, it should also interfere with the execution of a movement that is qualitatively different from that which is observed. An elegant confirmation of this hypothesis was demonstrated by Kilner et al. (2003), who required participants to make sinusoidal up-down or left-right movements with their right arm while concurrently observing either congruent or incongruent arm movements performed by another human or by a robotic arm. As predicted, the authors found that observing another human perform an incongruent movement led to significantly greater variance in participants' own movements than observing a congruent movement. Interestingly, this effect was not found during the observation of the robotic arm – this parallels early single cell data which showed that mirror neurons failed to respond during the observation of actions involving a mechanical device, such as a tool (Gallese et al., 1996).

The seemingly involuntary motor interference effect demonstrated by Kilner et al. (2003) has been shown in other behavioral tasks as well. A well-explored paradigm that is believed to reflect the visuomotor matching instantiated by the mirror system has been termed the 'automatic imitation' task (Heyes et al., 2005; Press et al., 2005). Typically in this task, participants perform actions faster and more accurately while concurrently observing the same (congruent) movement, relative to one that is different (incongruent) (Brass, Bekkering, & Prinz, 2001; Brass et al., 2000; Craighero, Bello, Fadiga, & Rizzolatti, 2002; Heyes et al., 2005; Stürmer et al., 2000; Vogt, Taylor, & Hopkins, 2003). Crucially, these effects are found in the absence of an explicit instruction to imitate even though the observed postures are entirely irrelevant to task requirements. In the first study to describe this effect, Stürmer et al. (2000) required participants to open or close their right hands in response to a go-stimulus, which could have been an image of the congruent or incongruent movement. However, the posture of the stimulus hand was irrelevant and participants selected their response based simply on its color, which at some point changed to red or blue. The key finding was that choice reaction times to initiate congruent movements were significantly faster than those to incongruent movements.

These automatic imitation effects have since been replicated when the requirement to actively discriminate the displayed hand is eliminated. In these simple detection tasks, participants exhibit the identical congruency effect, even when they perform a prespecified motor response to the onset of *any* visual stimulus within whole blocks of trials (Brass, Bekkering et al., 2001; Heyes et al., 2005; Vogt et al., 2003). Together, these behavioral data suggest that perceived gestures automatically activate representational structures also involved in the execution of those actions (Brass, Bekkering et al., 2001).

Most studies on automatic imitation have involved relatively simple, intransitive gestures, such as opening and closing a hand, or lifting or tapping a finger (Brass, Bekkering et al., 2001; Brass et al., 2000; Craighero et al., 2002; Heyes et al., 2005; Stürmer et al., 2000). Given the greater sensitivity of macaque mirror neurons to gestures that are more meaningful to the monkey (i.e., goal-directed actions), it is worth noting that two studies have also found similar congruency effects during the observation of transitive gestures (Craighero et al., 2002; Vogt et al., 2003). In both of these studies, participants were required to grasp an object that was positioned in one of two possible orientations, while they observed static images of a hand in a congruent or incongruent end position. For example, Craighero and colleagues (2002) required participants to grasp a clockwise or counterclockwise-oriented bar in response to a hand in the same relative to a different end position. In keeping with previous studies on automatic imitation, participants in this study initiated their movements faster in response to congruent compared to incongruent end-effector positions. The study by Craighero et al. (2002) was a 'Go/NoGo' task requiring participants to discriminate the displayed hand action prior to selecting their response, but similar findings were also found by Vogt et al. (2003) in a simple reaction time task, when no active visual discrimination of the target stimulus was required (Vogt et al., 2003).

The automatic imitation effects described above appear to be relatively invulnerable to participants' response strategies or beliefs about the type of stimulus they are observing (Press, Gillmeister, & Heyes, 2006). For example, a study by Press and colleagues (2006) exploited the finding that automatic imitation effects are most prominent during the observation of human, rather than robotic stimuli (Kilner et al., 2003; Tai, Scherfler, Brooks, Sawamoto, & Castiello, 2004). The authors presented their participants with silhouettes of a hand opening or closing, but prior to each block informed participants that the hand actions were either human or robotic in nature (when in fact they were identical). A post-test questionnaire confirmed that the instructions were effective in influencing participants' beliefs about the nature of the observed stimuli. Importantly, however, the experimental data found that the predicted automatic imitation effects occurred regardless of the type of stimuli that participants actually observed (human or robotic). Thus, the congruency effects did not vary as a function of participants' beliefs regarding the nature of the stimuli and imply a relative invulnerability of automatic imitation to strategic factors.

2.2 Electrophysiological and Neuroimaging Studies

Further evidence in favor of an automatic interference between observed and executed actions also come from studies involving transcranial magnetic stimulation (TMS). Stimulating the primary motor cortex of normal healthy observers leads to a measurable motor-evoked potential (MEP) from the muscles of the corresponding limb. In one study, MEPs were recorded while participants: (1) observed an experimenter grasping an object; (2) observed the object alone; (3) observed an experimenter tracing geometric figures with his arm; and (4) detected the dimming of a light (an attentional control task) (Fadiga, Fogassi, Pavesi, & Rizzolatti, 1995). Fadiga and colleagues (1995) found that the passive observation of goal-directed actions augmented the MEPs involving the corresponding muscles of the hand and arm, even though participants themselves were never required to initiate an action. Other authors have since replicated this finding with both meaningless and other goal-directed movements (Aziz-Zadeh, Maeda, Zaidel, Mazziotta, & Iacoboni, 2002; Fadiga et al., 1995; Strafella & Paus, 2000). This response priming is usually highly specific, such that only those muscles that would be involved in performing the action demonstrate an increased MEP to the observation of that action. The conclusion from these studies is that there is a covert motoric simulation of the observed action, that occurs even in the absence of having to perform an overt action (Fadiga, Craighero, & Olivier, 2005).

Additional electrophysiological evidence for this automaticity comes from electroencephalographic (EEG) and magnetoencephalographic (MEG) studies. Traditionally, EEG studies have noted a central mu rhythm within the 8–12 Hz range, which is present during motor rest and suppressed upon the initiation of a motor movement. Critically, subsequent studies have found that this mu rhythm suppression occurs, not only during action execution, but also during passive action observation (Altschuler, Vankov, Wang, Ramachandran, & Pineda, 1997; Cochin, Barthelemy, Lejeune, Roux, & Martineau, 1998; Cochin, Barthelemy, Roux, & Martineau, 1999; Gastaut & Bert, 1954). Similarly, MEG studies have documented the presence of a central 15–25 Hz rhythmic oscillation, which is enhanced less than 500 ms after median nerve stimulation (the 'post-stimulus rebound'), and which is abolished when participants manipulate an object. For example, Hari and colleagues (1998) recorded MEG activity while participants were: (1) idle; (2) manipulating a small object; and

(3) observing another individual performing the same task. As predicted, the post-stimulus rebound was diminished during action execution. Interestingly, however, it was also reduced when participants simply observed another individual perform the same task. Because the 15–25 Hz rhythm is thought to originate from the precentral cortex, these data are taken as indirect evidence that human primary motor cortex is active during both action observation and execution. Notably, the suppression of both the EEG mu rhythm and MEG post-stimulus rebound of the 15–25 Hz rhythm occurred in the absence of any active movements performed by participants. These data indicate that the human primary motor cortex is involuntarily activated during the observation as well as the execution of motor tasks, which may in turn facilitate peripheral activity in the corresponding limb musculature.

Few neuroimaging studies to date have investigated the role of top-down factors on mirror system activity. However, one functional magnetic resonance imaging (fMRI) study has examined the effect of 'context' during action observation (Iacoboni et al., 2005). In this study, one group of participants observed movie clips of a hand reaching toward a cup, and were required to infer the intention of the actor (to drink versus to clean up) from the appearance of objects in the background ('Explicit' task). A second group viewed the identical stimuli but was instructed to observe them passively ('Implicit' task). The authors argued that any area that is automatically activated during action observation should be insensitive to the context in which the action appears. Their results showed that an area of the posterior right IFG fulfilled this criterion, by being active regardless of the task that was required of participants. However, it should also be noted that the activity within left frontal areas, including the IFG, was greater in the Explicit relative to the Implicit task, suggesting that these left hemisphere areas were, in contrast, vulnerable to strategic influences.

An interesting question that has been largely neglected by the literature is the degree to which observed actions must be consciously attended in order to be processed. We recently used a binocular rivalry paradigm to examine this question (Cunnington, Lion, Chong, & Mattingley, 2007). We trained participants to recognize a series of gestures from American Sign Language, and then presented participants with these images in an fMRI environment. During the binocular rivalry paradigm, the hand gesture images were projected to one eye, while random dot textures were presented to the other. In alternating blocks, the hand gestures were either visible or entirely suppressed. When the gestures were visible, activity was localized to areas in the bilateral inferior and superior parietal lobe, and the left ventral premotor cortex. Critically, during the invisible condition, the bilateral activity within the IPL persisted, which suggests that the processing of observed hand gestures in the IPL occurs automatically even in the absence of perceptual awareness. These preliminary data imply that consciousness is not a precondition for the processing of observed gestures.

Thus, although few neuroimaging studies have directly investigated the automaticity of mirror system responses, those experiments by Iacoboni et al.

(2005) and Cunnington et al. (2007) provide early evidence that particular nodes of the mirror system may operate relatively independent of top-down factors such as strategy or awareness. Overall, the data reviewed in this section are consistent with the principle of ideomotor compatibility, by showing that perceptual events may influence the initiation of participants' motor responses by automatically activating their corresponding response image (Brass et al., 2000; Greenwald, 1970; Wohlschläger et al., 2003).

3 Controlled Processing in Macaque Mirror Neurons

If the responses of mirror neurons are indeed automatic, they should arise relatively independent of top-down control. Having considered the evidence for automatic processing within the human mirror system, we now address the susceptibility of the mirror neuron system to controlled processes. However, before considering this issue at the level of the mirror network in humans, we first discuss the vulnerability of mirror neurons at the cellular level to top-down modulation. Interestingly, one of the earliest accounts of macaque mirror neurons already alluded to their sensitivity to top-down control in the form of attention (Gallese et al., 1996). In this report, there is a brief reference to the fact that actions directed toward a food-related object always succeed in engaging mirror neurons but that those directed toward other three dimensional objects result in less consistent activation. In some cases, the responses to non-food items diminished within 'a few or even the first presentation' (Gallese et al., 1996). Given that non-food items would be less interesting to a monkey than food itself, a plausible explanation for this finding is that the monkey simply ceased attending to the action because it was no longer interested in the stimulus object. These anecdotal data provide indirect evidence that strategic factors such as attention might in fact exert an influence on individual mirror neurons.

More recent data build on this early finding by showing that the activity of individual mirror neurons can be altered by exposure and experience. A characteristic property of mirror neurons as first described was that they failed to respond to goal-directed actions when performed with a tool (Gallese et al., 1996). A possible explanation for this finding was that such actions lack familiarity to the animal. Evidence has since been found in favor of this account. Specifically, exposing monkeys to actions involving a tool over a prolonged period of time can sensitize some mirror neurons to acquire 'tool-responding' properties (Arbib & Rizzolatti, 1999; Ferrari, Rozzi, & Fogassi, 2005). Thus, the observation-execution matching process instantiated in mirror neurons may not be as intrinsic and inseparably linked as once thought, but may instead be a product of learned associations that are susceptible to voluntary control.

The vulnerability of mirror neurons to strategic factors is further suggested by findings that some mirror neurons only respond to an action when it occurs in a particular context. For example, Ferrari, Gallese, Rizzolatti, & Fogassi,

(2003) reported that F5 mirror neurons that responded to gestures performed live in the laboratory failed to respond when the identical gestures were presented as videotaped images, even if those actions were presented stereoscopically (Ferrari et al., 2003). This was quite a surprising finding, given that mirror neurons are capable of generalizing across multiple instances of an action (say, a precision grip), irrespective of factors such as the action's precise kinematic details, the distance from which the action is viewed, or the identity of the actor (e.g., a human or another conspecific). It is unlikely that the insensitivity of mirror neurons to videotaped gesture was caused by a general insensitivity of macaque neurons to such images, as other biologically-sensitive neurons in the macaque cortex (such as in the superior temporal sulcus, STS) are known to be capable of processing just such actions (Jellema, Baker, Wicker, & Perrett, 2000; Oram & Perrett, 1994). Instead, the sensitivity of F5 mirror neurons to the mode of presentation maybe due to the context in which these actions were perceived, with videotaped gestures failing to engage mirror neurons because of their irrelevance and limited naturalistic consequence to the monkey.

The findings of Ferrari et al. (2003), which show that mirror neurons are sensitive to the context in which an action is *observed* are complemented by other data which show mirror neurons are also sensitive to the context in which an action is *executed*. Fogassi and colleagues (2005) recorded from a population of parietal mirror neurons that responded to the execution of particular arm movements (e.g., grasping). Rather than respond whenever such movements were executed, these neurons were in fact differentially active depending on the context in which the action was performed (e.g., grasping to eat vs. grasping to place). This suggests that these parietal mirror neurons do not simply encode low level movement parameters such as its kinematics but can also be modulated by relatively sophisticated metacognitive functions such as the intention and purpose of an action.

Together, the broad conclusion from these single cell neurophysiological studies is that the responses of mirror neurons are not purely driven by bottom-up processes that run to completion following the observation or execution of a given stimulus. Rather, it appears that mirror neurons receive descending input from other areas of macaque cortex that are able to modulate their activity based on voluntary or strategic factors such as attention, observational learning, and context. In the following section, we consider evidence to suggest that a similar modulation also occurs within the human mirror system.

4 Controlled Processing Within the Human Mirror System

4.1 Behavioral Studies

Although the visuomotor priming studies reviewed in Section 2 show that observed actions spontaneously activate their corresponding motor representations, many of these studies were conducted in the absence of a secondary task or while attention was directed exclusively to the observed action. However, one of the defining characteristics of an automatic process is its invulnerability to capacity limitations and manipulations of attention. Given that the automaticity of the observation-execution matching process has not been explicitly challenged, the studies presented in Section 2 alone do not allow us to assess the independence of the visuomotor transformation process from selective processes.

Recently, however, Bach, Peatfield, and Tipper (2007) used a visuomotor priming task to determine if the perception-action link maybe modulated by spatial attention. The authors presented stationary images of individuals either kicking a ball or typing on a keyboard and participants were required to discriminate the color of a dot that appeared either near the head or the limb involved in the action (the foot for the kicking stimulus and the hand for the typing stimulus). Participants themselves were required to register their response by depressing a key with either their foot or finger. Although the observed actions were task-irrelevant, reaction times were nevertheless faster when participants responded with an effector that corresponded to the observed action (i.e., faster foot responses during the observation of kicking and faster hand responses for typing). *Prima facie* this would seem to be evidence in favor of the automaticity of the visuomotor matching process. Critically, however, this priming effect was found only when participants' spatial attention was directed toward the corresponding limb in the displayed image. In contrast, when attention was directed toward a neutral body part (i.e., the head), reaction times for the hand and foot responses were not differentially affected by the type of action presented on the display. The general conclusion from this study is therefore that spatial attention is necessary for an observed body site to prime an action involving the corresponding effector.

In addition to spatial attention, experience and prior exposure may also influence the visuomotor matching process. In a study by Heyes and colleagues (2005), the authors attempted to replicate the automatic imitation effects of Stürmer et al. (2000) across two groups of participants, each of which underwent separate periods of pre-test training. One group (the 'Incompatible Training' group) practised performing an incongruent movement to an observed action (e.g., hand closing in response to hand opening), while the other group (the 'Compatible Training' group) practised performing the corresponding congruent movements to the observed actions. In the main experiment, Heyes et al. (2005) found that automatic imitation effects were present only in the Compatible Training group and not the Incompatible training group. This suggests that learned associations between mismatched stimuli can attenuate the 'automatic imitation' effects. As such, it provides an interesting parallel with macaque data which showed that prolonged exposure to tool-use could sensitize a population of mirror neurons to respond to those actions (Ferrari et al., 2005). Together, these findings imply that the repeated observation of a particular stimulus and its response may result in a visual association that maybe instantiated at the level of mirror neurons and which may drive the apparently automatic imitation effects that are subsequently observed.

4.2 Electrophysiological and Neuroimaging Studies

As reviewed in Section 3, macaque mirror neurons can be sensitive to the context in which an action appears, as evidenced by their greater reactivity to live rather than videotaped gestures (Ferrari et al., 2003). A similar finding has been demonstrated in humans by Järveläinen, Schürmann, Avikainen, & Hari (2001), who presented participants with images of hand actions that were either presented live or on videotape. By using MEG, they compared the post-stimulus rebound within primary motor cortex during the observation of both types of actions. Importantly, they found that the rebound following the observation of a videotaped hand action, although present, was significantly lower in magnitude than that following the observation of a live action (Järveläinen et al., 2001). It appears, therefore, that the human mirror system may also be sensitive to the context of an observed action.

In addition to context, a series of PET studies by Grèzes, Costes, Decety, and colleagues (Decety et al., 1997; Grèzes, Costes, & Decety, 1998, 1999) showed that the activity of action observation areas is also sensitive to the explicit cognitive strategy that participants employ during action observation. In one study, they showed participants images of meaningful and meaningless hand gestures, while instructing half of their participants to observe these actions with no specific purpose, and the other half to observe the actions with the intention to imitate them offline (Grèzes et al., 1998). They found that, relative to the purposeless observation of action, observing with the intention to imitate resulted in greater bilateral activity in areas that included the IPL and premotor cortices. This was an early evidence that top-down effects, in the form of volitional strategy, can influence neural activity during action observation.

Apart from a susceptibility to strategic factors, a key criterion in evaluating the automaticity of a neural system is its invulnerability to capacity limitations. It is surprising for several reasons that few studies have investigated such limitations in the mirror neuron system. First, there are numerous examples in everyday life in which it would be beneficial to suppress the processing of an observed action, especially if it is task-irrelevant. Consider, for example, a basketball player whose primary task is to shoot a free throw. Skilled players are able to ignore the highly distracting gestures from the crowd, while they are engaged in the separate task of making the game-winning shot. For these players, selective attention would be critical to enhance the processing of behaviorally relevant stimulus (e.g., the location of the ring) while suppressing the processing of those that are irrelevant (e.g., the actions of spectators). In addition, selective attention plays a crucial role in the operation of many perceptual systems whose processing is limited by capacity demands (Lavie, 1995, 2000). For example, neural signals associated with the perception of biologically relevant stimuli, such as emotional faces, are reduced by engaging participants' attention in a separate task involving a high perceptual load (Pessoa, McKenna, Guttierez, & Ungerleider, 2002). There is also mounting evidence that the perception of more abstract biological stimuli such as point-light biological motion, appears to be at least partly dependent on selective mechanisms (Battelli, Cavanagh, & Thornton, 2003; Cavanagh, Labianca, & Thorntom, 2001; Thornton, Rensink, & Shiffrar, 2002). Finally, there is preliminary evidence in macaques that mirror neuron responses maybe reduced when the monkey no longer attends to the action or its goal (Gallese et al., 1996).

Recently, we used fMRI to directly examine the effect of capacity limits and selective attention on the activity of human mirror areas (Chong, Williams, Cunnington, & Mattingley, 2008). In this task, we aimed to determine whether cortical activity associated with action observation is modulated by the strategic allocation of selective attention. The predictions of our study were based on an influential account of selective attention, which claims that distractor stimuli will not be processed if attentional resources are exhausted by a demanding primary task (Lavie, 1995, 2000). In contrast, if the perceptual load of that task is low, any resources that are not involved will involuntarily 'spill over' to the perception of even task-irrelevant stimuli. Thus, in our study, we presented participants with images of reach-to-grasp hand actions, while they performed an attentionally demanding task at the fovea. Our localizer data showed that the observation of these gestures activated a network of areas within the IFG and IPL bilaterally. We then probed these regions-of-interest while participants observed the identical, but now task-irrelevant, hand actions and instead performed an easy (low attentional load) or difficult (high attentional load) visual discrimination task at the fovea. Our data showed that selective attention did in fact modulate the activity of the mirror system but interestingly this effect was not ubiquitous. Specifically, we found that the area that was most consistently suppressed under conditions of high attentional load was the left IFG, while the remaining fronto-parietal regions appeared to have been spared from this suppression and may therefore be considered to operate relatively automatically. This finding shows that the activity of the left IFG can be attenuated with a cognitive or perceptual load that limits the attentional resources available to the system, and is further evidence against a strong automaticity account of the mirror system.

Interestingly, these findings converge with the previously described study of Iacoboni and colleagues (2005), which examined the effect of context during action observation. Recall that the authors found that the activity in the right IFG did not differ as a function of participants' cognitive task (the 'Explicit' versus 'Implicit' interpretation tasks), and took this to imply that the right IFG operates automatically. However, the authors also found activity in left frontal areas, including the posterior left IFG, which was greater during the 'Explicit' interpretation task relative to the 'Implicit' interpretation task. In contrast to the right IFG, therefore, these left frontal areas do appear vulnerable to task instructions. Thus, although the experimental approach and motivation of this study differed significantly from those of our experiment, both sets of data complemented each other by indicating a susceptibility of the left IFG to top-down modulation – either by observational strategy (Iacoboni et al., 2005) or selective attention (Chong et al., 2008).

5 Implications of Controlled Processing on the Human Mirror System

To summarize, the apparent automaticity of the human mirror system has been demonstrated by a substantial body of behavioral, neurophysiological, and neuroimaging data, as reviewed in Section 2 (Brass, Bekkering et al., 2001; Craighero et al., 2002; Fadiga et al., 1995; Iacoboni et al., 2005; Press et al., 2006; Stürmer et al., 2000; Vogt et al., 2003). The behavioral data on automatic imitation, for example, show that the matching of an observed action with its motor representation proceeds despite being detrimental to task performance. However, few studies have manipulated the top-down input into the mirror system, and the automaticity of the visuomotor transformation process therefore remains largely unchallenged. As the data reviewed in Sections 3 and 4 indicate, the activity of the mirror system may indeed be modulated when a secondary task is imposed during action observation. Collectively, the data reviewed in the preceding sections allow us to draw two general conclusions. First, the activity of the mirror system is context-dependent and sensitive to different cognitive strategies that are implemented during action observation (Decety et al., 1997; Grèzes et al., 1998, 1999; Heyes et al., 2005; Iacoboni et al., 2005). Second, its activity is modulated by attentional load, and thus is susceptible to a capacity limit (Chong et al., 2008). In this section, we consider the effect of each of these factors on the mirror system, and the potential sources of top-down input into the mirror system.

5.1 The Effect of Prior Exposure and Strategy on the Mirror System

Data from Sections 3 and 4 emphasize the role of learned associations in modulating the formation of visuomotor cortical connections. These include the acquisition of tool-responding properties in previously tool-insensitive macaque mirror neurons (Ferrari et al., 2005) and the dissipation of automatic imitation effects following exposure to incompatible observed and executed actions (Heyes et al., 2005). Given that individual mirror neurons directly transform perceived actions into their corresponding motor representations, the question arises as to how this might occur. Several authors have postulated that the properties of mirror neurons are not innate; rather, that they emerge as a result of more general associative processes that occur over the course of normal development. For example, the ontogenetic evolution of the direct matching process maybe the result of a Hebbian learning process (Keysers & Perrett, 2004), or following the repeated reinforcement of observed and executed actions that become correlated over time – the Associative Sequence Learning model (Heyes, 2001). Such accounts could explain the findings of

Ferrari et al. (2005) and Heyes et al. (2005), and provide a basis for which the mirror system may adapt to novel goals or task demands.

The susceptibility of the mirror system to strategic control should not be completely surprising. From an anatomical point of view, by the time a perceived action reaches mirror areas in the IFG and IPL, it will have already undergone a significant amount of preliminary processing in earlier areas of the action processing stream. A current model of the mirror system suggests that the processing of an observed action begins in the superior temporal sulcus (STS), which encodes a simple visual description of the relevant action (Carr et al., 2003; Miall, 2003). This information is then sent to parietal mirror neurons, which extract kinesthetic information regarding the action, prior to that information being sent to the IFG, which decodes the goal of the observed action. If the action is to be imitated, efferent motor commands are sent via back-projections from the IFG to the STS, which compare the observed and intended motor acts in a forward model of imitation. Although the STS itself lacks motor (and therefore mirror) properties (C. Keysers, unpublished observations cited in Keysers & Perrett, 2004), it possesses very similar perceptual properties to mirror areas. In particular, it responds to articulated body movements and goal-directed hand actions, such as grasping, tearing, holding, or manipulating (Allison, Puce, & McCarthy, 2000; Bruce, Desimone, & Gross, 1981; Jellema & Perrett, 2003; Oram & Perrett, 1994; Oram & Perrett, 1996; Perrett et al., 1989; Perrett, Rolls, & Caan, 1982; Perrett et al., 1985; Puce & Perrett, 2003). Thus, prior to the action first reaching the mirror system, a relatively sophisticated representation of that action will have already been encoded in the STS.

In addition to the STS, there exist other more recently discovered areas that may provide perceptual information to the mirror system. For example, the extrastriate body area (EBA, Downing, Jiang, Shuman, & Kanwisher, 2001) and fusiform body area (FBA, Schwarzlose, Baker, & Kanwisher, 2005) both respond to the perception of human bodies and body parts, regardless of whether they depict an action. Furthermore, there is some suggestion that the EBA may also respond to the performance of simple movements such as pointing (Astafiev, Stanley, Shulman, & Corbetta, 2004), although this is controversial (Astafiev, Stanley, Shulman, & Corbetta, 2005; Peelen & Downing, 2005). As such, some authors have proposed a role for the EBA in distinguishing self-produced movements from those produced by others – information that would be critical during the processing of actions within the STS (Jeannerod, 2004). This proposition is verified by recent findings that the role of the EBA is to discriminate the individual causing an action, while the IFG is important in discriminating the actions themselves (Urgesi, Candidi, Ionta, & Aglioti, 2007).

Although the role of areas such as the EBA and FBA is an ongoing subject of investigation, the general conclusion that maybe drawn is that the representation of an observed action will have been heavily pre-processed prior to reaching the mirror system. In fact, the EBA, STS, IPL, and IFG may in turn be considered as part of an even larger network that is engaged in social cognition, and which includes areas such as the temporoparietal junction and medial prefrontal cortices

(Saxe, 2006; Saxe & Kanwisher, 2003). It has been proposed that the mirror system maybe modulated by areas that specialize in inferential processing of social stimuli (Nishitani, Avikainen, & Hari, 2004; Oberman et al., 2005) and that the temporal pole and/or medial prefrontal cortex enhances attention to social stimuli by modulating activity in the STS (Allison et al., 2000; Castielli, Frith, Happe, & Frith, 2002). While the precise functional connections between the mirror system and these other neural systems remain to be elucidated, the potential afferent and feedback connections arising from these networks could mutually inform the activity of the mirror system in different observational contexts.

5.2 Attentional Modulation of the Mirror System

Finally, we consider the role of attention in modulating the activity of the mirror system. The data reviewed in Section 4 suggest that the apparent automaticity in most mirror neuron studies is likely secondary to the absence of a secondary task. In this sense, the mirror system may operate in a similar manner to other closely allied neurobiological systems, such as those involved in the processing of emotionally neutral faces or biological motion, which may exhibit seemingly automatic properties until they are challenged by a competing cognitive load (Cavanagh et al., 2001; Pessoa et al., 2002; Thornton et al., 2002). On a broader level, this is also consistent with the general view that few cognitive processes, if any, occur entirely independently of selective processes (Kahneman & Chajczyk, 1983; Kahneman & Henik, 1981; Kahneman & Treisman, 1984; Logan, 1980; Treisman, 1960).

The role of selective processes in modulating mirror system activity is most evident when one considers the underlying automaticity of the observation-execution matching process. Clearly, if the visuomotor transformation of an observed action is as automatic as the data in Section 2 suggest, there should exist some mechanism to inhibit these imitative tendencies from taking place under normal conditions (Kinsbourne, 2005). Functional MRI data provide evidence that mirror areas are under just such inhibitory control. In one study, participants performed a predefined finger movement (lifting or tapping) that was either congruent or incongruent with an observed movement (Brass, Zysset, & von Cramon, 2001). The authors found that, during the execution of an Incongruent relative to a congruent movement, there was increased activity in prefrontal areas, including the frontopolar and middle frontal areas, in addition to the precuneus and anterior parietal areas. These data reveal the significance of these areas in inhibiting the automatic responses that would have otherwise been generated.

The significance of pre-frontal areas in inhibiting imitative responses becomes obvious when one considers neuropsychological patient data. Patients with echopraxia are characterized by their impulsive imitation of the actions of others, even when they are instructed to perform a separate task (Luria, 1966). Similarly, patients with 'imitation behavior' imitate the gestures of an experimenter even when these gestures are socially unacceptable or odd (Lhermitte, Pillon, & Serdaru, 1986). In contrast to patients with echopraxia, those with imitation behavior do not merely imitate movements at the level of gross actions, but at the level of the goal of the action. Interestingly, when the patients of Lhermitte were asked about the imitation behavior, they did not deny or disown their responses; rather, they felt that the gestures they observed included an instruction for them to be imitated, and that their responses were entirely acceptable. Although the precise loci of damage required to induce these syndromes remain uncertain, they have been associated with frontal lobe lesions since the time of their discovery. Specifically, echopraxia has been associated with prefrontal and mesial cortical lesions, while imitation behavior is usually associated with fronto-orbital lesions. Both echopraxia and imitation behavior may therefore represent instances of a 'release' of prefrontal cortical inhibitory inputs to the mirror system that prevent overt movement production in normal individuals.

It is worth emphasizing that the role of selective attention and inhibitory processes in the mirror system is incompletely explored. In particular, it is unclear whether selective mechanisms may differentially affect the stages at which an observed action is processed. In terms of the mirror system, the observation-execution matching process could be conceptualized as consisting of three components: the initial input to the mirror system (via areas such as the EBA and STS); the visuomotor transformation within the mirror system itself (within the IFG and IPL); and the overt execution of the observed action in the case of imitation (within motor and premotor areas). It remains for future investigations to determine the relative susceptibility of these stages to selective processes. Given that one of the defining hallmarks of an automatic function is its immunity to attentional manipulations, such investigations would be critical in testing the strength of the mirror system's automaticity.

6 Conclusions

Cognitive neuroscience has given rise to many significant advances since James' ideomotor theory in the 19th century, not least of which has been the discovery of the primate mirror neuron system. Broadly speaking, James' claim that the perception of an action activates a matching motor program was surprisingly prescient (James, 1890). Indeed, the existence of a putative human mirror system provides James' principle and current ideomotor theories with a plausible neuroanatomical substrate, and brings us closer to an understanding of how actions are matched with their corresponding motor representations. However, the seemingly trivial, yet nevertheless fundamental, capacity of humans to recognize and imitate the actions of others belies the complexity of the underlying mechanisms. As yet, we are only beginning to understand the basic response properties of the mirror neuron system, such as the automaticity with which it motorically encodes observed actions. The studies reviewed in this chapter indicate that the automaticity of this visuomotor matching process is not as strong as was once thought, and maybe influenced by controlled processes such as observational learning, strategy, and selective attention. Future studies should now seek to elucidate the specific stimulus features and top-down parameters that govern the operation of this system – research that will ultimately guide us toward uncovering the critical mechanisms that allow us to routinely navigate our complex social environments.

References

Allison, T., Puce, A., & McCarthy, G. (2000). Social perception from visual cues: The role of the STS region. *Trends in Cognitive Science, 4*(7), 267–278.

Altschuler, E. L., Vankov, A., Wang, V., Ramachandran, V. S., & Pineda, J. A. (1997). Person see, person do: Human cortical electrophysiological correlates of monkey see monkey do cells. *Society for Neuroscience Abstracts, 23*, 719.

Arbib, M. A. (2002). The mirror system, imitation, and the evolution of language. In K. Dautenhahn & C. L. Nehaniv (Eds.), *Imitation in animals and artifacts: Complex adaptive systems* (pp. 229–280). Cambridge, MA: MIT Press.

Arbib, M. A. (2005). From monkey-like action recognition to human language: An evolutionary framework for neurolinguistics. *Behavioral and Brain Sciences, 28*, 105–167.

Arbib, M. A., & Rizzolatti, G. (1999). Neural expectations. A possible evolutionary path from manual skills to language. In P. V. Loocke (Ed.), *The nature of concepts. Evolution, structure and representation* (pp. 128–154). New York: Routledge.

Astafiev, S. V., Stanley, C. M., Shulman, G. L., & Corbetta, M. (2004). Extrastriate body area in human occipital cortex responds to the performance of motor actions. *Nature Neuroscience, 7*(5), 542–548.

Astafiev, S. V., Stanley, C. M., Shulman, G. L., & Corbetta, M. (2005). Is the extrastriate body area involved in motor actions? *Nature Neuroscience, 8*, 125–126.

Aziz-Zadeh, L., Maeda, F., Zaidel, E., Mazziotta, J., & Iacoboni, M. (2002). Lateralization in motor facilitation during action observation: A TMS study. *Experimental Brain Research, 144*(1), 127–131.

Bach, P., Peatfield, N. A., & Tipper, S. P. (2007). Focusing on body sites: The role of spatial attention in action perception. *Experimental Brain Research, 178*, 509–517.

Bargh, J. A. (1992). The ecology of automaticity: Towards establishing the conditions needed to produce automatic processing effect. *American Journal of Psychology, 105*, 181–199.

Battelli, L., Cavanagh, P., & Thornton, I. (2003). Perception of biological motionin parietal patients. *Neuropsychologia, 41*, 1808–1816.

Brass, M., Bekkering, H., & Prinz, W. (2001). Movement observation affects movement execution in a simple response task. *Acta Psychologica, 106*, 3–22.

Brass, M., Bekkering, H., Wohlschläger, A., & Prinz, W. (2000). Compatibility between observed and executed finger movements: Comparing symbolic, spatial, and imitative cues. *Brain and Cognition, 44*, 124–143.

Brass, M., Zysset, S., & von Cramon, D. Y. (2001). The inhibition of imitative response tendencies. *NeuroImage, 14*, 1416–1423.

Bruce, C., Desimone, R., & Gross, C. G. (1981). Visual properties of neurons in a polysensory area in superior temporal sulcus of the macaque. *Journal of Neurophysiology,* 46, 369–384.

Buccino, G., Binkofski, F., & Riggio, L. (2004). The mirror neuron system and action recognition. *Brain & Language, 89*(2), 370–376.

Buccino, G., Vogt, S., Ritzl, A., Fink, G. R., Zilles, K., Freund, H. J., et al. (2004). Neural circuits underlying imitation learning of hand actions: An event-related fMRI study. *Neuron, 42*, 323–334.

Carey, D. P. (1996). 'Monkey see, monkey do' cells. *Current Biology, 6*, 1087–1088.
Carr, L., Iacoboni, M., Dubeau, M.-C., Mazziotta, J. C., & Lenzi, G. L. (2003). Neural mechanisms of empathy in humans: A relay from neural systems for imitation to limbic areas. *Proceedings of the National Academy of Sciences, USA, 100*, 5497–5502.
Castielli, F., Frith, C., Happe, F., & Frith, U. (2002). Autism, asperger syndrome and brain mechanisms for the attribution of mental states to animated shapes. *Brain, 125*, 1839–1849.
Cavanagh, P., Labianca, A. T., & Thorntom, I. M. (2001). Attention-based visual routines: Sprites. *Cognition, 80*(1–2), 47–60.
Chong, T.T.-J., Williams, M.A., Cunnington, R., & Mattingley, J.B. (2008). Selective attention modulates inferior frontal gyrus activity during action observation. *Neuroimage, 40*(1), 298–307.
Cochin, S., Barthelemy, C., Lejeune, B., Roux, S., & Martineau, J. (1998). Perception of motion and qEEG activity in human adults. *Electroencephalography and Clinical Neurophysiology, 107*, 287–295.
Cochin, S., Barthelemy, C., Roux, S., & Martineau, J. (1999). Observation and execution of movement: Similarities demonstrated by quantified electroencephalography. *European Journal of Neuroscience, 11*, 1839–1842.
Cohen, J. D., Dunbar, K., & McClelland, J. L. (1990). Automaticity, attention and the strength of processing: A parallel distributed processing account of the Stroop effect. *Psychological Review, 97*, 332–361.
Cole, J. D. (1999). *About face*. London: MIT Press.
Cole, J. D. (2001). Empathy needs a face. *Journal of Consciousness Studies, 8*, 51–68.
Coricelli, G. (2005). Two-levels of mental states attribution: From automaticity to voluntariness. *Neuropsychologia, 43*(2), 294–300.
Craighero, L., Bello, A., Fadiga, L., & Rizzolatti, G. (2002). Hand action preparation influences the responses to hand pictures. *Neuropsychologia, 40*, 492–502.
Cunnington, R., Lion, A., Chong, T., & Mattingley, J. B. (2007). Cortical responses to invisible hand gestures within the human mirror system. *Journal of Cognitive Neuroscience, 19*(Suppl), 139.
Decety, J., Grèzes, J., Costes, N., Perani, D., Jeannerod, M., Procyk, E., et al. (1997). Brain activity during observation of actions: Influence of action content and subject's strategy. *Brain, 120*, 1763–1777.
Downing, P. E., Jiang, Y., Shuman, M., & Kanwisher, N. (2001). A cortical area selective for visual processing of the human body. *Science, 293*, 2470–2473.
Fadiga, L., Craighero, L., & Olivier, E. (2005). Human motor cortex excitability during the perception of others' action. *Current Opinion in Neurobiology, 15*(2), 213–218.
Fadiga, L., Fogassi, L., Pavesi, G., & Rizzolatti, G. (1995). Motor facilitation during action observation: A magnetic stimulation study. *Journal of Neurophysiology, 73*(6), 2608–2611.
Ferrari, P. F., Gallese, V., Rizzolatti, G., & Fogassi, L. (2003). Mirror neurons responding to the observation of ingestive and communicative mouth actions in the monkey ventral premotor cortex. *European Journal of Neuroscience, 17*, 1703–1714.
Ferrari, P. F., Rozzi, S., & Fogassi, L. (2005). Mirror neurons responding to observation of actions made with tools in monkey ventral premotor cortex. *Journal of Cognitive Neuroscience, 17*(2), 212–226.
Fogassi, L., Ferrari, P. F., Gesierich, B., Rozzi, S., Chersi, F., & Rizzolatti, G. (2005). Parietal lobe: From action organization to intention understanding. *Science, 308*, 662–667.
Gallese, V. (2001). The 'shared manifold' hypothesis: From mirror neurons to empathy. *Journal of Consciousness Studies, 8*(5–7), 33–50.
Gallese, V. (2003a). The manifold nature of interpersonal relations: The quest for a common mechanism. *Philosophical Transactions of the Royal Society of London – Series B: Biological Sciences, 358*(1431), 517–528.
Gallese, V. (2003b). The roots of empathy: The shared manifold hypothesis and the neural basis of intersubjectivity. *Psychopathology, 36*, 171–180.

Gallese, V., Fadiga, L., Fogassi, L., & Rizzolatti, G. (1996). Action recognition in the premotor cortex. *Brain, 119*(Pt. 2), 593–609.
Gallese, V., & Goldman, A. (1998). Mirror neurons and the simulation theory of mind-reading. *Trends in Cognitive Sciences, 2*, 493–501.
Gallese, V., Rizzolatti, G., & Keysers, C., (2004). A unifying view of the basis of social cognition. *Trends in Cognitive Sciences, 8*(9), 396–403.
Gallese, V., & Metzinger, T. (2003). Motor ontology: The representational reality of goals, actions and selves. *Philosophical Psychology, 16*(3), 365–388.
Gastaut, H. J., & Bert, J. (1954). EEG changes during cinematographic presentation. *Electroencephalography and Clinical Neurophysiology, 6*, 433–444.
Greenwald, A. G. (1970). Sensory feedback mechanisms in performance control: With special reference to the ideo-motor mechanism. *Psychological Review, 77*, 73–99.
Greenwald, A. G. (1972). On doing two things at once: Time sharing as a function of idemotor compatibility. *Journal of Experimental Psychology, 94*, 52–57.
Grèzes, J., Costes, N., & Decety, J. (1998). Top-down effect of strategy on the perception of human biological motion: A PET investigation. *Cognitive Neuropsychology, 15*(6/7/8), 553–582.
Grèzes, J., Costes, N., & Decety, J. (1999). The effects of learning and intention on the neural network involved in the perception of meaningless actions. *Brain, 122*, 1875–1887.
Hari, R., Forss, N., Avikainen, S., Kerveskari, E., Salenius, S., & Rizzolatti, G. (1998). Activation of human primary motor cortex during action observation: A neuromagnetic study. *Proceedings of the National Academy of Sciences USA, 95*, 15061–15065.
Hasher, L., & Zacks, R. (1979). Automatic and effort full, processes in memory. *Journal of Experimental Psychology: General, 108*, 356–388.
Heyes, C. (2001). Causes and consequences of imitation. *Trends in Cognitive Sciences, 5*(6), 253–261.
Heyes, C., Bird, G., Johnson, H., & Haggard, P. (2005). Experience modulates automatic imitation. *Cognitive Brain Research, 22*, 233–240.
Hommel, B., Müsseler, J., Aschersleben, G., & Prinz, W. (2001). A theory of event coding (TEC): A framework for perception and action planning. *Behavioral and Brain Sciences, 24*, 849–937.
Iacoboni, M., Molnar-Szakacs, I., Gallese, V., Buccino, G., Mazziotta, J. C., & Rizzolatti, G. (2005). Grasping the intentions of others with one's own mirror neuron system. *PLoS Biology, 3*(3), 0525–0535.
James, W. (1890). *The Principles of Psychology*, Vol. 2. New York: Holt.
Järveläinen, J., Schürmann, M., Avikainen, S., & Hari, R. (2001). Stronger reactivity of the human primary motor cortex during observation of live rather than video motor acts. *NeuroReport, 12*(16), 3493–3495.
Jeannerod, M. (2001). Neural simulation of action: a unifying mechanism for motor cognition. *NeuroImage, 14*, S103–S109.
Jeannerod, M. (2004). Visual and action cues contribute to the self-other distinction. *Nature Neuroscience, 7*(5), 422–423.
Jellema, T., Baker, C., Wicker, B., & Perrett, D. (2000). Neural representation for the perception of the intentionality of actions. *Brain and Cognition, 44*(2), 280–302.
Jellema, T., & Perrett, D. I. (2003). Cells in monkey STS responsive to articulated body motions and consequent static posture: A case of implied motion? *Neuropsychologia, 41*, 1728–1737.
Kahneman, D., & Chajczyk, D. (1983). Tests of the automaticity of reading: Dilution of Stroop effects by color-irrelevant stimuli. *Journal of Experimental Psychology: Human Perception and Performance, 9*, 497–509.
Kahneman, D., & Henik, A. (1981). Perceptual organization and attention. In M. Kubovy & J. R. Pomerantz (Eds.), *Perceptual organization*. Hillsdale, NJ: Erlbaum.
Kahneman, D., & Treisman, A. M. (1984). Changing views of attention and automaticity. In R. Parasuraman & D. R. Davies (Eds.), *Varieties of attention* (pp. 29–62). Orlando, FL: Academic Press.

Keysers, C., & Perrett, D. I. (2004). Demystifying social cognition: A Hebbian perspective. *Trends in Cognitive Sciences, 8*(11), 501–507.
Kilner, J. M., Paulignan, Y., & Blakemore, S.-J. (2003). An interference effect of observed biological movement on action. *Current Biology, 13*, 522–525.
Kinsbourne, M. (2005). Overlapping brain states while viewing and doing. In S. Hurley & N. Chater (Eds.), *Perspectives on imitation: From neuroscience to social science. Volume 1: Mechanisms of imitation and imitation in animals* (Vol. 1, pp. 210–214). Cambridge, MA: Bradford.
Kornblum, S., & Lee, J. W. (1995). Stimulus-response compatibility with relevant and irrelevant stimulus dimensions that do and do not overlap with the response. *Journal of Experimental Psychology: Human Perception & Performance, 21*, 855–875.
Lavie, N. L. (1995). Perceptual load as a necessary condition for selective attention. *Journal of Experimental Psychology: Human Perception & Performance, 21*, 451–468.
Lavie, N. L. (2000). Selective attention and cognitive control: Dissociating attentional functions through different types of load. In S. Monsell & J. Driver (Eds.), *Attention and performance XVIII*. Cambridge, MA: MIT Press.
Lhermitte, F., Pillon, B., & Serdaru, M. D. (1986). Human autonomy and the frontal lobes: I. Imitation and utilization behavior. A neuropsychological study of 75 patients. *Annals of Neurology, 19*, 326–334.
Logan, G. D. (1980). Attention and automaticity in stroop and priming tasks: Theory and data. *Cognitive Psychology, 12*, 523–553.
Luria, A. R. (1966). *Higher cortical functions in man*. New York: Basic Books.
MacLeod, C. M., & Dunbar, K. (1988). Training and stroop-like interference: Evidence for a continuum of automaticity. *Journal of Experimental Psychology: Learning, Memory, and Cognition, 14*, 126–135.
Miall, R. C. (2003). Connecting mirror neurons and forward models. *NeuroReport, 14*(17), 2135–2137.
Nishitani, N., Avikainen, S., & Hari, R. (2004). Abnormal imitation-related cortical activation sequences in Asperger's syndrome. *Annals of Neurology, 55*, 558–562.
Oberman, L. M., Hubbard, E. M., McCleery, J. P., Altschuler, E. L., Ramachandran, V. S., & Pineda, J. A. (2005). EEG evidence for mirror neuron dysfunction in autism spectrum disorders. *Cognitive Brain Research, 24*, 190–198.
Oram, M. W., & Perrett, D. I. (1994). Responses of anterior superior temporal polysensory (STPa) neurons to 'biological motion' stimuli. *Journal of Cognitive Neuroscience, 6*, 99–116.
Oram, M. W., & Perrett, D. I. (1996). Integration of form and motion in the anterior superior temporal polysensory area (STPa) of the macaque monkey. *Journal of Neurophysiology, 76*, 109–129.
Pashler, H. (1998). *The psychology of attention*. Cambridge, MA: MIT Press.
Peelen, M., & Downing, P. (2005). Is the extrastriate body area involved in motor actions? *Nature Neuroscience, 8*(2), 125.
Perrett, D. I., Harries, M. H., Bevan, R., Thomas, S., Benson, P. J., Mistlin, A. J., et al. (1989). Frameworks of analysis for the neural representation of animate objects and actions. *Journal of Experimental Biology, 146*, 87–114.
Perrett, D. I., Rolls, E. T., & Caan, W. (1982). Visual neurones responsive to faces in the monkey superiore temporal cortex. *Experimental Brain Research, 47*, 329–342.
Perrett, D. I., Smith, P. A. J., Potter, D. D., Mistlin, A. J., Head, A. S., Milner, A. D., et al. (1985). Visual cells in the temporal cortex sensitive to face view and gaze direction. *Proceedings of the Royal Society of London Series B Biological Sciences, 223*(1232), 293–317.
Pessoa, L., McKenna, M., Guttierez, E., & Ungerleider, L. G. (2002). Neural processing of emotional faces requires attention. *Proceedings of the National Academy of Sciences of the United States of America, 99*(17), 11458–11463.
Posner, M. (1978). *Chronometric explorations of mind*. Hillsdale, NJ: Erlbaum.

Press, C., Bird, G., Flach, R., & Heyes, C. (2005). Robotic movement elicits automatic imitation. *Cognitive Brain Research, 25*, 632–640.

Press, C., Gillmeister, H., & Heyes, C. (2006). Bottom-up, not top-down, modulation of imitation by human and robotic models. *European Journal of Neuroscience, 24*, 2415–2419.

Prinz, W. (1990). A common coding approach to perception and action. In O. Neumann & W. Prinz (Eds.), *Relationships between perception and action: Current approaches* (pp. 167–201). Berlin: Springer-Verlag.

Prinz, W. (1997). Perception and action planning. *European Journal of Cognitive Psychology, 9*, 129–154.

Prinz, W. (2002). Experimental approaches to imitation. In A. Meltzoff & W. Prinz (Eds.), *The imitative mind: Development, evolution, and brain bases.* Cambridge, UK: Cambridge University Press.

Puce, A., & Perrett, D. I. (2003). Electrophysiology and brain imaging of biological motion. *Philosophical Transactions of the Royal Society of London B: Biological Science, 358*, 435–445.

Rizzolatti, G., & Craighero, L. (2004). The mirror-neuron system. *Annual Review of Neuroscience, 27*, 169–192.

Rizzolatti, G., Fadiga, L., Fogassi, L., & Gallese, V. (1999). Resonance behaviors and mirror neurons. *Archives Italiennes de Biologie, 137*, 85–100.

Rizzolatti, G., Fadiga, L., Matelli, M., Bettinardi, V., Paulesu, E., Perani, D., et al. (1996). Localization of grasp representations in humans by PET: 1. Observation versus execution. *Experimental Brain Research, 111*(2), 246–252.

Saxe, R. (2006). Uniquely human social cognition. *Current Opinion in Neurobiology, 16*, 235–239.

Saxe, R., & Kanwisher, N. (2003). People thinking about thinking people: The role of the temporo-parietal junction in 'theory of mind'. *NeuroImage, 19*, 1835–1842.

Schwarzlose, R. F., Baker, C. I., & Kanwisher, N. (2005). Separate face and body selectivity on the fusiform gyrus. *Journal of Neuroscience, 25*(47), 11055–11059.

Shiffrin, R. M., & Schneider, W. (1977). Controlled and automatic human information processing: II. Perceptual learning, automatic attending, and a general theory. *Psychological Review, 84*, 127–190.

Strafella, A. P., & Paus, T. (2000). Modulation of cortical excitability during action observation: a transcranial magnetic stimulation study. *NeuroReport, 11*, 2289–2292.

Stürmer, B., Aschersleben, G., & Prinz, W. (2000). Correspondence effects with manual gestures and postures: A study of imitation. *Journal of Experimental Psychology: Human Perception and Performance, 26*(6), 1746–1759.

Tai, Y. F., Scherfler, C., Brooks, D. J., Sawamoto, N., & Castiello, U. (2004). The human premotor cortex is 'mirror' only for biological actions. *Current Biology, 14*, 117–120.

Thornton, I., Rensink, R., & Shiffrar, M. (2002). Active versus passive processing of biological motion. *Perception, 31*, 837–853.

Treisman, A. M. (1960). Contextual cues in selective listening. *Quarterly Journal of Experimental Psychology, 12*, 242–248.

Urgesi, C., Candidi, M., Ionta, S., & Aglioti, S. M. (2007). Representation of body identity and body actions in extrastriate body area and ventral premotor cortex. *Nature Neuroscience, 10*(1), 30–31.

Vogt, S., Taylor, P., & Hopkins, B. (2003). Visuomotor priming by pictures of hand postures: Perspective matters. *Neuropsychologia, 41*, 941–951.

Wilson, M., & Knoblich, G. (2005). The case for motor involvement in perceiving conspecifics. *Psychological Bulletin, 131*(3), 460–473.

Wohlschläger, A., Gattis, M., & Bekkering, H. (2003). Action generation and action perception in imitation: An instance of the ideomotor principle. *Philosophical Transactions of the Royal Society of London – Series B: Biological Sciences, 358*, 501–515.

Embodied Perspective on Emotion-Cognition Interactions

Piotr Winkielman, Paula M. Niedenthal, and Lindsay M. Oberman

Abstract Emotions permeate social and non-social cognition. However, our understanding of the underlying mechanisms remains elusive. One reason is because traditional amodal or symbolic accounts of cognition view emotional information as equivalent to any other information. However, recent theories of embodied cognition suggest new ways to understand the processing of emotionally significant information. They suggest that both perceiving and thinking about such information involve perceptual, somatovisceral, and motoric reexperiences (embodiment) of the relevant emotion in the self. Consistent with this view, many studies show that processing of emotion recruits embodiments as reflected in psychological and psychological measures. Further, embodiment of emotion, even when induced by simple manipulations, such as facial expression, posture, or movement, can causally influence the processing of emotional information, including perception, learning, understanding, and use in language, judgment, and behavior. We review relevant studies and discuss potential neural mechanisms underlying embodiment and simulation. We especially highlight the importance of social context and flexible use of embodiment in emotional processing and discuss its importance for typical and atypical social functioning.

Keywords Emotion · Cognition · Embodiment · Mirroring · Social Psychology

1 Introduction

An extensive literature demonstrates interactions between emotion and cognition. Thus, emotions influence attention, perception, memory, reasoning, and decisions (Damasio, 1994; Eich, Kihlstrom, Bower, Forgas, & Niedenthal,

P. Winkielman
Department of Psychology, University of California, San Diego, California, San Diego, 9500 Gilman Drive, Mailcode 0109, La Jolla, CA 92093-0109, USA
e-mail: pwinkiel@ucsd.edu

J.A. Pineda (ed.), *Mirror Neuron Systems*, DOI: 10.1007/978-1-59745-479-7_11,

2000; Winkielman, Knutson, Paulus, & Trujillo, 2007). Emotions are also influenced by cognition. Beliefs about properties and causes of events shape the quality, intensity, and duration of emotion episodes (Ellsworth & Scherer, 2003). Even abstract symbols such as language that refer to emotional events can rapidly elicit an emotional response along with powerful physiological consequences (Phelps, O'Connor, Gateby, Grillon, Gore, & Davis, 2001).

Despite the many demonstrations, researchers are still far from understanding the mechanisms of the emotion-cognition interface. In this chapter we argue that progress can be offered by embodiment theories of cognition and emotion. Given the emphasis of the book, we relate the work on embodiment to the processes of mirroring, and highlight the relevance of these mechanisms to understanding of social cognition.

The chapter is structured roughly as follows. First, we locate embodiment theories in the general context of debates about mental representation. We then discuss possible neural mechanisms underlying embodiment, and related processes such as simulation and mirroring. Next, we review evidence for the embodiment account in several domains of emotion processing, such as perception, comprehension, learning, influence, concepts, and language. Finally, we conclude with some observations about the strengths and limitations of the embodied theories of emotion and cognition and raise some questions for future research.

2 Representing Emotion: Amodal and Modal Accounts

Traditionally, psychology and cognitive science were dominated by 'amodal' or symbolic accounts of information processes (e.g., Fodor, 1975; Newell, 1980). These accounts assume that information is initially encoded in various perceptual systems (vision, olfaction, audition, touch, etc), but then is transformed into a conceptual form functionally separated from its sensory origins. The resulting amodal symbols (concepts) bear no analogical relationship to the experienced event and it is these symbols that enter into high-level cognitive processes such as thought and language.

Applying the amodal view to processing of emotional information renders what individuals know about emotion equivalent to what they know about most other things. Just as people appear to know that *CARS* possess the features *engines*, *tires*, and *exhaust pipes*, they know that *ANGER* involves the experience of a *thwarted goal*, a *desire to strike out*, and even that it is characterized by *clenched fists*, *feeling of 'boiling inside,'* and a *willingness to strike*. In a way, the amodal view disposes of emotion as a topic of unique interest (Zajonc, 1980).

The last decade and a half witnessed a surge of interest in alternative models of representation clustered under the label *embodied cognition theories* (Barsalou, 1999, Clark, 1997; Prinz, 2002; Wilson, 2002). These theories assume that high-level cognitive processes are *modal*, that is., involve partial reactivations of states in modality-specific, sensory-motor systems. In such an embodied account,

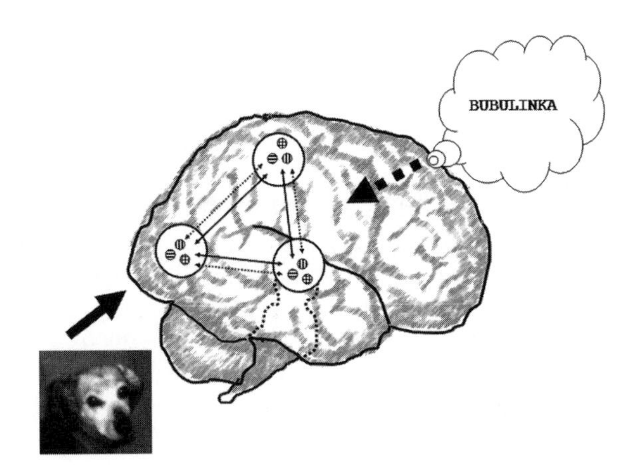

Fig. 1 A schematic illustration of activation patterns in visual (*left*), affective (*middle*), and somatosensory (*top*) systems upon perception of a dog. Later, thinking about the dog's name ('Bubulinka') reactivates parts of the original patterns

knowledge involves, in a sense, partially 're-experiencing' the event in sensory and motor modalities.

Recently, embodied accounts have been applied to understand processing of emotional information (Damasio, 1994; Decety & Jackson, 2004; Gallese, 2003; Niedenthal, Barsalou, Ric, & Krauth-Gruber, 2005; Niedenthal, Barsalou, Winkielman, Krauth-Gruber, & Ric, 2005; Niedenthal, 2007). One such application proposes that sensory-motor states triggered during the encounter with an emotion-eliciting stimulus (e.g., a dog) are captured and stored in modality-specific association areas (see Fig. 1). Later, during recovery of the experience in the consciousness (e.g., thinking about the dog), the original pattern of sensory-motor states that occurred during the encounter can be reactivated. Critically for such an account, the reactivation can be partial and involves a dynamic, online use of modality-specific information. That is, what gets reactivated depends on how selective attention is allocated and what information is currently relevant to the individual (Barsalou, 1999). For embodied cognition theories, using knowledge – as in recalling memories, drawing inferences, and making plans – is thus called 'embodied simulation' because parts of prior experience are reproduced in the originally-implicated neural systems *as if* the individual were there in the very situation (Gallese, 2003). For example, an embodied simulation of anger could involve simulating the experience of anger, including the activation of arm muscles that clench a fist, facial muscles that form a scowl, and an internal sensation of 'boiling inside.'

3 Mechanisms of Embodiment and Simulation

Many theories of embodied cognition suppose processes such as 're-experience' and 'simulation.' There is also a discussion about the role of 'mirroring.' An in-depth review of the proposed neural architecture supporting these processes is outside the scope of this chapter (see other contributions to this book).

However, let us indicate some areas of current debate. We will also come back to the neural instantiation issues when reviewing specific findings.

One debate concerns the relative role of central and peripheral mechanisms. Early embodiment theories (e.g., James, 1884), as well as some modern versions, (Zajonc & Markus, 1984) highlight the role of input from the autonomic nervous system. However, starting with Cannon's (1927) rebuttal of James-Lange emotion theory, critics have argued that bodily feedback is too undifferentiated and slow to support emotional experience. In fact, some of these criticisms are misplaced, as for example, facial musculature is refined and can respond quickly (Tassinary & Cacioppo, 2000). Further, recent research suggests that the autonomic feedback indeed contributes to emotional experience (Craig, 2002). More importantly, modern embodiment theories highlight that peripheral input works together with the brain's modality-specific systems, which can quickly simulate the necessary changes. For example, embodied states can be speedily and flexibly represented by 'as-if loops' linking the ventro-medial PFC, the limbic system, somatosensory, and motor cortex (Damasio, 1994; Goldman & Sripada, 2005).

Another imporant debate concerns the exact substrates of the '"simulation' mechanisms. Some researchers find it sufficient to assume that the brain represents information across a hierarchy of widely distributed associative areas, sometimes called *'convergence zones'* (Damasio, 1989). Those areas retain information about the modal (sensory-motor) features of the stimulus, with progressively 'higher' areas tuned to more abstract aspects of the representation. This way of representing information preserves its modal contents and allows sensory-motor representations to be selectively reactivated, via attentional mechanisms, whenever the perceiver needs to construct a simulation (Barsalou, 1999). Note that on this account, there is no anatomically unique 'simulation' system. In a way, the whole brain can function as a simulation machine, with different modality areas being recruited depending on the goals in a particular task (Grush, 2004).

One fascinating debate concerns the relation between the mechanisms of simulation and mirroring – a process that recreates the observer state in the perceiver. Some highlight that the later process might be supported by specialized *mirror neurons* or even an entire *mirror neuron system*, designed to map correspondences between the observed and performed actions and differentiating between goal-oriented vs. non-intentional action (Gallese, 2003). However, there is disagreement about the exact location of the mirror neurons, whether these neurons actually constitute a 'system' (in the sense of interconnected elements) and whether there actually are specialized neurons dedicated to mirroring or whether regular neurons can simply perform a mirroring function. Some of the original work in monkeys emphasized a unique role of neurons located in the inferior parietal and inferior frontal cortex, which discharge both when a monkey performs an action and when it observes another individual's action (Gallese, Keysers, & Rizzolatti, 2004). The implications of this work were quickly extended to humans. Some scientists argue that humans have a dedicated 'mirror neuron area,' located around the Broadmann area 44 (human homologue of the monkey F5 region). (Gallese et al., 2004). However, the empirical picture is a bit more complex, as we will discuss in more detail shortly. While there are some human studies that find activation in the putative mirror neuron area, which we discuss later in the chapter, there are also many studies suggesting that mirror-like responses, in the sense of an area's involvement in both perception and action, can be observed in variety of other regions of the brain. These may include a variety of emotion-related areas (insula, anterior cingulate), somato-sensory cortex, superior temporal sulcus, the extrastriate body area, or the dentate of the cerebellum (see Decety & Jackson, 2004 for a review). Of course, this could suggest that the mirror neurons are scattered throughout the brain, perhaps forming a distributed mirror neuron system. However, it could also suggest that there is no 'system' and mirroring is just a *function* that can be instantiated by many areas. In fact, one principle explaining these effects holds that neural coding generally tends to co-localize similar functions. For example, the neural representation of a visual image and linguistic concept of a 'leg' is partially co-localized with the neural representation of the physical leg because of Hebbian learning (seeing one's leg move and moving it at the same time). Thus, it is not surprising that the same area can be active during perception and action (Buccino et al., 2001). We will come back to these issues throughout the chapter.

4 Embodying Emotion

The idea of a tight connection between body and emotion is quite old. James (1884, p. 189) famously proposed that "the bodily changes follow directly the perception of the exciting fact, and that our feeling of the same changes as they occur is the emotion." Exactly a hundred years later, Zajonc and Markus (1984, p. 74) wondered "why do people who are angry squint their eyes and scratch their shoulders?" and proposed that sensory-motor actions are an intrinsic part of the representation of an emotional state.

These observations are confirmed by systematic investigation. Many studies find that merely thinking about emotional content can elicit incipient facial expressions and other physiological markers of emotional processing (Cacioppo, Petty, Martzke, & Tassinary, 1988; Adolphs, Damasio, Tranel, Cooper, & Damasio, 2000). In fact, even mere observation of emotional behavior tends to elicit covert mimicry, including matching facial expressions (Dimberg, Thunberg, & Elmehed, 2000; Wallbott, 1991) and emotional tone of voice (Neumann & Strack, 2000).

However, why do these phenomena occur? Do they play a causal role in emotion processing? Some views, like the associative account, emphasize that the role of previously established Stimulus-Response links (Lipps, 1907; Heyes, 2001). For example, people spontaneously smile when they observe another person smile because seeing and making a smile is frequently paired in the environment. Several studies have now clearly documented the role of experience in phenomena such as automatic imitation (e.g., Heyes, Bird, Johnson, & Haggard, 2005). This possibility makes some think that sensory-motor effects are causally inefficacious by-products of higher-order processes (Fodor & Pylyshyn, 1988). Others think that a system based on associative learning can play a causal role in the recognition and understanding of action and emotion, but doubt there is a need for assumptions beyond the standard associative approach (Heyes, 2001).

In contrast, theories of embodied cognition suggest that engagement of sensory-motor processes is part and parcel of the process of emotional perception, understanding, learning, and influence. On that account, the vicarious re-creation of the other's state provides information about the stimulus meaning and can go beyond the previously established associations. If so, manipulation (inhibition or facilitation) of somatosensory resources should influence the perception and understanding of emotional stimuli. Evidence for this interpretation has been now obtained in multiple domains.

5 Perceiving Emotional Information

A variety of behavioral and neuroscience studies provide evidence for the role of embodied simulation in emotion perception (for a review, see Goldman & Sripada, 2005). Those studies examined the involvement of somatosensory resources by manipulating and measuring both peripheral and central mechanism.

5.1 Peripheral Mechanisms

Focusing on the role of feedback from facial muscles, Niedenthal, Brauer, Halberstadt, and Innes-Ker (2001) examined the possibility that mimicry (reproducing the observed stimulus using one's own muscles) is causally involved in the perception of facial expression of emotion. Participants were asked to identify the point at which a morphed face changed from happy to sad and vice versa. During this task, some participants were free to move their faces naturally, whereas others were holding a pen sideways in their mouths, between their teeth and lips. This manipulation prevents facial mimicry and thus reduces somatic feedback that supports detection of change in the observed expressions. Participants whose facial movements were blocked by the pen detected the change in expression later in both directions (happy to sad and sad to happy) than those who were able to move their face freely, supporting the role of facial mimicry in recognition of facial expressions.

Oberman, Winkielman, and Ramachandran (2007) extended this research by adding several controls and, more importantly, examining the specificity of the mimicry-blocking effect. Note that the embodiment account predicts that recognition of a specific type of facial expressions should be impaired by blocking mimicry in the group of facial muscles used in production of this type of expression. Two experiments tested this hypothesis using four expressions (happy, disgust, fear, and sad) and four manipulation of facial mimicry (i) holding a pen sideways between the teeth, (ii) chewing gum, (iii) holding the pen just with the lips, and (iv) no task. Experiment 1 employed electromyography (EMG) and found that holding a pen sideways between the teeth selectively activates muscles involved in producing expressions of happiness. In contrast, the gum manipulation broadly activates several facial muscles, but only intermittently (the lip manipulation had no effect on EMG). Experiment 2 testing for accuracy of emotion discrimination found that the pen-biting manipulation selectively impaired recognition of happiness but had no effect on recognition accuracy for disgust, fear, and sad expressions. This finding suggests that recognition of a specific kind of facial expression involves selective recruitment of muscles used to produce that expression, as predicted by embodiment accounts.

5.2 Central Mechanisms

Several studies have investigated the role of central mechanisms underlying embodied simulation. In a pioneering study, Adolphs and colleagues (2000) asked 108 patients with a variety of focal brain lesions and 30 normal control participants to perform three visual emotion recognition tasks. In the first task, participants rated the intensity of basic emotional facial expressions. In the second task, participants matched a facial expression to its name. In the third task, participants sorted facial expressions into emotional categories. Though each task identified a slightly different group of regions, damage to primary and secondary somatosensory cortices impaired performance in all three tasks. This finding is consistent with the embodiment view on which emotion perception involves simulating the relevant state in the perceiver using somatosensory resources.

The above lesion study did not report a particularly critical role of the classic mirror neuron areas (BA 44) in the recognition of facial expressions. However, such suggestions have been made in the fMRI literature. Carr, Iacoboni, Dubeau, Mazziotta, & Lenzi (2003) asked participants to just observe or to observe and imitate emotional facial expressions. Compared to rest, both observation and imitation tasks activated a similar group of regions, including inferior frontal cortex (the mirror neuron area) as well as superior temporal cortex, insula, and amygdala.

Finally, there is some evidence for the selectivity of central mechanisms in embodied simulation of specific emotions. Wicker, Keysers, Plailly, Royet, Gallese, and Rizzolatti (2003) asked participants to inhale odors that generated strong feelings of disgust. The same participants then watched videos displaying other individuals expressing disgust. Results showed that the areas of the anterior insula and, to some extent, the anterior cingulate cortex were activated both when individuals experienced disgust themselves and when individuals observed disgust in others, presumably reflecting simulation. This interpretation is further supported by evidence that damage to the insula results in a paired impairment in the experience and recognition of disgust (Calder, Keane, Cole, Campbell, & Young, 2000).

6 Emotional Understanding and Empathy

Going beyond the perception of specific emotional stimuli such as facial expressions, the idea of embodied simulation can shed light on a more general process of emotional understanding and empathy. Social psychologists have long argued that empathy – 'putting oneself in some else's shoes' – can facilitate understanding (for review, see Batson, 1991). There is now evidence that this process might be supported by embodied simulation (Decety & Jackson, 2004). Much of the relevant data come from perceiver's reaction to another person's pain.

One early study assessed activity in areas related to the experience of pain with a precise technique for neural mapping – single cell recording. This study found activation of pain-related neurons when a painful stimulus was applied to the participant's own hand, and also when the patient watched the painful stimulus applied to the experimenter's hand (Hutchison, Davis, Lozano, Tasker, & Dostrovsky, 1999). This finding was extended by a recent fMRI study that revealed similar changes in pain-related brain regions (anterior cingulate and insula) of female participants while painful stimulation was applied to their own hand and to their partners' hand (Singer, Seymour, O'Doherty, Kaube, Dolan, & Frith, 2004). Further, the study showed that the change in relevant brain activations was related to the participants' level of empathy, suggesting the role of motivation to stimulate. Indeed, this interpretation is consistent with recent studies from the same lab which found an increase in activation of pain-related regions to the observation of a confederate receiving a painful stimulus, but only if the confederate had played fairly in a previous economic game (Singer, Seymour, O' Doherty, Klaas, Dolan, & Frith, 2006). This finding again highlights the goal- and context- dependent nature of simulation that is emphasized by modern embodiment theories. That is, the responses to another person are not simply automatic, but they are situated in a particular context that reflect the relationship with the person, or shared group memberships, and require active engagement of the perceiver in the process of constructing a simulation.

7 Social Functioning

If the ability to construct an embodied simulation is critical for emotion perception and understanding, one would expect that it would be related to social functioning. Evidence that this might indeed be a case comes from research on individuals with typical and atypical social functioning.

7.1 Typical Individuals

One early suggestion comes from a correlational study by Zajonc, Adelman, Murphy, and Niedenthal (1987) who found greater facial similarity in couples after 25 or more years of marriage than in the same couples at the beginning of marriage. Presumably, this effect occurs because married partners frequently mimic each other's facial expressions in order to empathize with each other. Consistently, the greater similarity in appearance was correlated with the quality of the marriage, and therefore presumably success in empathizing.

Experimental research in social psychology confirms that liking can enhance mimicry, and that mimicry can cause liking. Using facial electromyography (EMG) measures of mimicry, McIntosh (2006) found that observers who were first made to like a confederate mimicked the confederate's cheek movements more than observers who were first made not to like the confederate. Looking at the reverse of this process, Chartrand and Bargh (1999) found that participants liked the confederate more if she showed greater postural mimicry.

One way social relationships interface with mirroring is via modification of self-other overlap. For example, engaging in mimicry, or even in a simple exchange of touch, can reduce the psychological distance to the mental representation of the other (Smith & Semin, 2007). Or, in the reverse direction, enhancing self-other overlap by priming interdependent, rather than independent, self-construal can increase mimicry (Van Baaren, Maddux, Chartrand, de Bouter, & van Knippenberg (2003).

Again, the findings just discussed highlight the importance of social context for simulation and mirroring processes (Hess, Phillipot, & Blairy, 1999). Mimicry and simulation is not automatic in a strong sense, and is modified by several variables. In fact, there are also several contexts where mimicry is counterproductive (e.g., poker games) and where observers might even engage in countermimicry (e.g., watching political opponents).

7.2 Individuals with Autism

A link between embodiment and social functioning is also suggested by the literature on autism – a disorder characterized by severe deficits in social and emotional understanding. Several authors have suggested that these deficits could result from reduced imitative abilities (see Williams, Whiten, & Singh, 2004 for a review). In fact, there is now substantial evidence that individuals with ASD (autism spectrum disorder) have deficits in spontaneous imitation of both emotional and non-emotional stimuli (Hamilton, Brindley, & Frith, 2007). For example, McIntosh, Reichmann-Decker, Winkielman, and Wilbarger (2006) showed pictures of happy and angry facial expressions to adults with ASD and matched controls. In one condition, participants were simply asked to "watch the pictures as they appear on the screen." In another condition, participants were asked to "make an expression just like this one." Mimicry was measured by EMG, with electrodes placed over the cheek (smiling) and brow (frowning) regions. In the voluntary condition, there were no group differences, with ASD participants showing a normal pattern of voluntary mimicry (smile to a smile, frown to a frown). However, in the spontaneous condition, only typical participants mimicked with ASD showing no differential responses.

It has been proposed that the imitation deficits of ASD individuals result from impairments in their mirror neuron system (Oberman & Ramachandran, 2007) and their inability to spontaneously map the mental representation of the self to the representation of the other (Williams et al., 2004). Evidence consistent with these proposals has been obtained by several research groups using different techniques. First, there are reports of anatomical differences in the mirror neuron system. For example, Hadjikhani, Joseph, Snyder, and Tager-Flusberg (2006) found that ASD individuals have local decreases of gray matter in the mirror neuron system areas and that the cortical thinning of those areas was correlated with severity of ASD symptoms. Similarly, Villalobos, Mizuno, Dahl, Kemmotsu, and Muller (2005) found that ASD individuals have reduced functional connectivity between the primary visual cortex and area 44, the prefrontal mirror neuron area. Second, several studies observed functional differences in the activity of the mirror neuron system. Most of this research focused on simple, non-emotional gestures. Nishitani, Avikainen, and Hari, (2004) showed Asperger's Syndrome (AS) and control participants pictures of a woman performing orofacial gestures and asked them to imitate these gestures. Cortical activations were recorded using magnetoencephalography (MEG) – an electrophysiological technique that offers good temporal resolution. Compared to controls, the AS group showed weaker activations in inferior frontal lobe and primary motor cortex, suggesting a reduced mirror neuron activity. Focusing on spontaneous imitation, Oberman, Hubbard, McCleery, Ramachandran, and Pineda (2005) asked typical and ASD individuals to simply view videos of a person executing simple actions, or to perform the same actions. During these tasks, the experimenters recorded mu wave suppression – an EEG index of activity in primary motor cortex, and proposed to be indicative of activity in the premotor 'mirror neuron area' during the observation of action. The typically developing individuals showed mu wave suppression to both the execution and observation of action. However, individuals with ASD only showed mu wave suppression when performing their own actual movement, but not when observing movement (i.e., reduced mirror neuron activity).

Interestingly, consistent with social psychological literature on the role of self-other overlap, there is evidence that autistic impairment in spontaneous mirroring might relate to a deficit in mapping the representation of the observed action to the self. Theoret et al., (2005) asked typical and ASD groups to view videos of index finger and thumb movements that were directed either toward or away from the participants. During these tasks, the experimenters recorded motor-evoked-potentials (MEP) induced by Transcranial Magnetic Stimulation (TMS). In the typical group, both participant-directed and other-directed actions increased MEPs recorded from the participant's muscles, suggesting spontaneous mirroring. However, the ASD group showed increased MEPs (spontaneous mirroring) only when viewing actions directed toward the participant but not when viewing actions directed away from the participant. This suggests that ASD participants' mirroring failures might be due to reduction in self-other mapping.

Finally, a recent fMRI study investigated the role of mirror neurons in imitation of emotion stimuli in individuals with ASD and controls (Dapretto et al., 2005). Participants were asked to both imitate and observe emotional facial expressions. As compared to controls, ASD participants showed lower activation in a wide variety of regions, including visual cortices, primary motor, limbic, cerebellum, and the presumed 'mirror neuron region' (inferior frontal gyrus). Though the group differences in brain activations were fairly broad, one intriguing finding is a negative correlation of the activity in the mirror neuron region with severity of autism symptoms (measured by the Autism Diagnostic Observation Schedule (ADOS) and the Autism Diagnostic Interview (ADI)). Again, these findings suggest that deficits in social and emotional understanding in autism could be due to a reduction in spontaneous simulation.

8 Influence of Emotion on Complex Behavior

The embodiment perspective also sheds light on how emotions influence more complex behaviors (for review, see Winkielman et al., 2007). There is much evidence that stimuli such as emotional faces, evocative scenes, or valenced words can change subsequent behavior. For example, in one study participants were first exposed to a series of subliminal happy or angry faces and then asked to perform a set of consumption behaviors (pour and drink a novel beverage). The results showed that participants poured and drank more after being exposed to happy than angry faces, especially when they were thirsty (Winkielman, Berridge, & Wilbarger, 2005). Other studies have documented influence of incidental emotion stimuli on other behavior, including financial choices. For example, in one study, participants were first shown an erotic picture and then decided between making a small or large gamble (Knutson, Wimmer, Kuhnen, & Winkielman, 2008). Even though participants were fully informed about the incidental nature of the erotic picture and the random nature of gambles, they still preferred large gambles after erotic, rather than control pictures. But what are the mechanisms of such effects?

Many researchers treat affective priming in the same framework as any other type of 'cold' semantic priming. Affective primes activate valence-congruent material in semantic memory which then facilitates valence-congruent judgments and behaviors (Forgas, 2002). In contrast, the embodiment framework suggests that exposure to affective primes can elicit somatosensory reactions, which then bias and guide processing of subsequent stimuli (e.g., Niedenthal, Rohman, & Dalle, 2002). These considerations generate an interesting prediction regarding the impact of affective primes on behavior. Stimuli that trigger an embodied response should have greater impact on subsequent behavior than stimuli that are comparable in semantic aspects of valence but do not trigger an embodied response. This should be particularly true for behaviors that require some form of evaluative engagement with the stimulus, rather than just simply associative responding (we will return to this point later).

Winkielman and Gogolushko (under review) tested these predictions in a study that compared the impact of emotional faces and scenes versus emotional words on consumption behavior (pouring and drinking of a novel beverage). The priming stimuli were equated on valence and frequency but emotional faces and scenes were more likely to trigger a physiological response than emotional words (Larsen, Norris, & Cacioppo, 2003). In Experiment 1, both subliminal and supraliminal emotional facial expressions influenced consumption behavior in a valence-congruent way with happy primes leading to more pouring and drinking than angry primes. This effect was replicated in Experiment 2 with supraliminal pictures of high- and low-frequency emotional objects. In contrast, emotional words had no systematic effects on behavior in either study. There were also no differences between pictorial and word primes on more interpretive responses such as ratings of the beverage.

Similarly, in the earlier discussed study on erotic pictures and gambling, Knutson and colleagues (2008) were able to show that affective influence was mediated by the degree to which affective picture was able to activate the excitement and reward-related brain structures such as nucleus accumbens. In sum, these studies suggest that in order to influence a complex behavior (like consumption of a novel drink or a financial choice), an affective stimulus must not only be coldly processes regarding its valence implications – it must also elicit a 'hot' embodied response (for further discussion, see Winkielman et al., 2007).

9 Acquiring and Expressing Values, Preferences, and Attitudes

Over the last two decades, social psychologists have conducted a series of ingenious experiments that reveal the role of embodiment in both formation and expression of value as expressed in people's preferences and attitudes.

9.1 Attitude Formation

In an early demonstration of the role of embodiment, Wells and Petty (1980) instructed participants to nod their heads vertically or to shake their heads horizontally while wearing headphones under the pretext that the research was designed to investigate whether the headphones moved around if listeners danced while listening to music. While nodding or shaking their heads, participants then heard either a disagreeable or an agreeable message about a university-related topic. Later, they rated how much they agreed with the message. Results suggested that the earlier head movements moderated their judgments. Specifically, participants who had nodded while hearing the message were more favorable than participants who had shaken their heads.

Researchers have also shown that one can enhance an attitude toward an object by covertly inducing individuals to smile (Strack, Martin, & Stepper, 1988). In the study, participants were asked to rate different novel cartoons while holding a pencil between their front teeth, making it easier for the participants to smile. Other participants were instructed to hold a pencil between their lips without touching the pencil with their teeth, making it more difficult to smile. Results revealed that cartoons were evaluated as higher by individuals with facilitated smiling rather than inhibited smiling.

Cacioppo, Priester, and Berntson (1993) asked participants to view and evaluate neutral Chinese ideographs. During this task, the researchers manipulated the engagement of muscles involved in arm flexion and arm extension by having participants press against the bottom or top of the table. Arm flexion was associated with higher rating of ideographs than muscle extension, presumably because of differential association of these actions with evaluative outcomes.

9.2 Expression of Attitudes

One of the first studies that examined the role of embodied responses in attitude expressions was conducted by Solarz (1960). He asked participants to move cards with words that were mounted on a movable stage either toward or away from themselves. Participants responded faster with the pulling movement (typically associated with approach) to positive than to negative words and faster with the pushing movement (typically associated with avoidance) to negative than to positive words (see also Chen & Bargh, 1999).

9.3 Flexible Embodiment

Importantly, although findings like the ones we just discussed may suggest a relatively fixed link between valence and a specific muscle action or a specific direction of movement, this relationship is more complex. For example, Centerbar and Clore (2006) replicated the procedure of the Cacioppo et al. (1993) study with positively and negatively valenced ideographs and found that the impact of specific muscle movement on later evaluation depended on the initial stimulus valence. Thus, with initially negative stimuli, muscle extension (pushing away) led to more positive attitude than muscle flexion (pulling toward). Presumably, pushing away a bad stimulus is a more compatible action rather than pulling it toward oneself. As a result, this action might feel more fluent and pleasant (Winkielman, Schwarz, Fazendeiro, & Reber, 2003).

Further, the exact impact of action on valence depends on the meaning of the movement for the participant. For example, using a modified version of the Solarz /Chen and Bargh paradigm, Wentura, Rothermund, and Bak (2000) asked participants to respond to positive and negative words by either reaching out their hand to press a button or by withdrawing their hand from the button. Note that in this case pressing the button (i.e., approaching it) required an extension movement (away from the body). In contrast, *not* pressing a button (avoiding it) required a flexion movement, withdrawing the hand toward the body. Consistent with the primacy of functional meaning of movement (rather than a specific muscle movement), participants pressed the button faster for positive than for negative stimuli, but withdrew their hand faster for negative than positive stimuli. Similarly, Markman and Brendl (2005) demonstrated that the evaluative meaning of the movement does not depend on the relation to 'physical body,' but rather the relation to the more abstract representation of the 'self.' Specifically, they found that positive valence facilitates any motor action (push or pull) that brings the stimulus closer to the self, even when the self is represented as participants' name on a screen.

In summary, multiple studies suggest that bodily postures and motor behavior that are associated with positive and negative inclinations and action tendencies toward objects influence acquisition and expression of attitudes toward those objects. Thus, it seems that attitudes are in part grounded in embodied responses. Importantly, the link between these embodied responses and valence is flexible, and depends on features of the current situation, the initial value of the stimuli, and how participants interpret the meaning of the specific action. This flexibility is consistent with the modern theories of embodied cognition that view the use of somatosensory and motor responses as a constructive, on-line process, which dynamically utilizes relevant resources (Barsalou, 1999).

10 Linguistically Represented Emotion Knowledge

Thus far, most of our discussion has focused on relatively concrete emotional stimuli (e.g., faces or scenes). However, what about more symbolic forms of emotion knowledge? Can emotional words and sentences be embodied?

10.1 Emotion Concepts

Recent findings from studies of emotion concepts suggest that embodied simulation is also involved in representation of abstract emotion knowledge (Niedenthal, Winkielman, Mondillon, & Vermeulen, under review). In two experiments recently conducted in our laboratories, participants had to make judgments about whether concepts were associated with an emotion by providing a yes or no response. In the first study, the words referred to concrete objects (e.g., *PARTY*, *VOMIT*) previously rated by other participants as being strongly associated with the emotions of joy, disgust, and anger, or as being very neutral. In a second study, the stimuli were abstract concepts, specifically, adjectives that referred to affective states and conditions (e.g., *DELIGHTED*, *NAUSEATED*). During the judgment task, the activation of four different facial muscles was recorded with electromyography (EMG). Previous research shows that the activity in the region of two muscles, the *zygomaticus major* and the *orbicularis occuli* is elevated when an individual is smiling with happiness. The region of *corrugator supercilli* is activated when an individual is frowning with anger. And the *levator* muscle is largely activated when an individual makes the grimace of disgust (Tassinary & Cacioppo, 2000).

Results of both of our studies suggest that in processing emotionally significant concepts, individuals simulated the relevant, discrete, expressions of emotion on their faces. For example, the first study showed that in the very brief time (less than 3 seconds) it took participants to decide that, for example, *VOMIT* was related to an emotion, they activated disgust-related muscles on their faces. Similarly, the second study showed that when processing the concept *DELIGHTED* individuals activated happiness-related muscles. Importantly, these EMG effects did not simply reflect an automatic response to the word, but rather reflect a goal-dependent simulation of the word's referent (concrete objects for the first study and abstract emotional states for the second). Support for such a conclusion comes from an additional control condition of each study where participants were asked to simply judge (yes or no) whether the words were written in capital letters. In order to make such judgments these participants would not have to simulate the emotional meaning of the words, and indeed findings revealed that they showed no systematic activation of the facial musculature. This finding suggests that emotional simulation does *not* occur when the judgment can based on simpler features – a point that has been made in other research as well (Solomon & Barsalou, 2004; Strack, Schwarz, & Gscheidinger, 1985).

Further evidence for the embodiment of emotion concepts comes from extensions of research on the costs of switching processing between sensory modalities to the area of emotion. Previous research has shown that shifting from processing in one modality to another involves temporal processing costs (e.g., Spence, Nicholls, & Driver, 2001). For example, individuals take longer to detect the location of a visual stimulus after having just detected the location of an auditory, rather then another visual stimulus. For the present concerns it is of interest that similar 'switching costs' are also found when participants engage in conceptual tasks: Individuals are slower to verify that typical instances of object categories possess certain features if those features are processed in different modalities (Pecher, Zeelenberg, & Barsalou, 2003): They are slower to verify that a *BOMB* can be *loud* when they have just confirmed that a

LEMON can be *tart*, for example, than when they have just confirmed that *LEAVES* can be *rustling*. This provides support for the general assertion made by theories of embodied cognition that individuals simulate objects in the relevant modalities when they use them in thought and language.

Vermeulen and colleagues examined switching costs in verifying properties of positive and negative concepts such as *TRIUMPH* and *VICTIM* (Vermeulen, Niedenthal, & Luminet, 2007). Properties of these concepts were taken from vision, audition, and the affective system. Parallel to switching costs observed for neutral concepts, the study showed that for positive and negative concepts, verifying properties from different modalities produced costs such that reaction times were longer and error rates were higher than if no modality switching was required. Importantly, this effect was observed when participants had to switch from the affective system to sensory modalities, and vice-versa. In other words, verifying that a *VICTIM* can be *stricken* was less efficient if the previous trial involved verifying that a *SPIDER* can be *black* than if the previous trial involved verifying that an *ORPHAN* can be *hopeless*. And verifying that a *SPIDER* can be *black* was less efficient when preceded by the judgment that an *ORPHAN* can be *hopeless* than that a *WOUND* can be *open*. This provides evidence that affective properties of concepts are simulated in the emotional system when the properties are the subject of active thought.

10.2 Emotional Language

Finally, embodied cognition of language make the claim that even comprehension of complex sentences may involve embodied conceptualizations of the situations that language describes (Glenberg & Robinson, 2000; Zwaan, 2004). The first step in language comprehension, then, is to index words or phrases to embodied states that refer to these objects. Next, the observer simulates possible interactions with the objects. Finally, the message is understood when a coherent set of actions is created. Recently, Glenberg and colleagues provided some evidence for simulation of emotions in sentence comprehension (Havas, Glenberg, & Rinck, 2007). Motivation for this research was the hypothesis that if the comprehension of sentences with emotional meaning requires the partial reenactment of emotional bodily states, then simulation of congruent (or incongruent) emotions should facilitate (or inhibit) language comprehension. Participants had to judge whether the sentences described a pleasant or an unpleasant event, while holding a pen between the teeth (again, to induce smiling) or between the lips (to induce frowning). Reading times for understanding sentences describing pleasant events were faster when participants were smiling than when they were frowning. Those that described unpleasant events were understood faster when participants were frowning than when they were smiling. The same effect was observed in a second experiment in which participants had to evaluate if the sentences were easy or hard to understand. This research dovetails with other studies on comprehension of non-emotional language that have repeatedly demonstrated the role of embodied information (Zwaan, 2004).

11 Open Issues

We hope this chapter shows that theories of embodied cognition offer a fruitful approach to processing of emotional information in social and non-social cognition. The evidence supporting these theories comes from multiple domains, including emotion perception, learning, understanding, and emotional influence on behavior. Importantly, there are now many studies showing that embodiment can have a causal rather than just correlational relationship with processing of emotional information. Further, we now understand that embodiment is not a static, but a goal-dependent processes, with the exact impact of the particular sensory-motor input depending on dynamically contructed conceptualizations. There also has been remarkable progress in understanding the neural basis of embodiment and the specific peripheral and central mechanisms supporting processes of simulation. Finally, and perhaps most critically, the embodiment approach has been able to generate exciting and counterintuitive prediction across a variety of areas in psychology and neuroscience.

Of course, open issues remain. One theoretical challenge, for the embodiment account in general, not only of emotion, is representation and processing of abstract information. For example, how do people understand social concepts, such as indenture, legal concepts, such as eminent domain, or logical concepts, such as recursion? There are some promising attempts to solve this issue (Barsalou, 2003), even for such abstract domains as mathematics, though some of these solutions assume fairly powerful abilities to process metaphorical information (Lakoff & Núñez, 2000). A similar challenge applies to emotion concepts. For example, how do people understand the differences between shame, embarrassment, and guilt? Such understanding certainly involves the ability to simulate a relevant experience, but also requires the ability to connect the simulation to a more abstract knowledge about respective eliciting conditions and consequences (e.g., understanding that shame and guilt, but not necessarily embarrassment, involve norm violation, and that guilt, but not shame, implies recognition of responsibility). Recently in our laboratories we have begun to explore the idea that depending on their current goal, people represent an emotional concept (e.g., anger) in more modal or amodal fashion, and thus engage in different amount of perceptual simulation. These differences in representation can then change how the concept is used to interpret novel behaviors. For example, activation of a modal, rather than amodal, representation of anger should invite different inferences about the causes of a target's violent behavior, perhaps leading to different judgments of responsibility.

Another challenge for emotional embodiment theory is clarifying the difference between emotion and non-emotion concepts. That is, embodiment effects have been shown in a variety of modalities, including the motor system (Tucker & Ellis, 1998), visual, auditory, and gustatory system (Barsalou, 1999), so a legitimate question is whether there is anything unique about emotion concepts. One difference is that emotion concepts recruit a unique modality – internal representation of bodily state – and are tightly connected to motivation. A related difference is that emotion concepts organize information across modalities according to principles of subjective value, rather than relation to the objective external world. Finally, emotion concepts have a unique function and power in cognitive processing (e.g., they can prioritize processing according to perceivers' internal goals, interrupt ongoing processing streams, etc). In any case, exploring the differences between embodiment of emotional and non-emotional concepts, offers an exciting direction for future research.

Further, now that many emotion embodiment effects have been demonstrated, it is time to develop systematic models of their boundary conditions. For example, as we have discussed, there are now several demonstrations that peripheral manipulations of facial mimicry can influence emotion perception and judgment. However, there is little research directly contrasting inhibitory effects such as impairment in detection of happiness observed by Oberman et al. (2007) and Niedenthal, Brauer, Halberstadt, & Innes-Ker (2001) versus facilitatory effects, such as enhancement of cartoon judgments observed by Strack et al., (1988). One possible explanation is that forcing participants' face into a permanent smile (i.e., creating a constant level of muscular feedback) or immobilizing the face (i.e., removing muscular feedback) lowers participants' sensitivity to the presence versus absence of happy expressions by cutting out the differential information from the muscles. On the other hand, forming a half-smile creates a positivity bias – tendency toward happy responding to ambiguous stimuli. Incidentally, there is a similar debate in the cognitive literature which reports that similar embodiment manipulations can have both facilitatory and inhibitory effects, with subtle differences in timing and task demands leading either to resource priming or resource competition (Reed & McGoldrick, 2007).

Though challenges remain, it is clear that recent years have seen remarkable progress in understanding the nature of emotion. Emotion is no longer ignored by psychology and cognitive science, but in fact is one of the most studied topics. The embodiment account has inspired and is continuing to generate research that advances understanding of emotional perception, learning, comprehension, attitudes, prejudice, empathy, and even certain behavioral deficits.

Acknowledgments We appreciate discussions with Adam Aron, Larry Barsalou, Lisa Barrett, Thierry Chaminade, Vic Ferreira, Chris Frith, Celia Heyes, Dave Huber, Hal Pashler, Vilayanur Ramachandran, Cathy Reed, Norbert Schwarz, Gun Semin, Eliot Smith, and Robert Zajonc. This work is supported by NSF grant BCS-0350687 to Piotr Winkielman and Paula Niedenthal.

References

Adolphs, R., Damasio, H., Tranel, D., Cooper, G., & Damasio, A. (2000). A role for somatosensory cortices in the visual recognition of emotion as revealed by three-dimensional lesion mapping. *Journal of Neuroscience, 20*, 2683–2690.

Barsalou, L. W. (1999). Perceptual symbol system. *Behavioral and Brain Sciences, 22*, 577–660.

Barsalou, L. W. (2003). Abstraction in perceptual symbol systems. *Philosophical Transactions of the Royal Society of London: Biological Sciences, 358*, 1177–1187.

Batson, C. D. (1991). *The altruism question: Toward a social-psychological answer*. Hillsdale, NJ: Erlbaum.

Buccino, G., Binkofski, F., Fink, G. R., Fadiga, L., Fogassi, L., Gallese, V., et al. (2001). Action observation activates premotor and parietal areas in somatotopic manner: An fMRI study. *European Journal of Neuroscience, 13*, 400–404.

Cacioppo, J. T., Petty, R. E., Martzke, J., & Tassinary, L. G. (1988). Specific forms of facial EMG response index emotions during an interview: From Darwin to the continuous flow model of affect-laden information processing. *Journal of Personality and Social Psychology, 54*, 592–604.

Cacioppo, J. T., Priester, J. R., & Berntson, G. G. (1993). Rudimentary determinants of attitudes: II. Arm flexion and extension have differential effects on attitudes. *Journal of Personality and Social Psychology, 65*, 5–17.

Calder, A. J., Keane, J., Cole, J., Campbell, R., & Young, A. W. (2000). Facial expression recognition by people with Mobius syndrome. *Cognitive Neuropsychology, 17*, 73–87.

Cannon W. B. (1927). The James-Lange theory of emotions. *American Journal of Psychology, 39*, 115–124.

Carr, L., Iacoboni, M., Dubeau, M. C., Mazziotta, J. C., and Lenzi G. L. (2003). Neural mechanisms of empathy in humans: A relay from neural systems for imitation to limbic areas. *Proceedings of the National Academy of Science USA, 100*, 5497–5502.

Centerbar, D., & Clore, G. L. (2006). Do approach-avoidance actions create attitudes? *Psychological Science, 17*, 22–29.

Chartrand, T. L., & Bargh, J. A. (1999). The chameleon effect: The perception-behavior link and social interaction. *Journal of Personality and Social Psychology, 76*, 893–910.

Chen, S., & Bargh, J. A. (1999). Consequences of automatic evaluation: Immediate behavior predispositions to approach or avoid the stimulus. *Personality and Social Psychology Bulletin, 25*, 215–224.

Clark, A. (1997). *Being there: Putting brain body and world together again*. Cambridge, MA: MIT Press.

Craig, A. D. (2002). How do you feel? Interoception: the sense of the physiological condition of the body. *Nature Reviews Neuroscience, 3*, 655–66.

Damasio, A. R. (1989). Time-locked multiregional retroactivation: A systems-level proposal for the neural substrates of recall and recognition. *Cognition, 33*, 25–62.

Damasio, A. R. (1994). *Descartes' error: Emotion, reason, and the human brain*. New York: Grosset/Putnam.

Damasio, A. R., Grabowski, T. J., Bechara, A., Damasio, H., Ponto, L. L. B., Parvizi, J., et al. (2000). Subcortical and cortical brain activity during the feeling of self-generated emotions. *Nature Neuroscience, 3*,1049–1056.

Dapretto, M., Davies, M. S., Pfeifer, J. H., Scott, A. A., Sigman, M., Bookheimer, S.Y., et al. (2005). Understanding emotions in others: mirror neuron dysfunction in children with autism spectrum disorders. *Nature neuroscience, 9*(1), 28–30.

Decety, J., & Jackson, P. L. (2004). The functional architecture of human empathy. *Behavioral and Cognitive Neuroscience Reviews, 3*, 71–100.

Dimberg, U., Thunberg, M., & Elmehed, K. (2000). Unconscious facial reactions to emotional facial expressions. *Psychological Science, 11*, 86–89.

Eich, E., Kihlstrom, J., Bower, G., Forgas, J., & Niedenthal, P. (2000). *Cognition and emotion*. New York: Oxford University Press.
Ellsworth, P. C., & Scherer, K. R. (2003). Appraisal processes in emotion. In R. J. Davidson, H. Goldsmith, & K. R. Scherer (Eds.), *Handbook of affective sciences*. New York and Oxford: Oxford University Press.
Fodor, J. (1975). *The language of thought*. Cambridge, MA: Harvard University Press.
Fodor, J., & Pylyshyn, Z. (1988). Connectionism and cognitive architecture: A critical analysis. *Cognition*, *28*, 3–71.
Forgas, J. P. (2002). Feeling and doing: The role of affect in social cognition and behavior. *Psychological Inquiry*, *9*, 205–210.
Gallese, V. (2003). The roots of empathy: The shared manifold hypothesis and the neural basis of intersubjectivity. *Psychopathology*, *36*, 71–180.
Gallese, V., Keysers, C., & Rizzolatti, G. (2004). A unifying view of the basis of social cognition. *Trends in Cognitive Sciences*, *8*, 396–402.
Glenberg, A. M., & Robinson, D. A. (2000). Symbol grounding and meaning: A comparison of high-dimensional and embodied theories of meaning. *Journal of Memory and Language*, *43*, 379–401.
Goldman, A. I., & Sripada, C. S. (2005). Simulationist models of face-based emotion recognition, *Cognition*, *94*, 193–213.
Grush, R. 2004 The emulation theory of representation: motor control, imagery, and perception. *Behavioral Brain Sciences*, *27*, 377–396.
Hadjikhani, N., Joseph, R. M., Snyder, J., & Tager-Flusberg, H. (2006). Anatomical differences in the mirror neuron system and social cognition network in autism. *Cerebral Cortex*, *9*, 1276–1282.
Hamilton, A., Brindley, R. M., & Frith U. (2007). Imitation and Action Understanding in Autistic Spectrum Disorders: how valid is the hypothesis of a deficit in the mirror neuron system? *Neuropsychologia*, *45*, 1859–1868.
Havas, D. A., Glenberg, A. M., & Rinck, M. (2007). Emotion simulation during language comprehension. *Psychonomic Bullentin & Review*, *14*, 436–441.
Hess, U., Philippot, P., & Blairy, S., (1999). Facial mimicry: Facts and fiction. In P. Philippot, R. Feldman, & E. Coats (Eds.), *The social context of nonverbal behavior* (pp. 213–241). Cambridge, MA: Cambridge University Press.
Heyes, C. M. (2001). Causes and consequences of imitation. *Trends in Cognitive Sciences*, *5*, 253–261.
Heyes, C. M., Bird, G., Johnson, H., & Haggard, P. (2005). Experience modulates automatic imitation. *Cognitive Brain Research*, *22*, 233–240.
Hutchison, W. D., Davis, K. D., Lozano, A. M., Tasker, R. R., & Dostrovsky, J. O. (1999). Pain related neurons in the human cingulate cortex. *Nature Neuroscience*, *2*, 403–405.
Iacoboni, M., Koski, L., Brass, M., Bekkering, H., Woods, R. P., Dubeau, M. -C., et al. (2001). Re-afferent Copies of Imitated Actions in the Right Superior Temporal Cortex. *Proceedings of the National Academy of Science*, *98*, 13995–13999.
James W. (1884). What is an emotion? *Mind*, *9*, 188–205.
Knutson, B., Wimmer, G. E., Kuhnen, C. M., & Winkielman, P. (2008). Nucleus accumbens activation mediates the influence of reward cues on financial risk-taking. *NeuroReport*, *19*, 509–513.
Lakoff, G., & Núñez, R. (2000). *Where mathematics comes from: How the embodied mind erings mathematics into being*. New York: Basic Books.
Larsen, J. T., Norris, C. J., & Cacioppo, J. T. (2003). Effects of positive affect and negative affect on electromyographic activity over zygomaticus major and corrugator supercilii. *Psychophysiology*, *40*, 776–785.
Lipps, T. (1907). Das Wissen von fremden Ichen. In T. Lipps (Ed.), *Psychologische Untersuchungen (Band 1)* (pp. 694–722). Leipzig: Engelmann.

Markman, A. B., & Brendl, C.M. (2005), Constraining theories of embodied cognition, *Psychological Science*, *16*, 6–10.
McIntosh, D. N. (2006). Spontaneous facial mimicry, liking and emotional contagion. *Polish Psychological Bulletin*, *37*, 31–42.
McIntosh, D. N., Reichmann-Decker, A., Winkielman, P., & Wilbarger, J. (2006). When mirroring fails: Deficits in spontaneous, but not controlled mimicry of emotional facial expressions in autism. *Developmental Science*, *9*, 295–302.
Niedenthal, P.M., Winkielman, P., Mondillon, L., & Vermeulen, N. (under review). Embodiment of emotion concepts: Evidence from EMG measures.
Neumann, R., & Strack, F. (2000). "Mood contagion": The automatic transfer of mood between persons. *Journal of Personality and Social Psychology*, *79*, 211–223.
Newell, A. (1980). Physical symbol systems. *Cognitive Science*, *4*, 135–183.
Niedenthal, P. M. (2007). Embodying emotion. *Science*, *316*, 1002–1005.
Niedenthal, P. M., Barsalou, L. W., Winkielman, P., Krauth-Gruber, S., & Ric, F. (2005). Embodiment in attitudes, social perception, and emotion. *Personality and Social Psychology Review*, *9*, 184–211.
Niedenthal, P. M., Barsalou, L. W., Ric, F., & Krauth-Gruber, S. (2005). Embodiment in the acquisition and use of emotion knowledge. In L. Feldman Barrett, P. M. Niedenthal, & P. Winkielman (Eds.), *Emotion: Conscious and unconscious* (pp. 21–50). New York: Guilford.
Niedenthal, P. M., Brauer, M., Halberstadt, J. B., & Innes-Ker, Å. (2001). When did her smile drop? Facial mimicry and the influences of emotional state on the detection of change in emotional expression. *Cognition and Emotion*, *15*, 853–864.
Niedenthal, P. M., Rohman, A., & Dalle, N. (2002). What is primed by emotion words and emotion concepts? In J. Musch & K. C. Klauer (Eds.), *The psychology of evaluation: Affective processes in cognition and emotion*. Nahwah, NJ: Erlbaum.
Nishitani, N., Avikainen, S., & Hari, R. (2004). Abnormal imitation-related cortical activation sequences in Asperger's syndrome. *Annuls of Neurology*, *55*, 558–562.
Oberman, L. M., Hubbard, E. M., McCleery, J. P., Ramachandran, V. S., & Pineda, J. A. (2005). EEG evidence for mirror neuron dysfunction in autism. *Cognitive Brain Research*, *24*, 190–198.
Oberman, L. M., & Ramachandran, V. S. (2007). The simulating social mind: The role of simulation in the social and communicative deficits of autism spectrum disorders. *Psychological Bulletin*, *133*, 310–327.
Oberman, L. M., Winkielman, P., & Ramachandran, V.S. (2007). Face to Face: Blocking expression-specific muscles can selectively impair recognition of emotional faces. *Social Neuroscience*, *12*, 167–178.
Pecher, D., Zeelenberg, R., & Barsalou, L. W. (2003). Verifying different-modality properties for concepts produces switching costs. *Psychological Science*, *14*, 119–124.
Phelps, E. A., O'Connor, K. J., Gateby, J. J., Grillon, C., Gore, J. C., & Davis, M. (2001). Activation of the amygdala by cognitive representations of fear. *Nature Neuroscience*, *4*, 437–441.
Prinz, J. J. (2002). *Furnishing the mind: Concepts and their perceptual basis*. Cambridge, MA: MIT Press.
Reed, C. L., & McGoldrick, J. E. (2007). Action during body perception: Processing time affects self–other correspondences. *Social Neuroscience*, *2*, 1747.
Smith, E. R., & Semin, G. R. (2007). Situated social cognition. *Current Directions In Psychological Science*, *16*, 132–135.
Singer, T., Seymour, B., O'Doherty, J., Kaube, H., Dolan, R. J., Frith, C. D. (2004). Empathy for pain involves the affective but not sensory components of pain. *Science*, *303*, 1157–1162.
Singer, T., Seymour, B., O'Doherty, J., Klaas, E.S., Dolan, J.D., & Frith, C. (2006). Empathic neural responses are modulated by the perceived fairness of others. *Nature*, *439*, 466–469.
Solomon, K. O., & Barsalou, L. W. (2004). Perceptual simulation in property verification. *Memory & Cognition*, *32*, 244–259.

Solarz, A. K. (1960). Latency of instrumental responses as a function of compatibility with the meaning of eliciting verbal signs. *Journal of Experimental Psychology, 59*, 239–245.
Spence, C., Nicholls, M. E., & Driver, J. (2001). The cost of expecting events in the wrong sensory modality. *Perception & Psychophysics, 63*, 330–336.
Strack, F., Martin, L. L., & Stepper, S. (1988). Inhibiting and facilitating conditions of the human smile: A nonobtrusive test of the facial feedback hypothesis. *Journal of Personality and Social Psychology, 54*, 768–777.
Strack, F., Schwarz, N., & Gschneidinger, E. (1985). Happiness and reminiscing: The role of time perspective, mood, and mode of thinking. *Journal of Personality and Social Psychology, 49*, 1460–1469.
Tassinary, L. G., & Cacioppo, J. T. (2000). The skeletomuscular system: Surface electromyography. In J. T. Cacioppo, L. G. Tassinary, & G. G. Berntson (Eds.), *Handbook of psychophysiology*, (2nd ed., pp. 163–199). New York: Cambridge University Press.
Theoret, H., Halligan, E., Kobayashi, M., Fregni, F., Tager-Flusberg, H., & Pascual-Leone, A. (2005). Impaired motor facilitation during action observation in individuals with autism spectrum disorder. *Current Biology, 15*, R84–R85.
Tucker, M., & Ellis, R. (1998). On the relations between seen objects and components of potential actions. *Journal of Experimental Psychology: Human Perception and Performance, 24*, 830–846.
van Baaren, R. B., Maddux, W. W., Chartrand, T. L., de Bouter, C., & van Knippenberg, A. (2003). It takes two to mimic: Behavioral consequences of self-construals. *Journal of Personality and Social Psychology, 84*, 1093–1102.
Vermeulen, N., Niedenthal, P.M., & Luminet, O. (2007). Switching between sensory and affective systems incurs processing costs. *Cognitive Science, 31*, 183–192.
Villalobos, M. E., Mizuno, A., Dahl, B. C., Kemmotsu, N., & Muller, R. A. (2005). Reduced functional connectivity between V1 and inferior frontal cortex associated with visuomotor performance in autism. *Neuroimage, 25*, 916–25.
Wallbott, H. G. (1991). Recognition of emotion from facial expression via imitation? some indirect evidence for an old theory. *British Journal of Social Psychology, 30*, 207–219.
Wells, G. L., & Petty, R. E. (1980). The effects of overt head movements on persuasion: Compatibility and incompatibility of responses. *Basic and Applied Social Psychology, 1*, 219–230.
Wentura, D., & Rothermund, K., & Bak, P. (2000). Automatic vigilance: The attention-grabbing power of approach- and avoidance-related social information. *Journal of Personality and Social Psychology, 78*, 1024–1037.
Williams, J. H. G., Whiten, A., & Singh, T. (2004). A systematic review of action imitation in autistic spectrum disorder. *Journal of Autism & Developmental Disorders, 34*, 285–299.
Wilson, M. (2002). Six views of embodied cognition. *Psychonomic Bulletin & Review, 9*, 625–636.
Wicker, B., Keysers, C., Plailly, J., Royet, J. P., Gallese, V., & Rizzolatti, G. (2003). Both of us disgusted in My insula: the common neural basis of seeing and feeling disgust. *Neuron, 40*, 655–664.
Winkielman, P., Berridge, K. C., & Wilbarger, J. L. (2005). Unconscious affective reactions to masked happy versus angry faces influence consumption behavior and judgments of value. *Personality and Social Psychology Bulletin, 1*, 121–135.
Winkielman, P., Knutson, B., Paulus, M., Trujillo, J. L. (2007). Affective influence on judgments and decisions: Moving towards core mechanisms. *Review of General Psychology.*
Winkielman, P., Schwarz, N., Fazendeiro, T., & Reber, R. (2003). The hedonic marking of processing fluency: Implications for evaluative judgment. In J. Musch & K. C. Klauer (Eds.), *The psychology of evaluation: Affective processes in cognition and emotion.* (pp. 189–217). Mahwah, NJ: Lawrence Erlbaum.
Zajonc, R. B. (1980). Feeling and thinking: preferences need no inferences. *American Psychologist, 35*, 151–175.

Zajonc, R. B., Adelmann, P. K., Murphy, S. T., & Niedenthal, P. M. (1987). Convergence in the physical appearance of spouses. *Motivation and Emotion*, *11*, 335–346.

Zajonc, R. B., & Markus, H. (1984). Affect and cognition: The hard interface. In C. Izard, J. Kagan, & R. B. Zajonc (Eds.), *Emotions, cognition and behavior* (pp. 73–102). Cambridge, MA: Cambridge University Press.

Zwaan, R. A. (2004). The immersed experiencer: Toward an embodied theory of language comprehension. In B. H. Ross (Ed.), *The Psychology of Learning and Motivation*, *44*, 35–62.

Part V
Disorders of Mirroring

The Role of Mirror Neuron Dysfunction in Autism

Raphael Bernier and Geraldine Dawson

Abstract Autism spectrum disorders are characterized by impairments in the social, communication, and behavioral domains. Impairments in social cognition are considered a core symptom of the disorder. Dysfunction of an observation/execution matching system (the mirror neuron system) has been proposed to serve as a neural mechanism explaining the deficits in social cognition found in autism. Through direct matching of observed and executed behavior an individual can directly experience an internal representation of another's actions, and therefore, another's feelings, goals, and intentions. Dysfunction of this system would hinder this process. In this chapter, Bernier and Dawson review findings from behavioral research examining deficits in three areas of social cognition in individuals with autism: imitation, empathy, and theory of mind. They then discuss research implicating mirror neuron dysfunction in autism spectrum disorders; and explore the mirror neuron theory of autism. Autism research utilizing functional and structural brain imaging, electroencephalography, and transcranial magnetic stimulation to study the mirror neuron system is reviewed. Results from these studies suggest dysfunction of the mirror neuron system in both children and adults with autism. The mirror neuron theory of autism proposes that dysfunction of the execution/observation matching system interferes with the acquisition of internal representation of others' observed behavior, expressions, movements, and emotions. This precludes the individual with autism having an immediate, direct experience of the other through this internal representation. Social impairments, including impairments in imitation, empathy, and theory of mind, are hypothesized to cascade from this lack of immediate, experiential understanding of others in the social world.

Keywords Autism · Social cognition · Mirror neuron · Imitation · Theory of mind · Empathy

---

R. Bernier
Department of Psychiatry and Behavioral Sciences, University of Washington, Seattle, WA 98195
e-mail: rab2@u.washington.edu

J.A. Pineda (ed.), *Mirror Neuron Systems*, DOI: 10.1007/978-1-59745-479-7_12,

1 Introduction

Jimmy sits at the table opposite from his therapist. His therapist calls his name but he does not turn to look at her. He is instead focusing his attention on his fingers as he splays them in front of his eyes. On the third call of his name, Jimmy turns to look toward his therapist and he is rewarded with a cascade of bubbles floating through the air between them. His face erupts in a wide smile. His expression is not directed to his therapist or his mother sitting to the side, but he appears happy. As the last bubble settles on the tabletop and then dissolves in a soapy pop, Jimmy looks longingly at the bubble wand held by his therapist. She draws the wand up to her cheek and he follows willingly. For a split second his eyes stray and make contact with hers. She blows. Another cascade of bubbles rolls out of the wand's mouthpiece and is met by a squeal of delight from Jimmy. The splash of the final bubble's dissolution on Jimmy's nose results in quiet anticipation as Jimmy stares at the bubble wand. Again, his therapist brings it to her cheek and when their eyes meet for the briefest of moments, the task is repeated. Although the focus of today's treatment session is on eye contact, Jimmy has many areas of impairment beyond the social domain, including difficulties in spoken language, non-verbal communication, repetitive behaviors, restricted interests, and play. The other areas of impairment will have to wait. By working on Jimmy's basic social skills – his core impairment, his therapist will provide a foundation for addressing the other challenges facing Jimmy. These challenges will be many as Jimmy has autism.

Autistic disorder is a developmental disorder characterized by a triad of impairments involving the social, communication, and behavioral domains. Autistic disorder, along with Asperger syndrome, and Pervasive Developmental Disorder – Not Otherwise Specified (PDD-NOS), fall under the umbrella of Pervasive Developmental Disorders, commonly referred to as Autism Spectrum Disorders.

First described by Leo Kanner in 1943 and once believed to be a rare disorder, current estimates place the prevalence rate of autism spectrum disorders at 1 child in every 166 (Fombonne, 2003). Diagnoses of strictly defined Autistic Disorder prevalence rates are estimated at 1 in 1000 children (Fombonne, 2003). While the prevalence rate appears to be unrelated to socioeconomic status or ethnic background, males are affected between 3 and 4 times more often than females (Fombonne, 1999).

The etiology of autism is currently unknown but research indicates that there is a significant genetic component in the etiology of autism (Klauck, 2006; Veenstra-Vanderweele, Christian, & Cook, 2004). Indeed, heritability estimates are at 91–93% (Veenstra-Vanderweele & Cook, 2003; Bailey et al., 1995). Promising new research in the genetics of autism provides hope for breakthroughs in understanding the etiology. For example, studies examining endophenotypic traits, those traits that underlie the observable phenotype but that are present in individuals both with and without the disorder provide avenues for determining possible genes. Techniques such as quantitative trait locus analysis, previously used to help identify the chromosomes involved in reading disability are being utilized to study autism; and family members who possess traits similar to individuals with autism, but to a lesser degree, are participating in genetic studies to help understand what is termed as the broader autism phenotype. Through better understanding of the genes involved in autism, there can be a deeper understanding of the brain development and brain function of this neurological disorder.

At the core of these disorders in the autism spectrum are impairments in the social domain. These difficulties vary in both severity and form, and include impairments in eye contact and facial expressions, social interest, peer relationships, and social or emotional reciprocity (APA, 2000). Related to these diagnostic symptoms in the social domain, individuals with autism also have been found to have impairments in imitation (DeMeyer, Alpern, Barons, DeMyer, Churchill, Hingtgen et al., 1972; Williams, Whiten, & Singh, 2004; Dawson & Adams, 1984; Dawson & Lewy, 1989; Rogers & Pennington, 1991; Rogers, Bennetto, McEvoy, & Pennington, 1996; Williams, Whiten, Suddendorf, & Perrett, 2001), theory of mind (Baron-Cohen, Leslie, & Frith, 1985; Baron-Cohen, 1995), and empathy (Baron-Cohen & Wheelwright, 2004; Hobson, Ouston, & Lee, 1988; Sigman, Kasari, Kwon, & Yirmiya, 1992). Clearly, autism can be considered a disorder of social cognition. Recent research in non human primates and more recently in humans, highlights one possible neural mechanism for these impairments of social cognition, a dysfunction in mirror neurons. In this chapter, we will highlight findings from behavioral research examining deficits in three areas of social cognition in individuals with autism, including imitation, empathy, and theory of mind, review the research on implicating mirror neurons in autism spectrum disorders and explore the mirror neuron theory of autism.

2 Impairments of Social Cognition in Autism

2.1 Imitation

In the thirty-year history of imitation research in autism, imitative deficits in individuals with autism have consistently been observed (Williams et al., 2004). Several researchers have suggested that imitation deficits are one of the core impairments of autism (Dawson & Adams, 1984; Dawson & Lewy, 1989; Rogers & Pennington, 1991; Williams et al., 2001). Imitation impairments in autism were first reported by DeMeyer and colleagues (1972) following their study of nine children with autism (and three children with schizophrenia) and 12 age matched control subjects with mental retardation. On a large battery of tasks examining body movement imitation and imitation of actions on objects, the individuals with autism performed significantly poorer. Groups did not differ in performance on non-imitated actions on objects, which indicated that it was not motor deficits that accounted for the observed differences.

Despite the variability in imitation testing methodologies, sample characteristics, and control groups employed, 18 of 20 well-designed studies have found imitative deficits in autism (Williams et al., 2001). In the studies that have failed to find imitation deficits, the results were obscured by ceiling effects and tasks designed for infants that were administered to verbal individuals with autism (Charman & Baron-Cohen, 1994; Morgan, Cutrer, Coplin, & Rodrigue, 1989). Deficits in imitation include impairments of motor movements, facial expressions, actions involving imaginary objects, vocalizations, and imitation of the style in which actions are performed (Rogers, Hepburn, Stackhouse, & Wehner, 2003; Hertzig, Snow, & Sherman, 1989; Sigman & Ungerer, 1984; Loveland, Tunali-Kotoski, Pearson, Brelsford, Ortegon, & Chen, 1994; Hobson & Lee, 1999; Rogers et al., 1996; Dawson & Adams, 1984).

The imitation impairments in autism have been observed very early in life. Children with autism under six years of age show impaired imitation skills compared to children with mental retardation, developmental delay, and communication disorders as well as compared to typically developing children (Sigman & Ungerer, 1984; Stone, Lemanek, Fishel, Fernandez, & Altemeir, 1990; Charman, Swettenham, Baron-Cohen, Cox, Baird, & Drew, 1997, 1998; Stone, Ousley, & Littleford, 1997; Dawson, Meltzott, Osterling, & Rinaldi, 1998; Rogers et al., 2003; Aldridge, Stone, Sweeney, & Bower, 2000). Stone and colleagues (1990) examined imitation abilities in four groups of children (91 children total) between three and six years of age, including children with autism, mental retardation, hearing impairment, and language delay. They found imitative deficits only in the autism group and through discriminant function analysis determined that motor imitation differentiated the autism group from the other groups. In two studies, Stone and colleagues (1997) examined motor imitation abilities in younger children—children under the age of three and a half, and found imitation impairments in 18 toddlers with autism relative to 18 children with developmental delay and 18 typical children. The observed pattern of imitation impairments was similar across groups. They found that body movement imitation was more difficult than imitation of actions on objects and that non-meaningful actions were more difficult to imitate than meaningful actions. Additionally, as part of the second study, they found that body movement imitation was related to expressive language skills both currently and at 1 year follow up.

Dawson and colleagues (Dawson et al., 1998) also found imitation impairments in young children with autism relative to children with Down's Syndrome or typical development. In this battery of fifteen imitation items, of which ten were immediate imitation tasks and five were deferred imitation tasks, children imitated novel (touching elbow to panel) and familiar acts (banging blocks) as well as acts they could see themselves perform (hand opening/closing) and those they could not (mouth opening/closing). The children with autism successfully imitated significantly fewer tasks than both control groups. Moreover, for the autism group, degree of delayed imitation impairment was strongly correlated with a neuropsychological task that tapped the medial temporal lobe but not a task that tapped the dorsolateral prefrontal cortex.

Group differences between five-year old children with autism and children with developmental delay have also been found for both 'procedural' and 'gestural' imitation (Roeyers, Van Oost, & Bothuyne, 1998). While deficits in procedural imitation, which is the imitation of actions on novel objects, were observed, the gestural impairments—the imitation of meaningless gestures, were much greater for the autism group.

More recently, Rogers et al. (2003) examined manual, facial and object-oriented imitation in children less than three years of age. The sample consisted of 24 children with autism, 18 children with Fragile X syndrome, 20 children with other developmental disorders, and 15 children with typical development. The imitation battery consisted of 16 tasks including 7 manual tasks (pat elbow), 5 facial tasks (make a 'noisy kiss'), and 4 novel action on object tasks (turn car upside down and pat it). Compared to the chronological age, verbal ability and nonverbal ability matched control groups, the autism group showed significantly greater impairments on the overall measure of imitation and also on the subscales examining oral-facial imitation abilities, and imitation of novel actions on objects. Social cooperation and overall motor ability were also assessed and results indicated that the imitation impairments were not accounted for by a general motor impairment nor were they due to decreased social cooperation.

The imitation deficits observed in young children persist into school age (Hammes & Langdell, 1981; Ohta, 1987, Smith & Bryson, 1998; Green et al., 2002; Jones & Prior, 1985). In a study examining non-meaningful gestures and neurological signs, Jones and Prior (1985) found impairments of gestures relative to chronological age and mental age matched control children. Ohta (1987) also looked at gestures, both meaningful and non-meaningful, in ten year-old children with autism, and reported what are now termed reversal errors. These are errors in which the imitator replicates the gesture but in the perspective of how the gesture was observed. For example, if an experimenter facing a subject across the table presents a palm facing outward away from the body, the subject making the reversal error would direct her palm so that it faces herself. These reversal errors were also observed by Smith and Bryson (1998) in a study examining meaningful and non-meaningful gestures as well as single and multi step gesture sequences. They found greater impairment in imitating single step gestures in the children with autism as compared to the language impaired and typically developing controls, but found that the autism group made no more sequence errors than the other groups. However, the autism group made significantly more reversal errors.

Green and colleagues (2002) compared nine year-old children with Asperger Syndrome to children with Development Coordination Disorder (DCD) on tests of motor planning and gesture imitation. They found the children with Asperger Syndrome showed motor impairments that were sufficient to meet criteria for DCD. However, above and beyond the motor impairments, they found that the children with Asperger Syndrome performed significantly worse than the comparison group on the gesture imitation tasks.

Imitation impairments in autism have been found to persist into adolescence and adulthood (Rogers et al., 1996; Avikainen et al., 2003; Hobson & Lee, 1999; Bernier, Dawson, Webb, & Murias, 2007). Hobson and Lee (1999) found in a sample of 16 teens with autism and 16 controls, group matched on age and verbal ability, that individuals with autism were able to imitate, both immediately and after a delay, goal directed actions. However, the teens with autism failed to imitate the style in which the examiner performed the task with greater frequency than the control group. Moreover, once again, considerable reversal errors were noted in the autism group.

Avikainen et al. (2003) examined imitation in adults with Asperger Syndrome and High Functioning Autism. In this study, while sitting across from the experimenter, participants needed to imitate, as simultaneously as possible, grip type while placing a pen in one of two cups and in either a crossed or mirror image hand condition. In the crossed condition, participant used the same (right or left) hand as the experimenter. In the mirror image condition, the participant used the hand directly opposite (i.e., participant's left equaled experimenter's right). Typical adults will benefit from using the mirror image of the experimenter's actions and show fewer errors in this condition. The authors found that individuals with autism failed to capitalize on the mirror image of the experimenter's presentation of an action and demonstrated significantly more errors in this condition.

In a well designed study of adolescents, Rogers and colleagues (1996) tested two competing hypotheses regarding the imitation impairments in autism—the symbolic meaning and executive functioning hypotheses. A symbolic deficit posits that individuals should show differential impairment relative to the meaning of the gesture. That is, difficulties should be observed when the subject has to imitate movements with symbolic content. Following the executive functioning hypothesis, subjects should demonstrate greater impairments with sequenced tasks over single step tasks. To test these hypotheses, Rogers and colleagues examined imitation abilities varied by meaning and sequence in high functioning adolescents with autism and developmental delay. Tasks consisted of hand and face imitation tasks that were either single step or sequential. Additionally, the imitated movements were either meaningful (e.g., shaking hands over head as though a champion) or non meaningful (e.g., extending hand out with fingers extended and with thumb pointed upward). Motor tasks, memory tasks, and pantomime tasks were also administered. The autism group performed more poorly on the imitation and pantomime tasks with meaning aiding rather than hindering performance and some sequential tasks proving more difficult than single step tasks. Similar memory and motor task performance between groups indicated that the poorer performance is not due to motor difficulties or visual recognition memory impairments. Moreover, each participant attempted each imitation task so the poor performance was not a result of un-attempted trials. The authors concluded that there was no support for the symbolic meaning hypothesis and only partial support for the executive functioning hypothesis.

Using a similar paradigm also developed by Rogers, Cook, and Greiss-Hess (2005), Bernier and colleagues (2007) examined imitation ability in adults with autism. Fourteen adults with autism and 15 age, gender, and IQ matched controls were videotaped and coded by a rater blind to diagnostic status during the imitation of single and multi-step hand and face gesture, two-hand gestures, sequenced movements, and actions on objects. Across all types of imitative acts, the adults with autism demonstrated continued impairments in imitation ability compared to typical adults.

In summary, the research literature consistently points to a persistent and pervasive deficit in imitation ability in individuals with autism. This impairment is most marked by deficits in imitation of non-meaningful gestures and imitation of novel actions (Williams et al., 2004). Following earlier proposals by Rogers et al. (1991) and Hobson and Lee (1999), Williams recently concluded that the accumulated evidence suggests that a deficit in self-other mapping is at the core of the imitation impairment (Williams et al., 2004). Reversal errors and impairments in imitating the style of actions provide additional evidence for support of this hypothesis (Ohta, 1987; Smith & Bryson, 1998; Hobson & Lee, 1999).

Mirror neurons, in the role as an execution/observation matching system, are a mechanism for coordinating a representation of the self with another. A number of studies utilizing fMRI support this notion (Iacoboni et al., 1999; Decety, Chaminade, Grezes, & Meltzoff, 2002; Buccino, Vogt, Ritzl, Fink, Zilles, Freund et al., 2004; Iacoboni, 2005; Molnar-Szakacs, Iacoboni, Koski, & Mazziotta, 2005; Jackson, Brunet, Meltzoff, & Decety, 2006). As a result, dysfunction of the mirror neuron system has been proposed as an impairment that can explain the deficits in imitation found in autism (Williams et al., 2001).

2.2 *Empathy*

Empathy, the ability to experience the affective experience of others (while still recognizing this is the other's affective experience), is another important aspect of social cognition. Two related components of empathy: attention to others' affect and recognition of affect and expression have been found to be impaired in individuals with autism.

Since Leo Kanner's first description of autism in 1943 in which impairments in affective recognition were noted, a number of studies have shown that individuals with autism show atypical responses to others' affective displays. Children with autism attend differently to affective displays than children with developmental delay or typical development. For example, in experimental conditions in which an experimenter cries in response to a feigned injury, children with autism are more likely to continue playing with a toy and attend less to the crying experimenter compared to control children (Sigman, Kasari, Kwon, & Yirmiya, 1992; Charman et al., 1998). The responses appear to be stable and predictive of later responsiveness to affect (Dissanayake, Sigman, Kasari, & Bagley, 1996). In a similarly designed experiment, Dawson and colleagues found that children with autism attended less when the affective display was confusing such as a neutral expression as opposed to an expression of distress in response to a feigned injury (Dawson, Meltzoff, Osterling, & Rinaldi, 1998).

Several studies have also examined affect recognition abilities in individuals with autism. When asked to label emotions displayed by videotaped children telling stories and to label their own emotional response, children with autism performed significantly worse than age matched typically developing children (Yirmiya, Sigman, Kasari, & Mundy, 1992). In another study examining affect recognition, children with autism performed similarly to typical children in labeling emotionally ambiguous pictures but showed greater difficulty in discussing specific emotions of pride and embarrassment (Capps, Yirmiya, & Sigman, 1992). Adolescents and adults with autism and without mental retardation demonstrate greater impairments in matching emotional faces with prosodic voices compared to non-emotional pictures and related sounds, relative to typical adults (Hobson et al., 1988). When matching pictures of individuals with differing facial expressions with corresponding contexts, children with autism also demonstrated significant impairment (Hobson, 1986).

Imaging technology also indicates differential responding to affect recognition in individuals with autism at the neurological level. Using PET technology, Hall and colleagues (Hall, Szechtman, & Nahmias, 2003) found that in response to the matching of an emotional expression and a prosodic voice, adults with autism showed differential areas of regional blood flow in the brain from typical controls. Using ERP methodology, Dawson and colleagues (Dawson, Webb, Carver, Panagiotides, & McPartland, 2004) examined emotional expression processing in three year old children with autism, developmental delay, and typical development. De Haan and colleagues (De Haan, Belsky, Reid, Bolein, & Johnson, 2004) had previously shown that 6–7 month old typically developing infants show a differential ERP to a fear versus neutral face. During a passive viewing paradigm in which the children viewed pictures of faces showing a prototypic expression of fear a neutral expression, ERP activity was recorded via scalp electrodes. The children with autism failed to show differential ERP to the fear versus neutral face. Furthermore, longer latency at the early ERP component in response to the fear face was correlated with decreased attention to an experimenter feigning distress on a separate testing day. Importantly, latency of ERP response to the emotional expression was not correlated to non social attention, such as orienting to a non social sound. The authors concluded that the processing of emotional expressions is impaired in autism as early as three years of age and that abnormal facial expression processing is related to general impairments in social attention.

Empathy has also been measured through physiological measures such as heart rate. Children with autism did not show a change in heart rate in response to an experimenter's feigned injury and distress as control children with mental retardation did, despite similar increase in attention to the injured experimenter in both groups (Corona, Dissanayake, Arbelle, Wellington, & Sigman, 1998).

In an effort to examine the concept of empathy itself, Baron-Cohen and Wheelwright developed the Empathy Quotient. This instrument is a self report questionnaire consisting of 60 statements such as "seeing people cry doesn't upset me," to which respondents indicate their agreement with the statement on a four point scale. Using this measure with a sample of 90 high functioning adults with autism and 90 adult controls, the authors found that adults with autism spectrum disorders performed significantly worse than typical controls at empathic understanding (Baron-Cohen & Wheelwright, 2004).

Recent research suggests that the mirror neuron system plays a role in empathy (Gallese, Keysers, & Rizzolatti, 2004; Gallese, 2006). Functional imaging studies support this notion through findings of similar activation when experiencing disgust or observing others expressing disgust (Wicker et al., 2003); when observing or presenting facial expressions (Carr Iacoboni, Dubeau, Mazziotta, & Lenzi, 2003); when being touched or observing another being touched (Keysers et al., 2004); and when experiencing pain or observing a loved one experiencing pain (Singer et al., 2004). Given that consistent impairments in aspects related to empathy have been shown to be disrupted in autism and that the mirror neuron system may subserve empathy; this provides further support for a role of mirror neuron dysfunction.

2.3 Theory of Mind

The term 'theory of mind' refers to the awareness that others have beliefs and intentions that are different from our own and that these beliefs and intentions guide or direct others' behavior. The term theory of mind, first proposed in work with non-human primates, was adopted by developmental psychology to refer to the metarepresentational abilities that emerge in typically developing children (Premack & Woodruff, 1978; Leslie, 1987). This ability to impute mental states to others is a crucial aspect of social cognition because it allows for the understanding of others' beliefs, goals, and intentions and, importantly, allows for the prediction of what others will do in a given situation (Baron-Cohen et al., 1985).

By the age of four, this 'mentalizing' ability has been reliably demonstrated in children through experimental tasks. The basic theory of mind task requires that the child predicts a response based on a false belief. As shown in Fig. 1, the child observes two models. One of the models places an object in a hiding spot

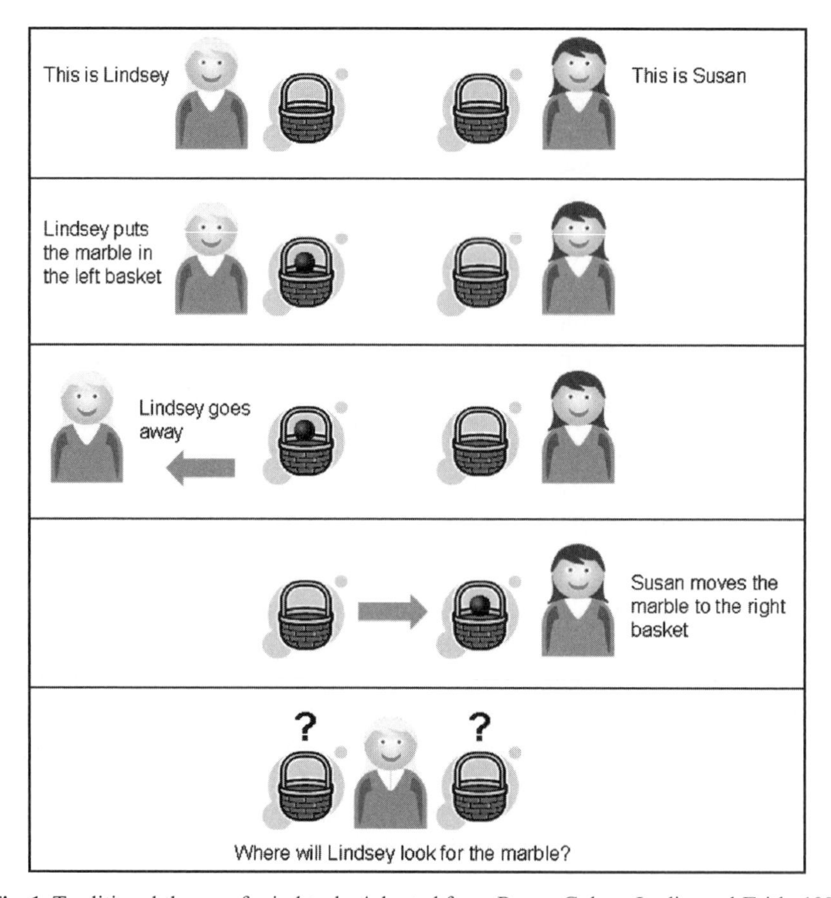

Fig. 1 Traditional theory of mind task. Adapted from Baron-Cohen, Leslie, and Frith, 1986

and then leaves. The second model moves the object to a new hiding spot. When the first model returns, the child must indicate where the first model will look for the object. In order to respond correctly, the child must be aware of the fact that the first model did not see the object being moved and therefore has the false belief that the object is where he/she originally placed it. Utilizing these experimental paradigms, Wimmer and Perner (1983) outlined the developmental progress of the ability to relate another's wrong belief to one's own knowledge. Prior to age four, performance on these tasks is inconsistent with some children showing this ability while others do not. However, between the ages of 4 and 6 years, this ability emerges and becomes consistent in typically developing children. This 'mentalizing' ability appears to be impaired in individuals with autism (Baron-Cohen, 1997).

Utilizing the classic theory of mind paradigm with dolls as agents, Baron-Cohen and colleagues demonstrated that 11 year-old children with autism failed to predict the doll's behavior based on the doll's belief (Baron-Cohen et al.,

1985). Instead, the children with autism consistently identified the location of the hidden object based on their knowledge, not the doll's belief. In contrast, chronologically age matched children with Down's Syndrome and 4 year-old typically developing children correctly predicted the doll's behavior based on the doll's belief. They, and not the children with autism, demonstrated a theory of mind ability. To further support their findings, the authors noted that the autism group's cognitive ability was higher than both control groups, highlighting the stricter control on the hypothesis that IQ might account for the theory of mind impairments. The authors reasoned that it was not general cognitive ability that accounted for these deficits.

Subsequent studies examining theory of mind abilities and children with autism have further explored the nature of this theory of mind deficit. For example, while children with autism demonstrate impaired performance on false belief tasks, they perform similarly to control groups on false photograph tasks (Charman & Baron-Cohen, 1992, Leekam & Perner, 1991). These are tasks that are modeled on the false belief paradigm. However, instead of false beliefs, drawings or photographs that do not conform to the presented situation are utilized. For example, a child is taught how to use a Polaroid camera. The child then takes a picture of an object placed on a table. While the photograph is developing, the object is moved to a different location on the table and the child is asked to predict in which location the object will be pictured in the photograph. In typically developing children, performance on this task is similar to performance on the false belief task (Zaitchik, 1990). In a replication of Zaitchik's (1990) paradigm, Leekam and Perner (1991) found similar performance on false belief and false photograph tasks for 3 and 4 year typically developing children, replicating Zaitchik's results. However, the adolescent participants with autism in this study performed significantly better, nearly perfectly, on the false photograph task, but showed false belief performance similar to the typically developing children. These studies suggest that in autism the deficit is specific to mental representations and not pictorial representations.

Studies have shown that not all individuals with autism show impairments in the traditional theory of mind tasks (Happe, 1994; Ozonoff, Pennington, & Rogers, 1991). However, performance on the traditional theory of mind tasks that were developed for young children does not indicate that these adults with autism that successfully perform the tasks have typical theory of mind ability (Rutherford, Baron-Cohen, & Wheelwright, 2002). To explore this, several 'advanced' theory of mind tasks have been developed. Baron-Cohen and colleagues developed the Reading the Mind in the Eyes task and examined performance on this measure in adults with autism and age matched adults with typical development or Tourette Syndrome (Baron-Cohen, Jolliffe, Mortimore, & Robertson, 1997; Baron-Cohen, Wheelwright, Hill, Raste, & Plumb, 2001). In this task, participants were asked to determine mental states by looking at photographs showing only people's eyes. Two control tasks were also administered. In one task, participants were also asked to determine the gender of the people depicted in the photographs, and in the second, participants were asked to determine the emotional expression from pictures of the whole face. The adults with autism performed significantly worse than the control groups on the mind in the eyes task while there were no differences in performance on the two control tasks.

Similar results were found on a task in which participants listened to a short dialogue and chose one of two adjectives to describe the speaker's mental state. High functioning adults with autism performed significantly worse than typical adults on this Reading the Mind in the Voice task (Rutherford, Baron-Cohen, & Wheelwright, 2002; Golan, Baron-Cohen, Hill, & Rutherford, 2006).

Another advanced theory of mind task is the Strange Stories Task (Happe, 1994). In this task participants are presented with a vignette in which people say things that they do not mean in the literal sense. For example, a story might describe a couple on a picnic during which it rains and one person says, "it's a perfect day for a picnic." High functioning individuals with autism and Asperger Syndrome perform significantly worse on this task compared to control groups (Happe, 1994; Jolliffe & Baron-Cohen, 1999).

Yet another related theory of mind task is the Faux Pas task. It is similar to the strange stories task in that participants are presented with short vignettes, but in this task participants are asked to identify the faux pas. A statement was considered a faux pas in the vignettes if the speaker said something without considering that it might not be something that the listener wanted to hear or know such as introducing someone with the incorrect name. While school aged children with typical development consistently, correctly identified the faux pas, the high functioning children with autism performed significantly worse, despite having performed successfully on traditional theory of mind tasks (Baron-Cohen, O'Riordan, Stone, Jones, & Plaisted, 1999).

Deficits in theory of mind ability appear to be specific to the nature of impairments observed in autism. Children with other developmental disorders, such as specific language impairment (Perner, Frith, Leslie, & Leekam, 1989), Down's Syndrome (Baron-Cohen, Leslie, & Frith, 1985), and Williams Syndrome (Karmiloff, Klirna, Bellugi, Grant, & Baron-Cohen, 1995), while possessing a range of cognitive and attentional deficits, do not show the theory of mind impairments noted in autism. Children with autism also perform significantly worse than children with psychiatric control groups diagnosed with dysthymia and conduct disorder on theory of mind tasks (Buitelaar, van der Wees, Swaab-Barneveld, & van der Gaag, 1999). The observed impairments on these tasks in individuals with autism also correlate with social functioning impairments and pragmatic language impairments (Baron-Cohen, 1991). Further, siblings of children with autism also perform significantly worse than age-, gender-, and verbal ability-matched children on theory of mind tasks, such as the Reading the Mind in the Eyes task (Dorris, Espie, Knott, & Salt, 2004).

Research suggests that theory of mind abilities are linked to a number of brain regions, including the mirror neuron system in humans. Given the role that mirror neurons play in the understanding of goals and the simulated representation of others, the mirror neuron system offers a mechanism for theory of mind abilities (Gallese & Goldman, 1998). However, other regions have been implicated as well, including the anterior cingulate cortex (Vogeley, Bussfield, Newen, Herrmann, Happe, Falkai et al., 2001; Gallagher & Frith, 2003), the medial frontal gyrus (Fletcher, Happe, Frith, Baker, Dolan, Frackowiak et al., 1995), the temporal lobe (Saxe & Kanwisher, 2003; Nieminen-von Wendt, Metsahonkala, Kulomaki, Aalto, Autti et al., 2003), and the frontal lobe (Stone, Baron-Cohen, & Knight, 1998) indicating that the mirror neuron system is just one component in the theory of mind circuitry (Frith & Frith, 1999; Brune & Brune-Cohrs, 2006).

3 Mirror Neurons and Autism

Mirror neurons, neurons that activate both when an individual is executing and observing an action, were first identified in monkeys using single electrode recording (Gallese, Fadiga, Fogassi, & Rizzolatti, 1996). In humans, evidence for the presence of a mirror neuron system has been supported indirectly through non-invasive techniques consisting of functional Magnetic Resonance Imaging (fMRI), Electroencephalogray (EEG), and Transcranial Magnetic Stimulation (TMS). The number of studies that consistently report an execution/observation matching system in humans provides convincing evidence of a mirror neuron system in humans (Rizzolatti & Craighero, 2004). Because mirror neurons provide a neurological mechanism for translating visual information about others into a direct translation of the self, they have been proposed to serve as a mechanism for self-other mapping and social cognition abilities, including imitation, empathy, and theory of mind. Because the core impairments in autism fall under the social domain, dysfunction of the mirror neuron system is a potential neural mechanism to account for the social impairments in autism.

Williams et al. (2001) first proposed that mirror neuron dysfunction underlies the imitation impairments in autism. Recently, in response to the interest in this area of autism, several studies have been conducted in individuals with autism (Altschuler, Vankov, Hubbard, Roberts, Ramachandran, & Pineda, 2000; Shenk, Hubbard, McCleery, Ramachandran, & Pineda, 2004; Theoret, Halligan, Kobayashi, Fregni, Tager-Flusberg, Pascual-Leone, 2005; Avikainen, Kulomaki, & Hari, 1999; Williams, Waiter, Gilchrist, Perrett, Murray, & Whiten, 2006; Oberman, Hubbard, McCleery, Altschuler, Ramachandran, & Pineda, 2005; Dapretto et al., 2006; Hadjikhani, Joseph, Snyder, & Tager-Flusberg, 2005; Bernier et al., 2007).

The first such study examining mirror neuron functioning in autism failed to find any differences between the autism group and controls (Avikainen et al., 1999). In the study, four adults with Asperger's Syndrome and one with autism were tested on a Theory of Mind task during magnetoencephalographic (MEG) recording. While manipulating an object with the hand or while viewing another person manipulating an object the same way, the oscillatory activity in the primary motor cortex was recorded during median nerve stimulation. Although the autism group showed impaired Theory of Mind abilities relative to the control group, there were no differences in MEG activity measured over the motor cortex between groups. The authors concluded that in autism spectrum disorders, the Theory of Mind deficits are not related to any impairment in mirror neuron activity in the motor cortex. The authors suggest examining activity over other regions of the brain, such as Broca's area, will be important for clarifying the functioning of the mirror neuron system in individuals with autism. In this small sample of individuals diagnosed without the rigorous gold standard autism diagnostic measures, mirror neuron functioning, as indexed using MEG, appears intact. Williams et al. (2001) propose that the small sample size and small effect size led to decreased power and an increased probability of type II error.

A recent study using transcranial magnetic stimulation (TMS) in individuals with autism found significantly lower excitability in the primary motor cortex as assessed through motor evoked potentials compared to controls during the observation of meaningless hand movements (Theoret et al., 2005). In ten high functioning adults with autism and ten age and gender matched controls, TMS was administered over the motor cortex. Motor evoked potentials were recorded from muscles in the index finger and thumb while participants viewed index and thumb movements directed toward or away from the observer. An interesting pattern emerged. In the control group, the observation of finger and thumb movements facilitated MEPs in the respective muscles regardless of the direction of the movement. For the autism group, when the hand was directed away from the observer there was no differential muscle MEP activation between thumb and finger movements. However, when the hand movement was directed toward the observer, the same pattern of differential facilitation of the MEPs seen in controls was observed. The authors rule out any potential underlying atypical motor cortex activation patterns by showing no differences between groups on a variety of neural motor comparisons. They conclude that a faulty self other representation system, potentially due to mirror neuron dysfunction, underlies these findings and the social abnormalities characteristic of autism.

A recent fMRI study provides stronger evidence for mirror neuron dysfunction in autism spectrum disorders (Dapretto et al., 2006). Nine male children with autism and 10 age and IQ matched control children imitated and observed facial expressions while undergoing fMRI. The school aged children either imitated or observed 80 faces expressing different emotions including anger, fear, happiness, sadness, or neutrality. Prior to conducting the scan, the participants imitated the faces and no differences in attending, task compliance, and imitation performance were found between the groups. While the typical children demonstrated mirror neuron system activity similar to that found in fMRI studies in adults, the autism group failed to show activation in the inferior frontal gyrus (Brodmann's area 44). Furthermore, the activity in this brain region that has consistently been identified as a core component of the mirror neuron system was inversely correlated with autism symptom severity.

A second fMRI study has recently been conducted in high functioning individuals with autism (Williams et al., 2006). Sixteen adolescent males with autism and 15 matched control teens executed and imitated simple finger movements, following the procedure developed by Iacoboni et al. (1999). Both groups showed right parietal lobe activity during imitation similar to that found in Iacoboni et al. (1999) study. The activation in the autism group was less extensive than the control group. However, the autism group showed less activation in the non-imitation execution condition in this region as well. The authors suggest possible poor thalamo-cortical connectivity which impacts proprioceptive feedback could account for this unusual finding. Additionally, the autism group showed greater activation than the control group for regions of the dorsal premotor and prefrontal cortices when imitating. The authors theorized that the individuals with autism were relying on additional strategies, such as visual experience and learning to guide the imitation. Both fMRI studies provide support for the hypothesis of an impaired mirror neuron system in individuals with autism.

A structural MRI study conducted in 14 high functioning adults with autism found that compared to matched controls, the individuals with autism showed decreases in gray matter in the mirror neuron system (Hadjikhani, Joseph, Snyder, & Tager-Flusberg, 2005). Additional regions of cortical thinning were also found in brain regions associated with other aspects of social cognition such as the superior temporal sulcus and medial prefrontal cortex among others. The amount of cortical thinning in the mirror neuron system regions was correlated with autism symptom severity as assessed by parent interview of early development. The authors suggested that early disruption of the mirror neuron system could impact the development of other areas of the brain involved in social cognition. This abnormal trajectory would thereby disrupt the behavioral presentation of the child with autism.

Four EEG studies focusing on the mu rhythm have also been conducted to examine mirror neuron activity in individuals with autism. The mu rhythm is an EEG wave oscillating in the 8–13 Hz frequency band recorded at centrally located scalp electrodes. At rest, the underlying cell assemblies fire synchronously but when executing an action, the underlying cells become active and fire asynchronously. This attenuates the mu rhythm amplitude. Importantly, when observing an action, attenuation of the mu rhythm is also observed. As a result, attenuation of the mu rhythm has been proposed to reflect mirror neuron activity (Muthukumaraswamy, Johnson, & McNair, 2004; Pineda, 2005).

In a preliminary study focusing on a child with autism and his fraternal twin without autism, attenuation of the mu wave was found for the execution of movement but not for the observation of movement (Altschuler et al.,

2000). In a second EEG study with a small sample size, individuals with high functioning autism failed to show mu wave suppression while watching others' movements but did show characteristic mu wave suppression while executing movements (Shenk et al., 2004). Recently, Oberman and colleagues published an EEG study reporting on 10 males with autism ranging from 6 to 47 years of age and 10 age and gender matched controls (Oberman et al., 2005). They recorded EEG activity from two electrodes (one in the left hemisphere and one in the right) while participants executed a hand movement (opening and closing the hand), watched a video of a moving hand, watched a video of a bouncing ball, or watched a static screen. They found attenuation of the mu rhythm during the execution and hand observation conditions for the control group but only during the execution condition for the autism group. They conclude that there is dysfunction of the mirror neuron system in individuals with autism.

Bernier and colleagues (2007) replicated Oberman et al.'s findings (2005) of decreased attenuation of the mu rhythm during the execution of biological movement but not during the observation of movement in individuals with autism and extended the findings through correlations with behaviorally assessed imitation skills. Following a paradigm developed by Muthukumaraswamy and colleagues (2004), 14 adults with autism and 15 age, gender, and IQ matched controls, were seated at a table facing a computer monitor. A wooden block that sent a signal to the EEG recording equipment when grasped, called a manipulandum, was placed on the table in front of the participants. While EEG data were being collected participants either executed or observed actions in four different conditions. During the observe conditions subjects simply watched a digital video of the experimenter grasping the manipulandum. During the execute condition, participants were verbally instructed to grasp the manipulandum on the table in front of them. During the imitate condition, participants imitated the experimenter grasping the manipulandum as shown on the digital video. The fourth condition was a resting condition in which participants sat quietly with their eyes open. The typical individuals demonstrated attenuation of the mu wave both when executing and observing the simple grasping motion of the manipulandum. The individuals with autism demonstrated similar levels of attenuation when executing the movement, but showed decreased levels of attenuation when observing the movement. In addition to replicating the findings by Oberman and adding to the support of a dysfunctional execution/observation matching system in individuals with autism, Bernier and colleagues compared the degree of EEG attenuation when observing actions to behaviorally assessed imitation skills in this sample of adults. As shown in Fig. 2, a positive correlation was found between imitation ability and level of attenuation. That is, as imitation ability increased so did the degree of mu rhythm attenuation when observing movements.

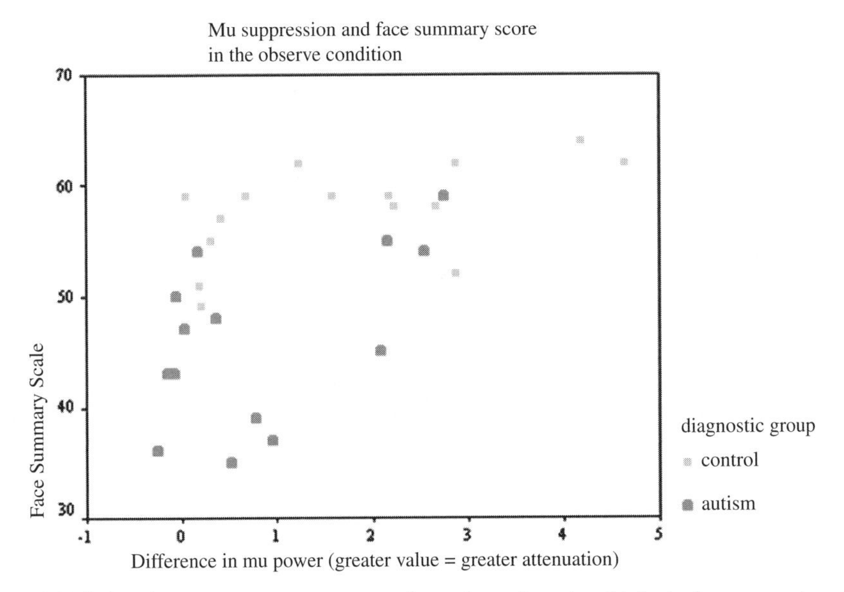

Fig. 2 Relation between mu wave attenuation when observing biological movement and behaviorally-assessed facial imitation ability

4 A Mirror Neuron Theory of Autism

The evidence from EEG, fMRI, and TMS studies suggests there is dysfunction of the mirror neuron system in children and adults with autism. The mirror neuron theory of autism simply proposes that dysfunction of the execution/observation matching system does not allow for the internal representation of others' observed behavior, expressions, movements, and emotions. This prevents the individual with autism from having an immediate, direct experience of the other through this internal representation. Social deficits, including deficits in imitation, empathy, and theory of mind, therefore cascade from this lack of immediate, experiential understanding of others in the social world (Williams et al., 2001; Iacoboni & Dapretto, 2006; Ramachandran & Oberman, 2006).

A disruption of the connections between the mirror neuron circuit and brain regions related to the inhibition of imitative acts could also account for another aspect of autism symptomology - echolalia and repetitive behaviors. Brass and Heyes report that portions of the anterior fronto-median cortex and the right temporo-parietal junction are activated when participants undergoing fMRI are asked to perform an action while observing a distinct action (Brass & Heyes, 2005). Portions of the right inferior parietal cortex show increased activation when participants undergoing a PET scan are imitated by another person (Decety, Chaminade, Grezes, & Meltzoff, 2002). Thus, Brass and Heyes conclude, the inhibition of imitative acts is related to self other differentiation, rather than inhibition specifically. Immediate and delayed echolalia, the echoing of verbal utterances of another person either immediately following the presentation or after some delay, is often reported in individuals with autism. An activated mirror neuron system that is ineffectively communicating with brain regions responsible for inhibition of imitation might result in these types of behaviors. Despite this, the mirror neuron theory is not able to account for all the symptoms observed in autism, such as an insistence on sameness or interests of unusual intensity (Fecteau, Lepage, & Theoret, 2006).

The precise nature of mirror neuron dysfunction has yet to be understood. The mirror neuron system could be functioning appropriately, yet receiving faulty input from upstream neural circuitry. For example, non human primate and human studies implicate the superior temporal sulcus in the perception of biological movement (Zilbovicius, Meresse, Chabane, Brunelle, Samson, & Boddaert, 2006) and there is evidence to suggest that there is a disruption biological motion processing, specifically the processing related to the superior temporal sulcus in individuals with autism (Dakin & Frith, 2005; Pelphrey, Morris, & McCarthy, 2005; Boddaert, Chabane, Gervaise, Good, Bourgeouis, Plumet et al., 2004; Waiter, Williams, Murray, Gilchrist, Perrett, & Whiten, 2004). Therefore, the reduced mirror neuron activity found in individuals with autism – as assessed through fMRI and EEG – could be a function of faulty input from the visual processing system, not mirror neuron dysfunction per se. However, in contrast to these conclusions, a study using MEG and MRI in adults with Asperger Syndrome and adult controls found normal activation of early cortical circuitry in the visual system, including the occipital, parietal and superior temporal sulcus regions (Nishitani, Avikainen, & Hari, 2004). According to this study, in which individuals imitated lip movements presented as still images, activation progressed from the occipital region through the superior temporal sulcus, through the inferior parietal lobe, to the inferior frontal lobe, and then to motor regions. The differences between the autism group and controls were in circuitry further downstream in the process, leading from the inferior parietal lobe (IPL) to Broca's area and the motor cortex. These differences included longer latencies between the IPL and inferior frontal regions and weaker activation in the inferior frontal lobe and the motor areas. The authors concluded that disruptions in the visual processing system could not account for mirror neuron system abnormalities. It is important to note that the stimuli used in this study were still images so it is difficult to determine if the typical activity observed in the STS would be found if the stimuli were moving images.

While a structural MRI study found differences in the cortical thickness of the mirror neuron system in individuals with autism (Hadjikhani et al., 2005), it is possible that these structural differences are the result of development following from faulty visual processing circuitry. Similarly, while two functional MRI studies of individuals with autism have found decreased activation when observing and imitating actions (Williams et al., 2006; Dapretto et al., 2006), it could be the result of disruptions in visual processing of biological motion and the signals sent to the mirror neuron system from this circuitry. Further research is necessary to clarify the role that the superior temporal sulcus and the processing of biological motion plays in observing and imitating actions, the mirror neuron system, and social deficits in autism.

Similarly, faulty attentional circuitry could impact the input to the mirror neuron system in autism. Research has shown that individuals with autism, even as young as 8 months old, demonstrate reduced attention to social information (Osterling & Dawson, 1994; Dawson, Meltzoff, Osterling, Rinaldi, & Brown, 1998; Dawson et al., 2004). This decreased social attention could result in subsequent dysfunction of differential activation of the mirror neuron system. The cerebellum and the amygdala, as well as other brain regions such as the fusiform gyrus and ventromedial prefrontal cortex, have also been implicated in autism and understanding the interconnections between these brain regions and the mirror neuron system will be necessary to understand the nature of the neural basis of autism.

The outstanding questions concerning the input to the mirror neuron system are the next step in better understanding the differential activity of this system in individuals with autism. The most parsimonious explanation is that disruptions to multiple circuits acting in concert result in the pattern of social impairments seen in autism.

5 Conclusion

Although many questions remain concerning the mirror neuron system in humans and in individuals with autism, the evidence compellingly indicates differential activity of the brain regions that comprise this execution/observation matching system. At a theoretical level, the mirror neuron theory of autism deserves significant consideration because of the role of the mirror neurons in self-other mapping. At an empirical level, support for this theory has grown through recent imaging and electrophysiological experiments linking the mirror neuron system to aspects of social cognition such as imitation, empathy, and theory of mind. Further, the growing number of studies demonstrating differential activity in this system in both children and adults with autism provide additional evidence to support this theory.

As well as providing important insight into the nature and conceptualization of autism, this theory has implications for early detection and intervention in autism. Early intensive behavioral intervention is successful and effective in children with autism, and the earlier the intervention begins, the greater the effectiveness (Dawson & Zanolli, 2003). However, in order for intervention to begin, children with autism need to be identified and diagnosed. Differential

EEG activation in the mu wave, specifically reduced attenuation when observing biological motion, may serve as a possible indicator for children with autism. Work is currently underway to examine whether the mu rhythm can be reliabily measured in young infants, and if so, whether abnormalities in mu activation are found in infants at risk for developing autism. If decreased attenuation is reliably observed in infants with autism, it is possible that it could be used as a marker in at risk children.

In the area of intervention, work is also being done related to the mirror neuron system. The mu rhythm can be trained through neurofeedback and it is possible that modification of the mu rhythm could impact mirror neuron functioning and therefore social cognitive abilities. Initial findings from pilot work suggest that individuals with autism can modify their mu rhythm through neurofeedback and that imitation skills can improve, at least temporarily (Pineda, 2006). Further work is clearly needed.

Possible pharmacological therapies are implicated by the mirror neuron theory of autism as well. It has been hypothesized that neuromodulators can impact mirror neuron activity (Ramachandran & Oberman, 2006). As a result, through the use of medications that can enhance the impact of neuromodulators or enhance their impact on the mirror neurons, it maybe possible to increase mirror neuron activity and the concomitant social cognitive abilities. Again, further work is necessary to explore this as a possible intervention for mirror neuron dysfunction.

Finally, direct intervention through stimulating and teaching imitative skills as a component of early behavioral intervention has been shown to be effective for improving imitation ability. Indeed, imitation skills are a core focus of virtually all early intervention programs (Rogers, 1998; Dawson & Osterling, 1994; Smith, Rogers, & Dawson, 2008). Imitation is considered a pivotal skill in early intervention because the ability to imitate others opens the door to many other forms of social learning and communication (Koegel, Koegel, Harrower, & Carter, 1999).

Our understanding of autism has progressed substantially since Leo Kanner first described the disorder in 1943. Advances in our understanding of the social, communication, and behavioral deficits have aided in our detection, treatment, and conceptualization of the disorder while technological advances in genetics and brain imaging have allowed for greater understanding of the deficits at a neural systems and molecular level. The mirror neuron system appears to play an important role in autism, but there is much work yet to be done to understand its nature and extent.

Acknowledgments The writing of this chapter was funded by grants from the National Institute of Child Health and Human Development (U19HD34565, P50HD066782, and R01HD-55741) and the National Institute of Mental Health (U54MH066399). Send correspondence to Raphael Bernier, Ph.D. at University of Washington, Box 357920, Autism Center, Seattle, WA 98195.

References

Aldridge, M., Stone, K., Sweeney, M., & Bower, T. (2000). Preverbal children with autism understand the intentions of others. *Developmental Science*, *3*, 294–301.

Altschuler, E., Vankov, A., Hubbard, E., Robers, E., Ramachandran, V., & Pineda, J. (2000). Mu wave blocking by observation of movements and its possible use as a tool to study theory of other minds. *Society of Neuroscience*, 68.1 (abstract).

American Psychiatric Association. (2000). *Diagnostic and statistical manual of mental disorders* (4th ed., Text Revision). Washington, DC: APA.

Avikainen, S., Kulomaki, T., & Hari, R. (1999). Normal movement reading in Asperger subjects. *Neuroreport*, *10*, 3467–3470.

Avikainen, S., Wohlschlager, A., Liuhanen, S., Hanninen, R., & Hari, R. (2003). Impaired mirror-image imitation in Asperger and high-functioning autistic subjects. *Current Biology*, *13*, 339–341.

Bailey, A., Le Couteur, A., Gottesman, I., Bolton, P., Simonoff, E., Yuzda, E., et al. (1995). Autism as a strongly genetic disorder: evidence from a British twin study. *Psychological Medicine*, *25*, 63–77.

Baron-Cohen, S. (1991). Precursors to a theory of mind: Understanding attention in others. In A. Whiten (Ed.), *Natural theories of mind*. Oxford: Blackwell.

Baron-Cohen, S. (1995). *Mindblindness: An essay on autism and theory of mind*. Cambridge, MA: MIT Press.

Baron-Cohen, S. (1997). Are children with autism superior at folk physics? *New Directions in Child Development*, *75*, 45–54.

Baron-Cohen, S., O'Riordan, M., Stone, V., Jones, R., & Plaisted, K. (1999). Recognition of faux pas by normally developing children and children with Asperger syndrome or high-functioning autism. *Journal of Autism and Developmental Disorders*, *29*, 407–418.

Baron-Cohen, S., Leslie, A., & Frith, U. (1985). Does the autistic have a "theory of mind"? *Cognition*, *21*, 37–46.

Baron-Cohen, S., Jolliffe, T., Mortimore, C., & Robertson, M. (1997). Another advanced test of theory of mind: evidence from very high functioning adults with autism or asperger syndrome. *Journal of Child Psychology and Psychiatry*, *38*, 813–22.

Baron-Cohen, S., & Wheelwright, S. (2004). The empathy quotient: Aninvestigation of adults with Asperger syndrome or high functionin gautism, and normal sex differences. *Journal of Autism and Developmental Disorders*, *34*, 163–175.

Baron-Cohen, S., Wheelwright, J., Hill, J., Raste, Y., & Plumb, I. (2001). The Reading the Mind in the Eyes? Test-Revised version: A study with normal adults, and adults with Asperger syndrome or high-functioning autism. *Journal of Child Psychology and Psychiatry*, *42*, 241–251.

Bernier, R., Dawson, G., Webb, S., & Murias, M. (2007). EEG mu rhythm and imitation impairments in individuals with autism spectrum disorder Brain & Congnition, *64*, 228-237.

Boddaert, N., Chabane, N., Gervaise, H., Good, C., Bourgeois, M., Plumet, M., et al. (2004). Superior temporal sulcus anatomical abnormalities in childhood autism: a voxel-based morphometry MRI study. *Neuroimage*, *23*, 364–369.

Brass, M., & Heyes, C. (2005). Imitation: Is cognitive neuroscience solving the corresponding problem? *Trends in Cognitive Sciences*, *9*, 489–495.

Brune, M., & Brune-Cohrs, U. (2006). Theory of mind-evolution, ontogeny, brain mechanisms and psychopathology. *Neuroscience & Biobehavioral Reviews*, *30*, 437–455.

Buccino, G., Vogt, S., Ritzl, A., Fink, G., Zilles, K., Freund, H., et al. (2004). Neural circuits underlying imitation learning of hand actions: an event-related fMRI study. *Neuron*, *42*, 323–334.

Buitelaar, J. K., van der Wees, M., Swaab-Barneveld, H., van der Gaag, R. J. (1999). Theory of mind and emotion-recognition functioning in autistic spectrum disorders and in psychiatric control and normal children. *Developmental Psychopathology*, *11*, 39–58.

Capps, L., Yirmiya, N., & Sigman, M. D. (1992). Understanding of simple and complex emotions in non-retarded children with autism. *Journal of Child Psychology and Psychiatry*, *33*, 1169–1182.

Carr, L., Iacoboni, M., Dubeau, M., Mazziotta, J., & Lenzi, G. (2003). Neural mechanisms of empathy in humans: a relay from neural systems of imitation to limbic areas. *Proceedings from the National Academy of Sciences in the USA*, *100*, 5497–5502.

Charman, T., & Baron-Cohen, S. (1992). Understanding drawings and beliefs: a further test of the metarepresentation theory of autism: a research note. *Journal of Child Psychology and Psychiatry*, *33*, 1105–1112.

Charman, T., & Baron-Cohen, S. (1994). Another look at imitation in autism. *Developmental Psychopathology*, *6*, 403–413.

Charman, T., Swettenham, J., Baron-Cohen, S., Cox, A., Baird, G., & Drew, A. (1997). Infants with autism: An investigation of empathy, pretend play, joint attention, and imitation. *Developmental Psychology*, *33*, 781–789.

Charman, T., Swettenham, J., Baron-Cohen, S., Cox, A., Baird, G., & Drew, A. (1998). An experimental investigation of social-cognitive abilities in infants with autism: Clinical implications. *Infant Mental Health Journal*, *19*, 260–275.

Corona, R., Dissanayake, C., Arbelle, S., Wellington, P., & Sigman, M. (1998). Is affect aversive to young children with autism? Behavioral and cardiac responses to experimenter distress. *Child Development*, *69*, 1494–1502.

Dakin, S., & Frith, U. (2005). Vagaries of visual perception in autism. *Neuron*, *48*, 597–507.

Dapretto, M., Davies, M., Pfeifer, J., Scott, A., Sigman, M., Bookheimer, S., et al. (2006). Understanding emotions in others: mirror neuron dysfunction in children with autism spectrum disorders. *Nature Neuroscience*, *9*, 28–30.

Dawson, G., & Adams, A. (1984). Imitation and social responsiveness in autistic children. *Journal of Abnormal Child Psychology*, *12*, 209–226.

Dawson, G., & Lewy, A. (1989). Arousal, attention, and socioemotional impairments of individuals with autism. In G. Dawson (Ed.), *Autism: Nature, diagnosis and treatment* (pp. 49–74). New York: Guilford.

Dawson, G., Meltzoff, A., Osterling, J., & Rinaldi, J. (1998). Neuropsychological correlates of early symptoms of autism. *Child Development*, *69*, 1276–1285.

Dawson, G., Meltzoff, A., Osterling, J., Rinaldi, J., & Brown, E. (1998). Children with autism fail to orient to naturally occurring social stimuli. *Journal of Autism and Developmental Disorders*, *28*, 479–485.

Dawson, G., & Osterling, J. (1994). Early Intervention in Autism. In M. Guralnick's (Ed.), *The Effectiveness of early intervention* (307–326). Baltimore, MD: Brookes Publishing Co.

Dawson, G., Toth, K., Abbott, R., Osterling, J., Munson, J., Estes, A., et al. (2004). Early social attention impairments in autism: Social orienting, joint attention, and attention to distress. *Developmental Psychology*, *40*, 271–283.

Dawson, G., Webb, S., Carver, L., Panagiotides, H., & McPartland, J. (2004). Young children with autism show atypical brain responses to fearful versus neutral facial expressions. *Developmental Science*, *7*, 340–359.

Dawson, G., & Zanolli, K. (2003). Early intervention and brain plasticity in autism. *Novartis Foundation Symposium*, *251*, 266–274.

De Haan, M., Belsky, J., Reid, V., Bolein, A., & Johnson, M. (2004). Maternal personality and infants' neural and visual responsivity to facial expressions of emotion. *Journal of Child Psychology and Psychiatry*, *45*, 1209–1218.

Decety, J., Chaminade, T., Grezes, J., & Meltzoff, A. (2002). A PET Exploration of the neural mechanisms involved in reciprocal imitation. *Neuroimage*, *15*, 265–272.

DeMeyer, M., Alpern, G., Barons, S., DeMyer, W., Churchill, D., Hingtgen, J., et al. (1972). Imitation in autistic, early schizophrenic, and non-psychotic subnormal children. *Journal of Autism and Child Schizophrenia*, *2*, 264–287.

Dissanayake, C., Sigman, M., Kasari, C., & Bagley, R. (1996). Long-term stability of individual differences in the emotional responsiveness of children with autism. *Journal of Child Psychology & Psychiatry & Allied Disciplines*, *37*, 461–467.

Dorris, L., Espie, C., Knott, F., & Salt, J. (2004). Mind-reading difficulties in the siblings of people with Asperger's syndrome: Evidence for a genetic influence in the abnormal development of a specific cognitive domain. *Journal of Child Psychology and Psychiatry*, *45*, 412–418.

Fecteau, S., Lepage, J., & Théoret, H. (2006). Autism spectrum disorder: seeing is not understanding. *Current Biology*, *16*, R131–R133.

Fletcher, P., Happé, F., Frith, U., Baker, S., Dolan, R., Frackowiak, R., et al. (1995). Other minds in the brain: a functional imaging study of 'theory of mind' in story comprehension. *Cognition*, *57*, 109–128.

Fombonne, E. (2003). Epidemiological surveys of autism and other pervasive developmental disorders: An update. *Journal of Autism and Developmental Disorders*, *33*, 365–382.

Fombonne, E. (1999). The epidemiology of autism: a review. *Psychological Medicine*, *29*, 769–786.

Frith, C., & Frith, U. (1999). Interacting minds—biological basis. *Science*, *286*, 1692–5.

Gallagher, H., & Frith, C. (2003). Functional imaging of 'theory of mind'. *Trends in Cognitive Science*, *7*, 77–83.

Gallese, V. (2006). Intentional attunement: a neurophysiological perspective on social cognition and its disruption in autism. *Brain Research*, *1079*, 15–24.

Gallese, V., Fadiga, L., Fogassi, L., & Rizzolatti, G. (1996). Action recognition in the premotor cortex. *Brain*, *119*, 593–609.

Gallese, V., & Goldman, A. (1998). Mirror neurons and the simulation theory of mind reading. *Trends in Cognitive Sciences*, *12*, 493–501.

Gallese, V., Keysers, C., & Rizzolatti, G. (2004). A unifying view of the basis of social cognition. *Trends in Cognitive Science*, *8*, 396–403.

Golan, O., Baron-Cohen, S., Hill, J., & Rutherford, M. (2006). The 'Reading the Mind in the Voice' test-revised: A study of complex emotion recognition in adults with and without autism spectrum conditions. *Journal of Autism and Developmental Disorders*, *37*, 1096–1106.

Green, D., Baird, G., Barnett, A., Henderson, L., Huber, J., & Henderson, S. (2002). The severity and nature of motor impairment in Asperger's syndrome: a comparison with specific developmental disorder of motor function. *Journal of Child Psychology and Psychiatry*, *43*, 655–68.

Hadjikhani, N., Joseph, R., Snyder, J., & Tager-Flusberg, H. (2005). Anatomical differences in the mirror neuron system and social cognition network in autism. *Cerebral Cortex*, *16*, 1272–1282.

Hall, G., Szechtman, H., & Nahmias, C. (2003). Sex differences in regional cerebral blood flow during the processing of facial emotion. *Brain and Cognition*, *51*, 164–166.

Hammes, J., & Langdell, T. (1981). Precursors of symbol formation and childhood autism. *Journal of Autism and Developmental Disorders*, *11*, 331–46.

Happe F. (1994). An advanced test of theory of mind: understanding of story characters' thoughts and feelings by able autistic, mentally handicapped, and normal children and adults. *Journal of Autism and Developmental Disorders*, *24*, 129–154.

Hertzig, M., Snow, M., & Sherman, M. (1989). Affect and cognition in autism. *Journal of the American Academy of Child and Adolescent Psychiatry*, *28*(2), 195–199.

Hobson, R. (1986). The autistic child's appraisal of expressions of emotion: A further study. *Journal of Child Psychology and Psychiatry*, *27*, 671–680.

Hobson, P., Ouston, J., & Lee, A. (1988). Emotion recognition in autism: coordinating faces and voices. *Psychological Medicine*, *8*, 911–923.

Hobson, R., & Lee, A. (1999). Imitation and identification in autism. *Journal of Child Psychology and Psychiatry*, *40*, 649–659.

Iacoboni, M., & Dapretto, M. (2006). The mirror neuron system and the consequences of its dysfunction. *Nature Review: Neuroscience*, *7*, 942–51.

Iacoboni, M., Woods, R., Brass, M., Bekkering, H., Mazziotta, J., & Rizzolatti, G. (1999). Cortical mechanisms of human imitation. *Science*, *286*(5449), 2526–2528.

Iacoboni, M. (2005). Neural mechanisms of imitation. *Current Opinion in Neurobiology*, *15*, 632–637.

Jackson, P. L., Brunet, E., Meltzoff, A. N., & Decety, J. (2006). Empathy examined through the neural mechanisms involved in imagining how I feel versus how you feel pain: An event-related fMRI study. *Neuropsychologia*, *44*, 752–761.

Jolliffe, T., & Baron-Cohen S. (1999). The Strange Stories Test: a replication with high-functioning adults with autism or Asperger syndrome. *Journal of Autism and Developmental Disorders*, *29*, 395–406.

Jones, V., & Prior, M. (1985). Motor imitation abilities and neurological signs in autistic children. *Journal of Autism and Developmental Disorders*, *15*, 37–46.

Kanner, L. (1943). Autistic disturbances of affective contact. *Nervous Child*, *2*, 217–250.

Karmiloff-Smith, A., Klirna, E., Bellugi, U., Grant, J., & Baron-Cohen, S. (1995). Is there a social module? Language, face processing and theory of mind in individuals with Williams syndrome. *Journal of Cognitive Neuroscience*, *7*, 196–208.

Keysers, C., Wickers, B., Gazzola, V., Anton, J., Fogassi, L., & Gallese, V. (2004). A Touching sight: SII/PV activation during the observation and experience of touch. *Neuron*, *42*, 335–346

Klauck, S. (2006). Genetics of autism spectrum disorder. *European Journal of Human Genetics*, *14*, 714–720.

Koegel, L. K., Koegel, R. L., Harrower, J. K., & Carter, C. M. (1999). Pivotal response intervention I: Overview of approach. *Journal of the Association for Persons with Severe Handicaps*, *24*, 174–185.

Leekam, S., & Perner, J. (1991). Does the autistic child have a metarepresentational deficit? *Cognition*, *40*, 203–218.

Leslie, A. M. (1987). Pretense and representation: The origins of "theory of mind". *Psychological Review*, *94*, 412–426.

Loveland, K., Tunali Kotoski, B., Pearson, D., Brelsford, K., Ortegon, J., & Chen, R. (1994). Imitation and expression of facial affect in autism. *Development and Psychopathology*, *6*, 433–444.

Molnar-Szakacs, I., Iacoboni, M., Koski, L., & Mazziotta, J. (2005). Functional segregation within pars opercularis of the inferior frontal gyrus: Evidence from fMRI studies of imitation and action observation. *Cerebral Cortex*, *15*, 986–994.

Morgan S., Cutrer P., Coplin J., & Rodrigue J. (1989). Do autistic children differ from retarded and normal children in Piagetian sensorimotor functioning? *Journal of Child Psychology and Psychiatry*, *30*, 857–864.

Muthukumaraswamy, S., Johnson, B., & McNair, N. (2004). Mu rhythm modulation during observation of an object-directed grasp. *Cognitive Brain Research*, *19*, 195–201.

Nieminen-von Wendt, T., Metsahonkala, L., Kulomaki, T., Aalto, S., Autti, T., Vanhala, R., et al. (2003). Changes in cerebral blood flow in Asperger syndrome during theory of mind tasks presented by the auditory route. *European Child & Adolescent Psychiatry*, *12*, 178–189.

Nishitani, N., Avikainen, S., & Hari, R. (2004). Abnormal imitation-related cortical activation sequences in Asperger's Syndrome. *Annals of Neurology*, *55*, 585–562.

Oberman, L., Hubbard, E., McCleery, J., Altschuler, E., Ramachandran, V., & Pineda, J. (2005). EEG evidence for mirror neuron dysfunction in autism spectrum disorders. *Cognitive Brain Research*, *24*, 190–198.

Ohta, M. (1987). Cognitive disorders of infantile autism: A study employing the WISC, spatial relationships, conceptualization, and gestural imitation. *Journal of Autism and Developmental Disorders*, *17*, 45–62.

Osterling, J., & Dawson, G. (1994). Early recognition of children with autism: A study of first birthday home videotapes. *Journal of Autism and Developmental Disorders, 24*, 247–257.
Ozonoff, S., Pennington, B., & Rogers, S. (1991). Executive function deficits in high-functioning autistic individuals: Relationship to theory of mind. *Journal of Child Psychology and Psychiatry, 32*, 1081–1105.
Pelphrey, J., Morris, J., & McCarthy, G. (2005). Neural basis of eye gaze processing deficits in autism. *Brain, 128*, 1038–1048.
Perner, J., Frith, U., Leslie, A., & Leekam, S. (1989). Exploration of the autistic child's theory of mind: knowledge, belief, and communication. *Child Development, 60*, 688–700.
Pineda, J. (2005). The functional significance of mu rhythms: Translating "seeing" and "hearing" into "doing." *Brain Research Reviews, 50*, 57–68.
Pineda, J. (2006, April). *Efficacy of Neurofeedback Training on Autism Spectrum Disorders* (poster). Presented at Cognitive Neuroscience Society, San Francisco CA, 8–11.
Premack, D., & Woodruff, G. (1978). Does the chimpanzee have a theory of mind? *Behavioral and Brain Sciences, 1*, 515.
Ramachandran, V. S., & Oberman, L. M. (2006, November). Broken mirrors: A theory of autism, *Scientific American.*
Rizzolatti, G., & Craighero, L. (2004). The mirror-neuron system. *Annual Review of Neuroscience, 27*, 169–192.
Roeyers, H., Van Oost, P., & Bothuyne, S. (1998). Immediate imitation and joint-attention in young children with autism. *Development and Psychopathology, 10*, 441–450.
Rogers, S. (1998). Empirically supported comprehensive treatments for young children with autism. *Journal of Clinical Child Psychology, 27*, 168–179.
Rogers, S., & Pennington, B. (1991). A theoretical approach to the deficits in infantile autism. *Development and Psychopathology, 3*, 137–162.
Rogers, S., Bennetto, L., McEvoy, R., & Pennington, B. (1996). Imitation and pantomime in high-functioning adolescents with autism spectrum disorders. *Child Development, 67*, 2060–2073.
Rogers, S., Hepburn, S., Stackhouse, T., & Wehner, E. (2003). Imitation performance in toddlers with autism and those with other developmental disorders. *Journal of Child Psychology and Psychiatry, 44*, 763–781.
Rogers, S., Cook, I., & Greiss-Hess, L. (2005). *Mature imitation task.* Unpublished coding manual, M.I.N.D. Institute, University of California – Davis.
Rutherford, M. D., Baron-Cohen, S., & Wheelwright, S. (2002). Reading the mind in the voice: A study with normal adults and adults with Asperger syndrome and high functioning autism. *Journal of Autism and Developmental Disorders, 32*, 189–194.
Saxe, R., & Kanwisher, N. (2003). People thinking about thinking people: fMRI studies of theory of mind. *Neuroimage, 19*, 1835–1842
Shenk, L., Hubbard, E., McCleery, J., Ramachandran, V., Pineda, J. (2004, April). EEG evidence for mirror neuron dysfunction in autism. *Cognitive Neuroscience Society* Abstract.
Singer, T., Seymour, B., O'Doherty, J., Kaube, H., Dolan, R., & Frith, C. (2004). Empathy for pain involves the affective but not the sensory components of pain. *Science 303*, 1157–1162
Sigman, M., Kasari, C., Kwon, J., & Yirmiya, N. (1992). Responses to the negative emotions of others by autistic, mentally retarded, and normal children. *Child Development, 63*, 796–807.
Sigman, M., & Ungerer, J. (1984). Attachment behaviors in autistic children. *Journal of Autism and Developmental Disorders, 14*(3), 231–244.
Smith, I., & Bryson, S. (1998). Gesture imitation in autism I: Nonsymbolic postures and sequences. *Cognitive Neuropsychology, 15*, 747–770.

Smith, M., Rogers, S., & Dawson, G. (2008). *The early start Denver model: A comprehensive early intervention approach for toddlers with autism, third edition*. Austin, TX: Pro Ed Corporation, Inc.

Stone, V., Baron-Cohen, S., & Knight, R. (1998). Frontal lobe contributions to theory of mind. *Journal of Cognitive Neuroscience, 10*, 640–656.

Stone, W., Lemanek, K., Fishel, P., Fernandez, M., & Altemeier, W. (1990). Play and imitation skills in the diagnosis of autism in young children. *Pediatrics, 64*, 1688–1705.

Stone, W., Ousley, O., & Littleford, C. (1997). Motor imitation in young children with autism: What's the object? *Journal of Abnormal Child Psychology, 25*, 475–485.

Theoret, H., Halligan, E., Kobayashi, M., Fregni, F., Tager-Flusberg, H., & Pascual-Leone, A. (2005). Impaired motor facilitation during action observation in individuals with autism spectrum disorder. *Current Biology, 15*, R84–R85.

Veenstra-Vanderweele, J., Christian, S., Cook, E. (2004). Autism as a paradigmatic complex genetic disorder. *Annual Review of Genomics and Human Genetics, 5*, 379–405.

Veenstra-Vanderweele, J., & Cook, E. (2003). Genetics of Childhood Disorders: XLVI. Autism, Part 5: Genetics of Autism. *Journal of Academy of Child and Adolescent Psychiatry, 42*, 116–118.

Vogeley, K., Bussfield, P., Newen, A., Herrmann, S., Happe, F., Falkai, P., et al. (2001). Mind reading: neural mechanisms of theory of mind and self-perspective. *Neuroimage, 14*, 170–181.

Waiter, G., Williams, J., Murray, A., Gilchrist, A., Perrett, D., & Whiten, A. (2004). A voxel-based investigation of brain structure in male adolescents with autistic spectrum disorder. *Neuroimage, 22*, 619–625.

Wicker, B., Keysters, C., Plailly, J., Royet, J., Gallese, V., & Rizzolatti, G. (2003). Both of us disgusted in my insula: the common neural basis of seeing and feeling disgust. *Neuron, 40*, 655–664.

Williams, J., Whiten, A., Suddendorf, T., & Perrett, D. (2001). Imitation, mirror neurons and autism. *Neuroscience and Biobehavioral Reviews, 25*, 287–295.

Williams, J., Whiten, A., & Singh, T. (2004), A systematic review of action imitation in autistic spectrum disorder. *Journal of Autism and Developmental Disorders, 34*(3), 285–299.

Williams, J., Waiter, G., Gilchrist, A., Perrett, D., Murray, A., & Whiten, A. (2006). Neural mechanisms of imitation and 'mirror neuron' functioning in autism spectrum disorder. *Neuropsychologia, 44*, 610–621.

Wimmer, H., & Perner, J. (1983). Beliefs about beliefs: Representation and constraining function of wrong beliefs in young children's understanding of deception. *Cognition, 13*, 103–128.

Yirmiya, N., Sigman, M., Kasari, C., & Mundy, P. (1992). Empathy and cognition in high-functioning children with autism. *Child Development, 63*, 150–160.

Zaitchik, D. (1990). When representation conflict with reality: The preschooler's problem with false belief and false photographs. *Cognition, 35*, 41–68.

Zilbovicius, M. Meresse, I., Chabane, N., Brunelle, F., Samson, Y., & Boddaert, N. (2006). Autism, the superior temporal sulcus and social perception. *Trends in Neurosciences, 29*, 359–366.

Synaesthesia for Pain: Feeling Pain with Another

Melita J. Giummarra and John L. Bradshaw

Abstract In this chapter, we establish a theoretical framework for the rare manifestation of *synaesthesia for pain*; that is, *the sensation in one part of the body (pain) produced by stimulus (pain) observed or imagined in another*. We first describe mirror neuron systems (MNS) and the implications these systems have on imitation, behavioral mimicry, communication, socialization, and empathy. Behavioral and emotional contagion may signify disinhibition of behavioral mirroring that is closest to synaesthesia for pain. We describe cases of synaesthesia for pain, including one case of mirrored pain with hyper-algesia, eight cases of mirrored pain in a phantom limb following amputation, one case of mirrored pain in the stump following amputation, and one case of mirrored pain following traumatic childbirth. We then explore the mechanisms that may underlie the experience of pain when observing, or even thinking about, another in pain, including mirror neurons, empathy, and motor systems, the Autonomic Nervous System and visceral mechanisms, and the potential role of sensitization to pain and hypervigilance to pain cues. We conclude that synaesthesia for pain is most likely a consequence of disinhibited activation, or central sensitization, of a fundamentally adaptive system for the empathic perception of pain in another.

Keywords Mirror neurons · Synaesthesia · Pain · Sensitization · Empathy · Behavioral mimicry

---

M.J. Giummarra
Experimental Neuropsychology Research Unit, School of Psychology, Psychiatry and Psychological Medicine, Monash University, Clayton VIC 3800, Australia; National Ageing Research Institute, Parkville, VIC, Australia
e-mail: melita.giummarra@med.monash.edu.au

J.A. Pineda (ed.), *Mirror Neuron Systems*, DOI: 10.1007/978-1-59745-479-7_13,

1 Introduction

Mirror neuron systems (MNS) are activated equally during the observation and execution of any given goal-directed action (see 2, below), and may underlie action-understanding, imitation, communication, socialization, and empathy. Behavioral mimicry (Chameleon effect) may also play a key bioevolutionary role in enhancing social communication and cohesion (Yabar, Johnston, Miles, & Peace, 2006), and may be a behavioral correlate of MNS activity. In this chapter, we establish a theoretical framework for the rare manifestation of **synaesthesia for pain**; that is, *the sensation in one part of the body (pain) produced by a stimulus (pain) observed or imagined in another*. We first describe mirror neuron systems and the implications MNS have on imitation, communication, and empathy. We then describe behavioral and emotional contagion, the manifestation of disinhibited behavioral mirroring closest to synaesthesia for pain. Finally, we describe some cases of synaesthesia for pain and explore the mechanisms that may underlie the experience of pain when observing, or even thinking about, another in pain, including mirror neurons, empathy and motor systems, and the potential role of sensitization to pain and hypervigilance to pain cues.

2 Mirror Neuron Systems

2.1 Action Understanding

Mirror neurons—initially identified in the macaque in area F5 of the ventral premotor cortex (vPMC) (Rizzolatti et al., 1988)—fire not only when the monkey executes a goal-directed action, but also when another monkey or the human experimenter is observed to perform the same goal-directed motor action. Activation of the MNS through observing another perform a goal-directed action leads to activity in the same supraspinal and peripheral neural networks associated with personally performing the observed action, leading to sub-threshold activity in the homologous motor systems (Buccino, Binkofski, & Riggio, 2004; Fadiga, Craighero, & Olivier, 2005; Rizzolatti, Craighero, & Fadiga, 2002). MNS activity is likely to play a key role in action-understanding (Fadiga et al., 2005), such that to understand the goal of another person's behavior, we must not only match it against our own motor system, but also covertly imitate the other's action (Buccino et al., 2004; Rizzolatti et al., 2002).

Skill acquisition can be facilitated through imitation (Heyes, Bird, Johnson, & Haggard, 2005) or even through repeated imagination of the sequence of actions required to perform a new activity (Wohldmann, Healy, & Bourne, 2007). When observing another perform an action there appears to be covert imitation through the activation of representations of movement in the MNS, which probably underlies the ease with which we learn through imitation (Wilson, 2006). When no overt imitation is required, a series of mechanisms prevents the observer from emitting a motor behavior that mimics the observed one (Rizzolatti & Arbib, 1998); this is perhaps analogous to the inhibitory mechanisms that normally prevent 'acting out' during rapid-eye-movement sleep (Siegel, 2006). Spinal cord inhibition selectively blocks the motoneurons involved in the execution of the observed action, although sometimes, particularly when the observed action is of particular interest, the premotor cortex system will allow a brief prefix of the movement (Rizzolatti & Arbib, 1998). The strength of activity in the MNS (particularly in the pars opercularis in the inferior frontal gyrus and the rostral posterior parietal cortex) is greater during *execution* than *observation* of an action, and even greater during *imitation* (that is, both observation *and* performance) of an action (Iacoboni et al., 1999). Covert 'imitation' of another's observed experience, through MNS activity, evokes multiple motor-based representations of the observed situation, including "what another person *is* doing, what another person *ought* to do, and what you yourself *intend* to do" (Wilson, 2006, p. 215), and is thus an adaptive tool for predicting the behavior associated with a particular environmental situation.

MNSs have now been identified for other domains of perception and perceptual-understanding, including emotion (Ruby & Decety, 2004), disgust (Wicker et al., 2003), touch (Keysers et al., 2004), and pain (Jackson, Meltzoff, & Decety, 2005; Morrison, Lloyd, & Di Pellegrino, 2004; Singer et al., 2004). MNSs potentially have a key communicative role in not only interpreting and understanding another person's goals, but also their emotional state.

2.2 *Communication and Speech Perception*

Mirror neuron systems likely played a key role in the evolution of communication (Hamzei et al., 2003). The primate vPMC is the homologue of Broca's region in humans. Broca's region is not only responsible for the motor control of oro-laryngeal movements during language production and during sign language (Corballis, 2003), but is also active during the observation of dynamic and still images depicting movements of the hands and face (Binkofski & Buccino, 2004; Hamzei et al., 2003), suggesting that verbal language may have evolved from gestural communication rather than vocalization, although see Bradshaw (2003). Furthermore, production and perception of speech excites the motor cortices of the hands (Flöel, Ellger, Breitenstein, & Knecht, 2003). MNSs were likely involved in the development of verbal communication considering their involvement in the interpretation of goal-directed nonverbal activity and gestural communication (Corballis, 2002; 2003; Rizzolatti & Arbib, 1998).

2.3 Empathy

Another important element of communicative behavior is interpreting and understanding the meaningful behavior of others and empathically processing the associated affective and motivational aspects of another's experience (Avenanti, Bueti, Galati, & Aglioti, 2005). As with action observation, the observation or imagination of another person in a particular emotional state automatically activates a representation of the other's actions, goals, intentions, and emotional state in largely the same neural networks that are involved in experiencing the respective emotional state, and triggers associated autonomic and somatic responses (Gallese, 2001; Preston & de Waal, 2002; Singer, 2006). Feeling empathy for another is, however, a complex, multi-level process where the empathizer is in an isomorphic state elicited by the observation or imagination of another person's affective state, but where one is nevertheless aware that the other person is the source of one's own affective state (de Vignemont & Singer, 2006). By evoking the other's inner state, the observer is able to better predict the other's future actions, to analyze important environmental properties, and subsequently perhaps to employ cooperative, pro-social behavior (de Vignemont & Singer, 2006). Gallese (2001, 2003) proposes that empathy occurs through a shared manifold, through the MNS, at three levels: (i) the *phenomenological level*, responsible for the sense of similarity, of being an individual within a larger social community, whereby the actions, emotions, sensations experienced by others become implicitly meaningful to us; (ii) the *functional level*, characterized in terms of '*as if*' processes to create models of others; and (iii) the *sub-personal level*, at the level of activity in a series of mirror matching neural circuits, and the activity of these neural circuits is, in turn, tightly coupled with multilevel changes within bodily states.

3 Behavioral Mimicry and Emotional Contagion

We have highlighted the importance of MNSs in action understanding, imitation, communication, and empathy. We now review related findings on imitative behavior, behavioral mimicry, and emotional contagion. While much of this research has not been analyzed with respect to MNSs, we propose that they are intimately linked and provide an important insight to the evolutionary role of empathic processes within MNSs. In particular, the enormous literature on behavioral and emotional mimicry has important implications for other domains of perceptual mimicry, particularly synaesthesia for pain.

3.1 Clinical Imitative Behavior

Imitation Behavior (IB) is a form of utilisation behavior (the disinhibited, stereotyped, and situationally inappropriate use of an object) (Lhermitte,

Pillon, & Serdaru, 1986). Patients with IB automatically imitate the gestures and behavior of others even if not asked to do so. When asked to stop, they either continue to imitate or recommence imitation when their attention is diverted (De Renzi, Francesca, & Stefano, 1996). Rather than perform an action identical to the observed one, patients with IB perform largely similar actions that achieve the same goal. IB is associated with lesions to the frontal cortex, particularly the inferior, mediobasal (Lhermitte et al., 1986) and upper medial and lateral frontal cortex (De Renzi et al., 1996), as with utilisation behavior, and is thus likely caused by impairment of the inhibitory action of the frontal lobe on the parietal lobe, thereby releasing parietal lobe activity. While IB was first described prior to the discovery of the MNS in humans, recent MNS research suggests that patients with IB have disinhibited activity in the fronto-parietal mirror neuron system that is active when both performing and observing goal-directed behavior in others (Archibald, Meateer, & Kerns, 2001). Lesions in this fronto-parietal circuit would result in the patient performing actions that achieve the same goal as that observed in the actions of others, not necessarily mirroring the specific actions, as is the case in IB.

3.2 Behavioral Mimicry and Emotional Contagion

Behavioral imitation also occurs readily in normal populations: the perception of another's behavior automatically increases the likelihood of engaging in that behavior oneself (Chartrand & Bargh, 1999). This *Chameleon effect* leads one to adopt the posture, gestures, mannerisms, and speech patterns of an interaction partner (Lakin, Jefferis, Cheng, & Chartrand, 2003). Chartrand & Bargh (1999) proposed that perception causes the execution of similar behaviors, and such perception creates smoother interactions, shared feelings of empathy and rapport, and greater social liking (Chartrand & Bargh, 1999). Having a goal to affiliate with an interaction partner increases non-conscious mimicry, and failure to affiliate results in a further increase in mimicry (Lakin & Chartrand, 2003). Accordingly, people mimic in-group members more than out-group members (Yabar et al., 2006); for example, Christians mimic other Christians more than non-Christians. Mimicry may have been evolutionarily advantageous by increasing the opportunity for food sharing, mating, and predator avoidance, leading to further selection and retention of the tendency to mimic others during social encounters (Chartrand & Bargh, 1999; Lakin & Chartrand, 2003; Lakin et al., 2003; Yabar et al., 2006). Furthermore, appropriate displays of sexual arousal and attraction in the animal kingdom generally can generate 'chameleon' mirror-type responses in the other, enhancing the likelihood of successful mating, and reproduction.

Aspects of human behavior that are rapidly and spontaneously mimicked include speech patterns, such as the accents, rates, and rhythms of the speech of interaction partners (Lakin et al., 2003; Yabar et al., 2006); facial and bodily expressions of inner emotional states, for example, fear, pain, laughter, smiling, affection, embarrassment, discomfort, sadness, and disgust (Bavelas, Black, Lemery, & Mullett, 1987); as well as some physiological processes, for example, yawning, or vomiting (Platek, Critton, Myers, & Gallup, 2003). Thus, when an interaction partner is sad and crying, we automatically produce tears (Singer, 2006), whereas when we hear or see laughter, smiling, and happiness—as makers of television comedy programs well know—we are more likely to laugh, smile, or feel happy (Hatfield, Cacioppo, & Rapson, 1994).

Common facial expressions of emotion are universally produced and recognized (Russell, 1994), despite vast cultural differences; for example, facial muscular patterns and discrete emotions in Western cultures are also evident in isolated preliterate societies such as the highlands of New Guinea (Ekman & Friesen, 1971). Deceptive facial expressions differ from genuine, spontaneous facial expressions, and may be controlled by the cortical-pyramidal and subcortical-extrapyramidal systems respectively (Rinn, 1984). Deceptive facial expressions may exhibit a mix of emotional expressions with 'leakage' of genuine emotions, which most often occurs around the eyes due to poor differentiated control of the musculature in the upper compared with the lower face. Faked expressions also show a decreased rate of blinking, possibly due to increased cognitive activity required to fake emotional expression (Hill & Craig, 2002). Faked expressions of pain show a greater frequency and intensity of pain-related facial actions compared with genuine pain expressions, together with an increased frequency and intensity of non-pain-related facial actions (Hill & Craig, 2002). While a bioevolutionary perspective would argue that one should be adept at detecting deceptive emotional expression, people have great difficulty in determining the veracity of facial expressions of pain (Hill & Craig, 2004).

Facial mimicry is already apparent in one-month-old infants who may smile, stick out their tongues, and open their mouths when they see someone else doing the same (Meltzoff & Moore, 1977). Adults also spontaneously and subconsciously imitate the emotional expressions of others, which can affect autonomic arousal in accordance with the respective emotion embodied (for a summary, see Effron, Niedenthal, Gil, & Droit-Volet, 2006). The tendency for couples to grow to resemble each other may be due to many years of mimicking the other's facial expressions, leading to similar facial lines (Zajonc, Adelmann, Murphy, & Niedenthal, 1987). Highly empathic people exhibit more facial mimicry (measured with electromyography; EMG) and are more likely to report feelings that reflect their muscle reactions compared with people with low empathy (Sonnby-Borgström, 2002; Sonnby-Borgström, Johnsson, & Svensson, 2003).

Primates generally tend not only to read the emotion displayed in facial expressions but also to follow the direction of another's gaze (Emery, 2000; Myowa-Yamakoshi, Tamonaga, Tanaka, & Matsuzawa, 2003). From the age of three months, infants follow the direction of an adult's gaze and exhibit biased attention to objects previously so-cued (Reid, Stiano, Kaufman, &

Johnson, 2004). Adults show a strong predisposition to imitate another's oculomotor behavior, even when it is detrimental to task performance (Ricciardelli, Bricolo, Aglioti, & Chelazzi, 2002). Brain structures involved in oculomotor control may process information about another individual's gaze direction, directly translating it into a matching oculomotor command. The actual release of this command would manifest itself as gaze-following behavior (Ricciardelli et al., 2002). Instinctive gaze-following is evolutionarily significant as it cues attention to objects of interest in one's environment such as potential sources of food or predators/danger.

The embodiment of another's emotional state through facial mimicry is likely to be a manifestation of emotional empathy and the processes of interpretation of the emotional states of others. MNSs are vital for understanding, through neural replication, the inner states of others, particularly the transmission of emotion through facial expression (Iacoboni & Dapretto, 2006). Observation and purposeful imitation of facial expressions activates a similar network of brain areas, including Broca's area, bilateral dorsal, and ventral premotor areas, right superior temporal gyrus, supplementary motor area, posterior temporo-occipital cortex, and cerebellar areas, with greater activity during imitation compared with mere observation (Carr, Iacoboni, Dubeau, Mazziotta, & Lenzi, 2003). Evidently, to empathize one must invoke the representation of the actions associated with the emotions that are witnessed. This empathic resonance occurs through communication between action representation networks and limbic areas.

4 Empathic Perception of Another's Painful Experience

Pain is a perceptual experience with complex sensory, affective, and cognitive qualities (Jackson et al., 2005; Morrison et al., 2004; Singer et al., 2004). Humans express pain experience through verbal and nonverbal vocalization, and behavioral expression (e.g., facial expressions, gestures, alterations to body posture, and guarding or limited use of the affected body part). Facial expression of pain following an acute injury may play an evolutionary role in promoting survival by eliciting empathic, helping behaviors in others. The survival of onlookers is also enhanced, as another's expression of pain alerts them to a potential immediate or object-related threat (Williams, 2002), and is thus associated with activation of the supplementary motor area triggering motor plans that would facilitate action, such as escape (Saarela et al., 2007). However, interpretation of expressions of pain differs according to the gender of the person in pain, such that male expression of pain elicits increased activity in emotion related areas—including amygdala, perigenual anterior cingulate cortex (ACC), and primary somatosensory cortex (SI)—which may elicit a threat-related response from the observer, whereas for female expression of pain there is decreased activity in these same areas, suggesting that the defensive response is inhibited, and instead promotes helping behaviors (Simon, Craig, Miltner, & Rainville, 2006). Overall, it is clearly advantageous for one not only to effectively communicate the experience of pain, but also to rapidly and accurately understand another's pain. Understanding another's experience appears to be mediated by a MNS for pain, and the associated mechanisms underlying behavioral mimicry.

When observing another person in pain, we not only consciously comprehend that the other is in pain, but we also automatically interpret the experience throughout the same cortical networks, or 'pain matrix,' that mediate personal experience of pain. The pain matrix includes the secondary somatosensory cortex (SII), thalamus, insular regions, the ACC, and the movement related areas such as the cerebellum and supplementary motor area (Jackson, Rainville, & Decety, 2006; Singer et al., 2004). In particular, the anterior insula cortex (AIC) and the ACC are both involved in the personal and empathic experience of pain (Jackson et al., 2005; Morrison et al., 2004; Singer et al., 2004). These same areas are also active during feelings of 'emotional pain,' such as social rejection (Eisenberger, Lieberman, & Williams, 2003) or frustration (Abler, Walter, & Erk, 2005), and during the anticipation of pain (Sawamoto et al., 2000). The AIC is involved in integrating autonomic and visceral information, and the ACC is active during tasks related to affect, repose selection, interoception, attention, conflict, autonomic arousal as well as processing pain unpleasantness and affect (Saarela et al., 2007). The level of activity in the ACC strongly correlates with an observer's ratings of the intensity of another's pain (Jackson et al., 2005; Saarela et al., 2007). The ACC receives projections from the superior temporal areas, which play a role in semantic visual processing, and is probably involved in the affective components of the 'pain matrix' during empathy for pain. The ACC has extensive outputs to the premotor and motor areas (Morrison et al., 2004), and there are similar motor-action responses in homologous muscles when a painful stimulus is personally experienced, and when it is observed to be experienced by another (Avenanti et al., 2005). These findings suggest that there is empathic inference of the sensory qualities of another's pain, automatic embodiment of the observer's motor system, and empathic activity in the autonomic nervous system's 'flight' mechanism during empathy for pain.

While the ACC is active during perception of both self and other pain, activation sites follow a clear caudo-rostral organization based on the target of the pain (self or other) (Jackson et al., 2006). Perception of pain in the *self* is associated with caudal and ventral activation (Brodman Area 24; BA24) of the ACC; whereas perception of pain in *others* is represented in two distinct clusters within more rostral regions of the ACC (perigenual BA24/BA33; subcallosal BA32/BA25). Jackson, Brunet, Meltzoff, and Decety (2006; Jackson, Rainville, & Decety 2006) argue that empathy for another's pain does not, and ideally should not, lead to a complete merging of representations of the self and other, otherwise the mere observation of others in pain may lead to heightened emotional distress and actual perception of pain, as is indeed reported in the cases below.

4.1 Synaesthetic Experience: Case Summaries

There are few published cases of sensory synaesthetic experiences. Here we summarize one published case of synaesthesia for touch, and eleven cases of synaesthesia for pain.

4.1.1 Synaesthesia for Touch

Blakemore Bristow, Bird, Frith, and Ward (2005) used fMRI to investigate the structures activated in the perception of touch in a subject reported to have synaesthesia for touch (C), and a group of control subjects. Activations in the somatosensory cortex—which has previously been identified as having MNS properties (Keysers et al., 2004)—and the left pre-motor cortex were significantly higher in C, compared with controls, when she observed touch. The AIC was bilaterally activated in C but there was no evidence of such activation in the non-synaesthetic group. The findings from this study suggest that the mirror system for touch is over-active in C.

4.1.2 Synaesthesia for Pain

Mirror Pain in Association with Hyperalgesia

Bradshaw and Mattingley (2001) described the first patient who felt mirrored pain when he observed his wife in pain. This man, who also had hyperalgesia, felt pain that was immediate and intense and appeared to be qualitatively similar to his own hypersensitivity to touch when he saw his wife hurt herself.

Mirror Pain in the Phantom Limb Following Amputation

We have identified eight amputees in our phantom limb research (N = 280) who experience pain in the phantom limb or stump that is triggered by observing, thinking about, or inferring that another person is in pain (Giummarra, Georgiou-Karistianis, Gibson, Chou, & Bradshaw, 2006). These cases reported incidences of mirrored pain, together with other triggers, when asked "does anything trigger or cause your phantom limb sensations or phantom limb pain to emerge or change?" Similar experiences have not been identified in prior phantom pain research, thus the identification of the present cases emerged without specific probing.

The participants who reported experiencing mirrored pain included seven males and one female, aged 26–88 years old ($M = 54$ years; $SD = 19$) who had been missing a limb on average for 8 years ($SD = 8$; range: one month to 25 years). Six had suffered traumatic limb loss and two suffered complications of vascular disease leading to amputation. There were five above-knee, two below-knee, and one bilateral lower limb amputee (right below-knee/left above-knee). All but one amputee experienced pain in the limb prior to amputation, ranging from 13 hours to three years (M = 258 days; SD = 401). Anxiety (M = 6.6; SD = 2.8) and depression (M = 4.6, SD = 4.9) scores were normal in all but two cases (both had suggestive anxiety and one had probable depression).

All cases reported phantom limb sensations and phantom pain. Five had experienced phantom sensations constantly since the limb loss, while one perceived phantom sensations a few times a day, one perceived only occasional phantom sensations, and one reported only having a general awareness of a phantom limb. These cases reported that their phantom pain was triggered or modified when:

(a) *Observing another person hurt themselves or in pain* (n = 6, cases: AR, NW, BM, NN, SM, SW), such as seeing someone with an injured, painful leg. For example, AR reported that when he observes someone cut or hurt themselves he feels pain, initially as a 'cringe' in the stump, and then in the phantom.
(b) *Observing circumstances associated with being in pain* (n = 3, cases: RB, LD, SM), such as seeing blood, accidents, war, and traumatic events on television. For example, RB's father had recently had a quadruple bypass and when looking at the sutured wounds on his father's chest RB suddenly experienced strong, painful 'electric' impulses in his phantom foot. He found the phantom pain to be so intense that he had to look away from his father's chest for relief.
(c) *Thinking about others in pain* (n = 2, cases: BM, NN). For example, NN reported that when he hears a 'gruesome story,' or observes another person in pain, "instead of my testicles retracting, my [phantom] foot goes crazy" suggesting that his reaction to another's pain is one of the increased phantom pains instead of the expected cremasteric reflex. NN explained that it is not the similarity between the other's apparent pain and his personal experiences of pain prior to amputation that triggers the phantom pain, but rather the 'gruesome aspect' of it.

These cases also reported various other triggers of their phantom pain, including (a) *emotional triggers* (n = 4, cases: AR, NN, LD, SM) such as talking about the phantom, tiredness or being 'run down,' loud noises, fright, surprise, 'strong emotions' and stress; (b) *physical triggers and referred sensations* (n = 3, cases: NW, BM, SW), such as clothes rubbing on the stump, strenuous activity or work, or leaning toward the amputated side; (c) *environmental triggers* such as feeling cold (n= 1, case BM), and (d) *stimulation of the genitals* during toileting (n = 2, cases: AR, NN). BM's experience of phantom pain triggered by leaning toward the amputated side is analogous to the amelioration of left hemi-neglect—that is, the ability to perceive stimuli in the previously neglected hemi-space when attention is re-oriented toward that side (Mattingley, 2002). Furthermore, one case (LD) also reported that he feels phantom pain more often at night time, often associated with spontaneous movements in the phantom and amputated limb, and he felt that he must move the phantom to relieve his pain. These symptoms, consistent with a diagnosis of Restless Legs Syndrome (Allen et al., 2003), warrant further research.

Mirror Pain in the Stumps Following Amputation

One additional amputee (GD) reported mirrored pain in his stumps when observing activities that he associated with pain (walking barefoot). He was a 79 year-old man who, in 2002, became a bilateral lower limb amputee due to complications from diabetes and vascular disease. GD had multiple amputations within a three-month period: first of three toes, then the right leg below the knee, and finally the left leg below the knee. In the two months leading up to the amputations, he reported being in 'extreme pain.' At interview, four years following amputation, GD's mood was within normal ranges (HADS Anx = 0, HADS Dep = 2) and he wore bilateral prostheses most of the time, which caused him no pain.

GD did not perceive phantom sensations or pain, but did report some puzzling neuropathic sensations within his stumps. He explained that as a child he had always been troubled by foot spurs, which made walking barefoot (prior to the amputations) extremely painful. Now, as a bilateral lower limb amputee, *GD reports tingling, stinging pain in his stumps whenever he sees another person running or walking barefoot, whether it be on the television or in real life*.

Mirror Pain Following Traumatic Childbirth

CB, a 36 year-old woman, is a specialist medical practitioner. She experienced a long and painful childbirth, with prolonged foetal distress, which she in turn found to be distressing, and resulted in emergency caesarean section delivery. CB reports that *when she is told of another person's traumatic experience, she experiences shooting pains from the groin that radiate down the legs* (Giummarra & Bradshaw, 2007).

5 Mechanisms Underlying Synaesthesia for Pain

Each of the above cases has one main thing in common: they have all endured intense, traumatic, or chronic pain. Second, they have all (except the patient with hyperalgesia about whom there is limited information because he had died) experienced empathic pain in the lower limbs or genital region. We propose that the primary mechanism underlying empathically felt pain is the disinhibition of MNSs for pain. Such disinhibition may be caused by central sensitization to pain, selective attention, and/or hypervigilance to pain cues.

5.1 Mirror Neurons and Empathy for Pain

While all people may exhibit MNS activity in the pain matrix when observing another in pain, this rarely manifests in actual perception of pain. The cases described here suggest that elements of the pain matrix may be reorganized, disinhibited, or sensitized to pain cues, leading to empathically perceived pain (Giummarra, Gibson, Georgiou-Karistianis, & Bradshaw, 2007) in the same way that the case of synaesthesia for touch exhibited higher levels of activity in the mirror system for touch (somatosensory cortex, left premotor cortex, and AIC) (Blakemore et al., 2005). Furthermore, the levels of activity in MNS for action are higher during performance than observation and higher still during imitation (performance and observation) (Iacoboni & Dapretto, 2006). Together, these findings suggest that understanding through mimicry not only allows the sensory and affective qualities of the other's experience (pain) to be embodied and mapped onto the observer's body schema (Schwoebel, Boronat, & Coslett, 2002), but also permits the development of the associated empathic motivational state of readiness for action (Avenanti et al., 2005). Motor systems are inherently involved in processing pain, particularly for preparation for action (McCabe et al., 2003). From an evolutionary perspective, pain cues are important for survival, as reviewed above, and even pain-free volunteers have significant difficulty in disengaging attention from a threatening cue of impending pain (van Damme, Crombez, Eccleston, & Goubert, 2004).

At present, MNSs have primarily been found to be active during the performance and observation of meaningful activities, performed by the upper limbs, that are involved in gestural communication and goal-directed, operant behavior. However, most people would generally be able to identify with feeling that one's leg muscles become tense at the moment their favorite football team takes a penalty kick or when observing another walk dangerously close to a cliff's edge. There is now growing evidence from imaging and psychophysical research that the MNS is indeed involved in processing activities performed by the lower limbs. Buccino et al. (2001; Riggio, Melli, Binkofski, Gallese, & Rizzolatti, 2005), using fMRI, found somatotopically organized activation of the premotor cortex during observation not only of hand and mouth movements but also of foot movements. More recently, Bach and Tipper (2006) showed that perceiving a highly skilled athlete kick a soccer ball affects similar motor behaviors in the observer. These same authors also found that observing actions in the upper limbs (e.g., typing) and lower limbs (kicking a ball) facilitates more fluent responses in the compatible motor system; for example, kicking a ball facilitated more fluent foot-key responses compared with finger-key responses (Bach & Tipper, 2007). Evidently, the MNS is not only restricted to hand actions but also includes a rich repertoire of lower body actions.

While MNSs for pain are the most likely systems to explain empathically perceived pain, little is known about the mechanisms that weaken the integrity of the pain matrix during observation of another person in pain. Propensity to empathize with others may differentiate these cases from people who do not experience mirrored pain, in the same way that high empathizers demonstrate a higher degree of facial mimicry (Sonnby-Borgström, 2002; Sonnby-Borgström et al., 2003). Traumatic painful experience alone may increase one's propensity to exhibit empathy for others in pain. What is apparently lacking in people who report synaesthesia for pain is the typical distinction between self and other, which is usually mediated by activity in the caudal and rostral regions of the ACC, respectively, together with activity in the temporo-parietal junction (Jackson, Rainville, & Decety, 2006). Perhaps through peripheral or central sensitization to pain there is a greater degree of activity in the ACC that evokes representations of pain in the self when perceiving pain in another. The primary mechanisms leading to mirrored pain—through MNSs—potentially include the disinhibition of the autonomic nervous system (and the involvement of visceral reflexes), peripheral and central sensitization to pain, and hypervigilance to pain cues.

5.1.1 Autonomic Nervous System and Visceral Mechanisms

In the cases described above, mirrored pain was perceived to originate in the grain, stump, and/or phantom limb, which had been pre-sensitized, by trauma, to pain. The experience of mirror pain only in the lower limbs or genital regions suggests heightened activity of the autonomic nervous system, as a result of embodiment of the other's fight or flight dilemma, which is more likely to involve (a) the lower limbs in readiness for flight from danger and (b) the uro-genital system through the increased reflex urge to empty the bladder or bowels for more efficient escape from danger. Cortical reorganization of the lower limb onto the homuncular-adjacent genitals is also associated with heightened phantom pain (Flor, 2002). The genitals are highly involved in emotional *perceptual* processes (Both, Everaerd, & Laan, 2003; van der Velde & Everaerd, 2001), while the face—localized close to the upper limbs in the somatosensory homunculus—plays an emotionally *expressive* role.

5.2 *Sensitization and Hypervigilance to Pain*

5.2.1 Sensitization to Pain

Processing an acute painful stimulus starts by stimulating nociceptors that respond to noxious mechanical, chemical, or thermal stimuli which then activate peripheral C-fibers (Suzuki & Dickenson, 2005). The first synapse in the transmission of noxious information is in the superficial dorsal horn of the spinal cord (Suzuki & Dickenson, 2005). *Peripheral sensitization* to pain is characterized by an increase in pain sensitivity and excitability of the nociceptors at the site of injury or inflammation. An increase in sensitivity of neurons in the dorsal horn and supraspinal regions following, and outlasting, an ongoing barrage of nociceptive insult results in *central sensitization* (Ji, Kohno, Moore, & Woolf, 2003). Peripheral sensitization to pain is an adaptive response to nociceptive stimulation signaling tissue or nerve injury, and produces short-term hypersensitivity to low threshold stimulation to protect the site of injury and promote tissue-repair (Ji et al., 2003; Zusman, 2002). Central sensitization to pain can, however, lead to hypersensitivity to painful stimuli (hyperalgesia), expansion of receptive fields, and reduced pain threshold with low threshold, non-painful sensory fibers activating high threshold nociceptive neurons (allodynia) (Ji et al., 2003). Peripheral nerve injury with neuropathic pain is characterized by persistent hypersensitivity to pain and is associated with high rates of ectopic activity in sensory fibers eliciting central sensitization (Ji et al., 2003). People with a greater pain history tend to show lower tolerance for pain and exhibit hypervigilance to painful stimuli (Rollman, Abdel-Shaheed, Gillespie, & Jones, 2004); that is, the perceptual tendency to focus attention on threatening stimuli.

5.2.2 Attention and Hypervigilance to Pain Cues

Processing pain draws on attentional resources (Veldhuijzen, Kenemans, de Bruin, Olivier, & Volkerts, 2006). Areas of the pain matrix—such as the posterior parietal cortex, the ACC, dorsolateral prefrontal cortex, and thalamus—also belong to attentional networks (Bantick et al., 2002). Selective attention to relevant stimuli activates descending, facilitatory pain-modulation systems involved in sensitization of dorsal horn neurons (Zusman, 2002). While distraction from a noxious stimulus reduces pain intensity scores and associated activity in the pain matrix (Bantick et al., 2002), directing attention toward a painful stimulus increases its perceived intensity and unpleasantness (Miron, Duncan, & Bushnell, 1989). During distraction from painful stimuli, the ACC—which has a pivotal role in executive processes, motivation, allocation of attentional resources, premotor functions, and error detection—shows increased activation (Bantick et al., 2002). Pain related activity (in the medial prefrontal cortex and cerebellum) is modulated by the demand level of cognitive tasks (Wiech et al., 2005) and, accordingly, pain ratings are significantly lower when performing a high load compared with low load attentional task (Veldhuijzen et al., 2006). While attentional, motivational, and cognitive functions associated with a physiological state of injury may be processed through shared structures, these processes also draw on resources implemented in at least partly different cortical areas (Hsieh, Belfrage, Stoneelander, Hansson, & Ingvar, 1995).

The descending brainstem pathways implicated in central sensitization to pain include fibers from limbic 'emotional' structures (e.g., amygdala) and attentional networks (such as the ACC) (Zusman, 2002). Negative emotions such as depressive and anxious feelings, pain catastrophizing and somatic awareness extend the attentional demand of painful stimuli (Arntz, Dreessen, &

Merckelbach, 1991; Veldhuijzen et al., 2006), and can trigger neuropathic pain, such as phantom limb pain (Katz, 1992; Melzack, Isreal, Lacroix, & Schultz, 1997; Wilkins, McGrath, Finley, & Katz, 2004). Mechanisms related to fear might contribute to greater attention to pain stimuli, leading to hypervigilance for pain-related stimuli (Sullivan, Martel, Tripp, Savard, & Crombez, 2006). Early and/or frequent pain experiences may sensitise one—both neurally (with central sensitization) and behaviorally (with hypervigilance)—to rate experienced pain as more intense (Rollman et al., 2004). Furthermore, anticipation of pain appears to pre-cue the pain matrix, resulting in changes in activity of cortical networks for nociception (Porro et al., 2002); in particular, the ACC and the parietal operculum/posterior insula show increased activity during anticipation of pain, and may be the neural circuit that conveys somatosensory information to the limbic system relating painful events with the relevant affective states (Sawamoto et al., 2000). The expectation of painful stimulation even amplifies the perceived unpleasantness of otherwise innocuous stimuli (Sawamoto et al., 2000). Pain perception is transiently heightened when subjects observe images of human pain concomitant with receiving pain stimulation, most likely due to the automatic generation of an emotionally empathic reaction (Godinho, Magnin, Frot, Perchet, & Garcia-Larrea, 2006).

High levels of catastrophizing (i.e., rumination, magnification of the threat value of stimuli, and helplessness behaviors) are associated with more intense pain experience—together with greater activation of ACC—more pronounced displays of pain behavior, heightened emotional distress and greater disability (Sullivan et al., 2006). Thus individuals with high levels of catastrophizing experience more intense pain than low catastrophizers, and also rate the pain of others to be more intense. Preferential processing of, and attention to, pain behavior might underlie the relationship between catastrophizing and perception of heightened pain in others.

6 Summary

Synaesthesia for pain is most likely a consequence of disinhibited activation or central sensitization, of a fundamentally adaptive system for the empathic perception of pain in another. The known cases of synaesthesia for pain exhibit experiences associated with central sensitization to pain, including (a) hyperalgesia (Bradshaw & Mattingley, 2001); (b) long duration painful experience (e.g., foot spurs throughout child- and adulthood); (c) traumatic amputation, often precipitated by vascular or mechanical pain; and (d) painful distressing childbirth. What is not clear is why more people who have had intensely painful experiences have not reported synaesthesia for pain. In particular, amputees have been the subject of much pain research and have all endured some degree of traumatic, vascular, or mechanical pain, whether they consciously recall it or not; however, there are no prior reports of synaesthesia for pain in amputees.

There are many unanswered questions about why and how cases of synaesthesia for pain differ from other people—normal and clinical—who do not report synaesthesia for pain; including: First, how common is synaesthesia for pain, and in which clinical (e.g., burns victims) and non-clinical populations does it predominantly manifest? Second, how do the implications of synaesthesia for pain differ across populations? For example, mothers with mirrored pain who have experienced a more traumatic childbirth (a) may exhibit greater empathy for pain and intimacy with their infant, promoting the child's survival, or alternatively (b) may exhibit greater catastrophizing and maternal distress, leading to over-protection following normal painful experiences in the infant. Third, how does mirrored pain relate to one's level of empathy for another, or perspective-taking (theory of mind), and suggestibility (for example, on hypnotisability scales)? Fourth, is there a familial aspect? Fifth, are there gender differences? Finally, research is required to specifically explore (a) whether synaesthesia for pain involves MNS activity in the pain matrix and motor systems; and (b) whether people with dysfunctional MNSs, such as those with autism, are less likely to exhibit empathy for pain, or synaesthesia for pain. Further understanding of synaesthesia for pain has major implications for management of clinical pain as well as understanding the mechanisms underlying empathy for pain.

References

Abler, B., Walter, H., & Erk, S. (2005). Neural correlates of frustration. *NeuroReport, 16*(7), 669–672.

Allen, R. P., Picchietti, D., Hening, W. A., Trenkwalder, C., Walters, A. S., & Montplaisir, J. (2003). Restless legs syndrome: Diagnostic criteria, special considerations, and epidemiology. A report from the restless legs syndrome diagnosis and epidemiology workshop at the National Institutes of Health. *Sleep Medicine, 4*(2), 101–119.

Archibald, S. J., Meateer, C. A., & Kerns, K. A. (2001). Utilization behavior: Clinical manifestations and neurological mechanisms. *Neuropsychology Reviews, 11*(3), 117–130.

Arntz, A., Dreessen, L., & Merckelbach, H. (1991). Attention, Not Anxiety, Influences Pain. *Behaviour Research and Therapy, 29*(1), 41–50.

Avenanti, A., Bueti, D., Galati, G., & Aglioti, S. M. (2005). Transcranial magnetic stimulation highlights the sensorimotor side of empathy for pain. *Nature Neuroscience, 8*(7), 955–960.

Bach, P., & Tipper, S. P. (2006). Bend it like Beckham: Embodying the motor skills of famous athletes. *The Quarterly Journal of Experimental Psychology, 59*(12), 2033–2039.

Bach, P., & Tipper, S. P. (2007). Implicit action encoding influences personal-trait judgements. *Cognition, 102*, 151–178.

Bantick, S. J., Wise, R. G., Ploghaus, A., Clare, S., Smith, S. M., & Tracey, I. (2002). Imaging how attention modulates pain in humans using functional MRI. *Brain, 125*(2), 310–319.

Bavelas, J. B., Black, A., Lemery, C. R., & Mullett, J. (1987). Motor mimicry as primitive empathy. In N. Eisenberg & J. Strayer (Eds.), *Empathy and its development* (pp. 317–338). New York: Cambridge University Press.

Binkofski, F., & Buccino, G. (2004). Motor functions of the Broca's region. *Brain and Language, 89*(2), 362–369.

Blakemore, S.-J., Bristow, D., Bird, G., Frith, C., & Ward, J. (2005). Somatosensory activations during the observation of touch and a case of vision-touch synaesthesia. *Brain, 128*(7), 1571–1583.

Both, S., Everaerd, W., & Laan, E. (2003). Modulation of spinal reflexes by aversive and sexually appetitive stimuli. *Psychophysiology, 40*, 174–183.

Bradshaw, J. L. (2003). Gesture in language evolution: Could I but raise my hand to it! *Behavioral and Brain Sciences, 26*(2), 213–214.

Bradshaw, J. L., & Mattingley, J. B. (2001). Allodynia: A sensory analogue of motor mirror neurons in a hyperaesthetic patient reporting instantaneous discomfort to another's perceived sudden minor injury? *Journal of Neurology, Neurosurgery and Psychiatry, 70*(1), 135–140.

Buccino, G., Binkofski, F., Fink, G. R., Fadiga, L., Fogassi, L., Gallese, V., et al. (2001). Action observation activates premotor and parietal areas in a somatotopic manner: An fMRI study. *European Journal of Neuroscience, 13*(2), 400–404.

Buccino, G., Binkofski, F., & Riggio, L. (2004). The mirror neuron system and action recognition. *Brain and Language, 89*(2), 370–376.

Buccino, G., Riggio, L., Melli, G., Binkofski, F., Gallese, V., & Rizzolatti, G. (2005). Listening to action-related sentences modulates the activity of the motor system: A combined TMS and behavioural study. *Cognitive Brain Research, 24*(3), 355–363.

Carr, L., Iacoboni, M., Dubeau, M.-C., Mazziotta, J. C., & Lenzi, G. L. (2003). Neural mechanisms of empathy in humans: A relay from neural systems for imitation to limbic areas. *Proceedings of the National Academy of Sciences of the United States of America, 100*(9), 5497–5502.

Chartrand, T. L., & Bargh, J. A. (1999). The chameleon effect: The perception-behaviour link and social interaction. *Journal of Personality and Social Psychology, 76*(6), 893–910.

Corballis, M. C. (2002). *From hand to mouth : The origins of language*. Princeton: Oxford Princeton University Press.

Corballis, M. C. (2003). From mouth to hand: Gesture, speech and the evolution of right-handedness. *Behavioral and Brain Sciences, 26*, 199–260.

De Renzi, E., Francesca, C., & Stefano, F. (1996). Imitation and utilisation behaviour. *Journal of Neurology, Neurosurgery & Psychiatry, 61*(4), 396–400.

de Vignemont, F., & Singer, T. (2006). The empathic brain: How, when and why? *Trends in Cognitive Sciences, 10*(10), 435–441.

Effron, D. A., Niedenthal, P. A., Gil, S., & Droit-Volet, S. (2006). Embodied temporal perception of emotion. *Emotion, 6*(1), 1–9.

Eisenberger, N. I., Lieberman, M. D., & Williams, K. D. (2003). Does rejection hurt? An fMRI study of social exclusion. *Science, 302*, 290–292.

Ekman, P., & Friesen, W. V. (1971). Constants across cultures in the face and emotion. *Journal of Personality and Social Psychology, 17*(2), 124–129.

Emery, N. J. (2000). The eyes have it: The neuroethology, function and evolution of social gaze. *Neuroscience and Biobehavioural Reviews, 24*(6), 581–604.

Fadiga, L., Craighero, L., & Olivier, E. (2005). Human motor cortex excitability during the perception of others' action. *Current Opinion in Neurobiology, 15*(2), 213–218.

Flöel, A., Ellger, T., Breitenstein, C., & Knecht, S. (2003). Language perception activates the hand motor cortex: Implications for motor theories of speech perception. *European Journal of Neuroscience, 18*, 704–708.

Flor, H. (2002). Phantom limb pain. In V. S. Ramachandran (Ed.), *Encyclopedia of the human brain* (Vol. 3). New York: Academic Press.

Gallese, V. (2001). The 'shared manifold' hypothesis - From mirror neurons to empathy. *Journal of Consciousness Studies, 8*(5–7), 33–50.

Gallese, V. (2003). The roots of empathy: The shared manifold hypothesis and the neural basis of intersubjectivity. *Psychopathology, 36*(4), 171–180.

Giummarra, M. J., & Bradshaw, J. L. (2007). *Synaesthesia for pain following painful labour*. Australia, Clayton: Monash University.

Giummarra, M. J., Georgiou-Karistianis, N., Gibson, S. J., Chou, M., & Bradshaw, J. L. (2006, 25–31 August). *The menacing phantom: What triggers phantom limb pain and why?* Paper presented at the Australasian Winter Conference on Brain Research, Queenstown, New Zealand.

Giummarra, M. J., Gibson, S. J., Georgiou-Karistianis, N., & Bradshaw, J. L. (2007). Central mechanisms in phantom limb perception: the past, present and future. *Brain Research Reviews, 54*(1), 219–232.

Godinho, F., Magnin, M., Frot, M., Perchet, C., & Garcia-Larrea, L. (2006). Emotional modulation of pain: Is it the sensation or what we recall? *The Journal of Neuroscience, 26*(44), 11454–11461.

Hamzei, F., Rijntjes, M., Dettmers, C., Glauche, V., Weiller, C., & Buchel, C. (2003). The human action recognition system and its relationship to Broca's area: An fMRI study. *Neuroimage, 19*(3), 637–644.

Hatfield, E., Cacioppo, J. T., & Rapson, R. L. (1994). *Emotional contagion*. Cambridge, MA: Press Syndicate of the University of Cambridge.

Heyes, C., Bird, G., Johnson, H., & Haggard, P. (2005). Experience modulates automatic imitation. *Cognitive Brain Research, 22*, 233–240.

Hill, M. L., & Craig, K. D. (2002). Detecting deception in pain expressions: the structure of genuine and deceptive facial displays. *Pain, 98*(1–2), 135–144.

Hill, M. L., & Craig, K. D. (2004). Detecting deception in facial expressions of pain: Accuracy and training. *Clinical Journal of Pain, 20*(6), 415–422.

Hsieh, J. C., Belfrage, M., Stoneelander, S., Hansson, P., & Ingvar, M. (1995). Central representation of chronic ongoing neuropathic pain studied positron emission tomography. *Pain, 63*(2), 225–236.

Iacoboni, M., & Dapretto, M. (2006). The mirror neuron system and the consequences of its dysfunction. *Nature Reviews Neuroscience, 7*, 942–951.

Iacoboni, M., Woods, R. P., Brass, M., Bekkering, H., Mazziotta, J. C., & Rizzolatti, G. (1999). Cortical mechanisms of human imitation. *Science, 286*(5449), 2526–2528.

Jackson, P. L., Brunet, E., Meltzoff, A. N., & Decety, J. (2006). Empathy examined through the neural mechanisms involved in imagining how I feel versus how you feel pain. *Neuropsychologia, 44*(5), 752–761.

Jackson, P. L., Meltzoff, A. N., & Decety, J. (2005). How do we perceive the pain of others? A window into the neural processes involved in empathy. *NeuroImage, 24*, 771–779.

Jackson, P. L., Rainville, P., & Decety, J. (2006). To what extent do we share the pain of others? Insight from the neural bases of pain empathy. *Pain, 125*(1–2), 5–9.

Ji, R. R., Kohno, T., Moore, K. A., & Woolf, C. J. (2003). Central sensitization and LTP: do pain and memory share similar mechanisms? *Trends in Neurosciences, 26*(12), 696–705.

Katz, J. (1992). Psychophysiological contributions to phantom limbs. *Canadian Journal of Psychiatry, 37*, 282–298.

Keysers, C., Wicker, B., Gazzola, V., Anton, J. L., Fogassi, L., & Gallese, V. (2004). A touching sight: SII/PV activation during the observation and experience of touch. *Neuron, 42*, 335–346.

Lakin, J. L., & Chartrand, T. L. (2003). Using nonconscious behavioural mimicry to create affiliation and rapport. *Psychological Science, 14*(4), 334–339.

Lakin, J. L., Jefferis, V. E., Cheng, C. M., & Chartrand, T. L. (2003). The chameleon effect as social glue: Evidence for the evolutionary significance of nonconscious mimicry. *Journal of Nonverbal Behaviour, 27*(3), 145–162.

Lhermitte, F., Pillon, B., & Serdaru, M. (1986). human autonomy and the frontal lobes: I. Imitation and utilization behavior: A neurpsychological study of 75 patients. *Annals of Neurology, 19*(4), 326–334.

Mattingley, J. B. (2002). Visuomotor adaptation to optical prisms: A new cure for spatial neglect? *Cortex, 38*, 277–283.
McCabe, C. S., Haigh, R. C., Ring, E. F. J., Halligan, P., Wall, P. D., & Blake, D. R. (2003). A controlled pilot study of the utility of mirror visual feedback in the treatment of complex regional pain syndrome (type 1). *Rheumatology, 42*(1), 97–101.
Meltzoff, A. N., & Moore, M. K. (1977). Imitation of facial and manual gestures by human neonates. *Science, 198*, 75–78.
Melzack, R., Isreal, R., Lacroix, R., & Schultz, G. (1997). Phantom limbs in people with congenital limb deficiency or amputation in early childhood. *Brain, 120*, 1603–1620.
Miron, D., Duncan, G. H., & Bushnell, M. C. (1989). Effects of attention on the intensity and unpleasantness of thermal pain. *Pain, 39*(3), 345–352.
Morrison, I., Lloyd, D., & Di Pellegrino, G. (2004). Vicarious responses to pain in anterior cingulate cortex: Is empathy a multisensory issue? *Cognitive, Affective, & Behavioural Neuroscience, 4*(2), 270–278.
Myowa-Yamakoshi, M., Tamonaga, M., Tanaka, M., & Matsuzawa, T. (2003). Preference for human direct gaze in infant chimpanzees (Pan troglodytes). *Cognition, 89*, B53–B64.
Platek, S. M., Critton, S. R., Myers, T. E., & Gallup, G. G. (2003). Contagious yawning: the role of self-awareness and mental state attribution. *Cognitive Brain Research, 17*, 223–227.
Porro, C. A., Baraldi, P., Pagnoni, G., Serafini, M., Facchin, P., Maieron, M., et al. (2002). Does anticipation of pain affect cortical nociceptive systems? *The Journal of Neuroscience, 22*(8), 3206–3214.
Preston, S. D., & de Waal, F. B. M. (2002). Empathy: Its ultimate and proximate bases. *Behavioral and Brain Sciences, 25*, 1–72.
Reid, V. M., Stiano, T., Kaufman, J., & Johnson, M. H. (2004). Eye gaze cueing facilitates neural processing of objects in 4-mont-old infants. *NeuroReport, 15*(16), 2553–2555.
Ricciardelli, P., Bricolo, E., Aglioti, S. M., & Chelazzi, L. (2002). My eyes want to look where your eyes are looking: Exploring the tendency to imitate another individual's gaze. *NeuroReport, 13*(17), 2259–2264.
Rinn, W. E. (1984). The neuropsychology of facial expression: A review of the neurological and psychological mechanisms for producing facial expressions. *Psychological Bulletin, 95*(1), 52–77.
Rizzolatti, G., & Arbib, M. A. (1998). Language within our grasp. *Trends in Neuroscience, 21*(5), 188–194.
Rizzolatti, G., Camarda, R., Fogassi, L., Gentilucci, M., Luppino, G., & Matelli, M. (1988). Functional organization of inferior area 6 in the macaque monkey. II. Area F5 and the control of distal movements. *Experimental Brain Research, 71*, 491–507.
Rizzolatti, G., Craighero, L., & Fadiga, L. (2002). The mirror system in humans. In M. I. Stamenov & V. Gallese (Eds.), *Mirror neurons and the evolution of brain and language* (Vol. 42, pp. 37–59). Amsterdam: John Benjamins Publishing Company.
Rollman, G. B., Abdel-Shaheed, J., Gillespie, J. M., & Jones, K. S. (2004). Does past pain influence current pain: biological and psychosocial models of sex differences. *European Journal of Pain, 8*(5), 427–433.
Ruby, P., & Decety, J. (2004). How would you feel versus how do you think she would feel? A neuroimaging study of perspective taking with social emotions. *Journal of Cognitive Neuroscience, 16*(6), 988–999.
Russell, J. A. (1994). Is there universal recognition of emotion from facial expression? A review of the cross-cultural studies. *Psychological Bulletin, 115*(1), 102–141.
Saarela, M., Hlushchuk, Y., de C Williams, A. C., Schürmann, M., Kalso, E., & Hari, R. (2007). The compassionate brain: Humans detect intensity of pain from another's face. *Cerebral Cortex, 17*(1), 230–237.

Sawamoto, N., Honda, M., Okada, T., Hanakawa, T., Kanda, M., Fukuyama, H., et al. (2000). Expectation of pain enhances responses to nonpainful somatosensory stimulation in the anterior cingulate cortex and parietal operculum/posterior insula: An event-relate functional magnetic resonance imaging study. *Journal of Neuroscience, 20*(19), 7438–7445.

Schwoebel, J., Boronat, C. B., & Coslett, H. B. (2002). The man who executed "imagined" movements: Evidence for dissociable components of the body schema. *Brain and Cognition, 50*(1), 1–16.

Siegel, J. M. (2006). The stuff dreams are made of: Anatomical substrates of REM sleep. *Nature Neuroscience, 9*(6), 721–722.

Simon, D., Craig, K. D., Miltner, W. H. R., & Rainville, P. (2006). Brain responses to dynamic facial expressions of pain. *Pain, 126*(1–3), 309–318.

Singer, T. (2006). The neuronal basis and ontogeny of empathy and mind reading: Review of literature and implications for future research. *Neuroscience and Biobehavioural Reviews, 30*, 855–863.

Singer, T., Seymour, B., O'Doherty, J., Kaube, H., Dolan, R. J., & Frith, C. (2004). Empathy for pain involves the affective but not sensory components of pain. *Science, 303*(5661), 1157–1162.

Sonnby-Borgström, M. (2002). Automatic mimicry reactions as related to differences in emotional empathy. *Scandinavian Journal of Psychology, 43*, 433–443.

Sonnby-Borgström, M., Johnsson, P., & Svensson, O. (2003). Emotional empathy as related to mimicry reactions at different levels of information processing. *Journal of Nonverbal Behaviour, 27*(1), 3–23.

Sullivan, M. J. L., Martel, M. O., Tripp, D. A., Savard, A., & Crombez, G. (2006). Catastrophic thinking and heightened perception of pain in others. *Pain,* 123(1–2), 37–44.

Suzuki, R., & Dickenson, A. (2005). Spinal and supraspinal contributions to central sensitization in peripheral neuropathy. *Neurosignals, 14*(4), 175–181.

van Damme, S., Crombez, G., Eccleston, C., & Goubert, L. (2004). Impaired disengagement from threatening cues of impending pain in a cross-modal cueing paradigm. *European Journal of Pain, 8*, 227–236.

van der Velde, J., & Everaerd, W. (2001). The relationship between involuntary pelvic floor muscle activity, muscle awareness and experienced threat in women with and without vaginismus. *Behaviour Research and Therapy, 39*, 395–408.

Veldhuijzen, D. S., Kenemans, J. L., de Bruin, C. M., Olivier, B., & Volkerts, E. R. (2006). Pain and attention: Attentional disruption or distraction? *The Journal of Pain, 7*(1), 11–20.

Wicker, B., Keysers, C., Plailly, J., Royet, J. P., Gallese, V., & Rizzolatti, G. (2003). Both of us disgusted in my insula: The common neural basis of seeing and feeling disgust. *Neuron, 40*(3), 655–664.

Wiech, K., Seymour, B., Kalisch, R., Stephan, K. E., Koltzenburg, M., Driver, J., et al. (2005). Modulation of pain processing in hyperalgesia by cognitive demand. *NeuroImage, 27*, 59–69.

Wilkins, K. L., McGrath, P. J., Finley, G. A., & Katz, J. (2004). Prospective diary study of nonpainful and painful phantom sensations in a preselected sample of child and adolescent amputees reporting phantom limbs. *Clinical Journal of Pain, 20*(5), 293–301.

Williams, A. C. D. (2002). Facial expressions of pain: An evolutionary account. *Behavioral and Brain Sciences, 25*(4), 439–488.

Wilson, M. (2006). Covert imitation: How the body schema acts as a prediciton device. In G. Knoblich, I. M. THornton, M. Grosjean & M. Shiffrar (Eds.), *Human body perception from the inside out* (pp. 211–228). New York: Oxford University Press.

Wohldmann, E. L., Healy, A. F., & Bourne, L. E. (2007). Pushing the limits of imagination: Mental practice for learning sequences. *Journal of Experimental Psychology – Learning, Memory and Cognition, 33*(1), 254–261.

Yabar, Y., Johnston, L., Miles, L., & Peace, V. (2006). Implicit behavioural mimicry: Investigating the impact of group membership. *Journal of Nonverbal Behaviour, 30*(3), 97–113.

Zajonc, R. B., Adelmann, K. A., Murphy, S. T., & Niedenthal, P. M. (1987). Convergence in the physical appearance of spouses. *Motivation and Emotion, 11*, 335–346.

Zusman, M. (2002). Forebrain-mediated sensitization of central pain pathways: 'non-specific' pain and a new image for MT. *Manual Therapy, 7*(2), 80–88.

Part VI
Alternative Views

Mirroring, Mindreading, and Simulation

Alvin I. Goldman

Abstract What is the connection between mirror processes and mindreading? The paper begins with definitions of mindreading and of mirroring processes. It then advances four theses: (T1) mirroring processes in themselves do not constitute mindreading; (T2) some types of mindreading ('low-level' mindreading) are based on mirroring processes; (T3) not all types of mindreading are based on mirroring ('high-level' mindreading); and (T4) simulation-based mindreading includes but is broader than mirroring-based mindreading. Evidence for the causal role of mirroring in mindreading is drawn from intention attribution, emotion attribution, and pain attribution. Arguments for the limits of mirroring-based mindreading are drawn from neuroanatomy, from the lesser liability to error of mirror-based mindreading, from the role of imagination in some types of mindreading, and from the restricted range of mental states involved in mirroring. 'High-level' simulational mindreading is based on enactment imagination, perspective shifts or self-projection, which are found in activities like prospection and memory as well as theory of mind. The role of cortical midline structures in executing these activities is examined.

Keywords Cortical midline structures · Emotion · Folk psychology · Intention · Mentalizing · Mindreading · Mirror neurons · Mirroring processes · Pain · Simulation · Self-projection · Social cognition · Theory of mind

1 Introduction

Mirror systems are well established as a highly robust feature of the human brain (Rizzolatti, Fogassi, & Gallese, 2004; Gallese, Keysers, & Rizzolatti, 2004; Iacoboni et al., 1999; Rizzolatti & Craighero, 2004). Mirror systems

A.I. Goldman
Department of Philosophy, Center for Cognitive Science, Rutgers, The State University of New Jersey, New Brunswick/Piscataway, NJ
goldman@philosophy.nutgers.edu

J.A. Pineda (ed.), *Mirror Neuron Systems*, DOI: 10.1007/978-1-59745-479-7_14,

and mirroring processes are found in many domains, including action planning, sensation, and emotion (for reviews, see Keysers & Gazzola, 2006; Gallese et al., 2004; Goldman, 2006). Since mirroring commonly features an interpersonal matching or replication of a cognitive or mental event, it is a social interaction. It involves two people sharing the same mental-state type, although activations in observers are usually at a lower level than endogenous ones, commonly below the threshold of consciousness. There is strong evidence that mirror systems play pivotal roles in empathy and imitation (Iacoboni et al., 1999; Rizzolatti, 2005; Iacoboni, 2005; Gallese, 2005; Decety & Chaminade, 2005). Indeed, in a minimal sense of the term, 'empathy' might simply *mean* the occurrence of a mirroring process. In this paper, however, I shall focus on the connection between mirroring processes and another category of social cognition, namely mindreading or mentalizing.

By 'mindreading' I mean the attribution of a mental state to self or other. In other words, to mindread is to form a judgment, belief, or representation that a designated person occupies or undergoes (in the past, present, or future) a specified mental state or experience. This judgment may or may not be verbally expressed. Clearly, not all judgments about other people are acts of mindreading. To judge that someone makes a certain facial expression, or performs a certain action, or utters a certain sound is not to engage in mindreading, because these are not attributions of mental states. To attribute a mental state, the judgment must deploy a mental concept or category. Thus, if 'empathize' simply means 'echo the emotional state of another,' empathizing is not sufficient for mindreading. A person who merely echoes another's emotional state may not *represent* the second person at all, and may not represent her *as* undergoing that emotional state (a species of mental state). It is certainly possible that mirroring processes are responsible for acts of mindreading as well as for imitation or empathy, but this involvement in mindreading does not 'logically' follow from their role in either imitation or empathy. A connection between mirroring and mindreading must be considered separately.

2 Definitional Issues

The phrase 'mirroring process' can have either a wide sense or a narrow sense. In the wide sense, it refers to an interpersonal process that spans both sender and receiver. In a narrower sense, it refers to an intrapersonal process that includes only a receiver (a single individual). Unless otherwise specified, I shall understand 'mirroring process' in the narrow sense. To define 'mirroring process' in the narrow sense, we first need a definition of 'mirror neuron' or 'mirror system.' Rizzolatti, Fogassi, and Gallese (2004) offer the following definition of mirror neurons:

> Mirror neurons are a specific class of neurons that discharge both when the monkey performs an action and when it observes a similar action done by another monkey or the experimenter. (2004: 431)

This is a good definition of *action* mirror neurons but not mirror neurons in general. We do not want to restrict mirror neurons or mirroring processes to action-related events; they should equally be allowed in the domains of touch, pain, and emotion, for example. So let me propose a more general definition:

> Mirror neurons are a class of neurons that discharge both when an individual (monkey, human, etc.) undergoes a certain mental or cognitive event *endogenously* and when it observes a *sign* that another individual undergoes or is about to undergo the same type of mental or cognitive event.

One type of 'sign' to which an observer's mirror neuron might respond is a behavioral manifestation of the mental event in question; another example is a facial expression. A third type of sign is a stimulus that can be expected to produce the mental event in question. Thus, an observer's pain mirror neurons discharge when he sees a sharp knife being applied to someone else's body.

The above definition of mirror neurons can be extended essentially unchanged to mirror *systems* or *circuits*:

> Mirror systems are neural systems that get activated both when an individual undergoes a certain mental or cognitive event endogenously and when he observes a sign that another individual is undergoing, or is about to undergo, the same type of mental or cognitive event.

What counts as an 'endogenous' occurrence varies from one type of event to another. For present purposes I will not try to characterize endogenousness any further.

Even given the definitions of mirror neurons and mirror systems, it is not trivial to produce a definition of a *mirroring process*. One cannot say there is a mirroring process whenever a mirror neuron or mirror system discharges, or is activated. When a mirror neuron or system is endogenously activated, this is not a mirroring event. Only when a mirror neuron or system is activated in the *observation* mode is there a mirroring process. What exactly do we mean by 'observation mode'? Must the observer perceive a genuine behavioral manifestation or expression (etc.) of an endogenous mirror event? I think not. If a good imitation of a pained expression triggers an observer's pain mirror-neuron to fire, the process in the observer is still a mirroring process although the imitation is not a genuine expression, or sign, of an endogenous mirror event.

It would also be incorrect to say, as a simple definition, that every non-endogenous activation of a mirror neuron or mirror system is a mirroring process. An important non-endogenous mode of activation is imagination-generated activation. Motor imagery, for instance, is the result of imagining the execution of a motor act, and the generation of motor imagery uses largely the same mirror circuits as used in the endogenous generation of an action (M. Jeannerod, personal communication). However, we would not consider the process of creating motor imagery a type of mirroring process. In light of these points, let us define a mirroring process as follows:

> Neural process N is an mirroring process if and only if (1) N is an activation of a mirror neuron or mirror system, and (2) N results from observing something that is normally a behavioral or expressive manifestation (or a predictive sign) of a matching mirror event in another individual.

3 Four Theses about Mirroring Processes and Mindreading

With these clarifications and definitions in hand, let me now present the main theses I wish to advance and defend in this chapter.

(T1) Mirroring processes in themselves do not constitute mindreading.
(T2) *Some* acts of mindreading ('low-level' mindreading) are caused by, or based on, mirroring processes.
(T3) Not *all* acts of mindreading (in particular, not 'high-level' mindreading) are based on mirroring.
(T4) Simulation-based mindreading is broader than mirroring-based mindreading; some simulation-based mindreading (the 'low-level' type) involves mirroring and some of it (the 'high-level' type) does not.

In this section I will defend thesis T1, and in succeeding sections theses T2, T3, and T4.

An act of mindreading consists of a belief or judgment about a mental state. So, if a mirroring process in itself were to *constitute* mindreading (as opposed to merely *cause* it), the 'receiving' mirroring event would itself have to *be*, or *include*, a judgment or attribution of a mental state. In particular, it would have to be an attribution to a third person, presumably the originator of the mirroring process. Is there any reason to suppose that belief 'constitution' of this kind generally holds of mirroring processes?

Our definition of mirror neurons and mirror systems says that they are neural units that serve as substrates of one and the same cognitive event type, whether activated endogenously or observationally. This presumably implies that tokens of a mirror unit have substantially the same functional properties in whichever mode they are activated. If they had sharply different functional properties under different modes of activation, nobody would regard them as tokens of a mirroring type.[1] Having the same neuroanatomical location is not

[1] The functional properties of two mirror tokenings need not be identical, however. First, it is taken for granted that mirror discharges in execution and observation mode are not perfectly identical (for a review, see Csibra, 2007). In observation mode, the frequency or amplitude of firing may not coincide with that of the execution mode. Thus, the 'strength' of two tokenings may diverge slightly, with implications of slight differences in functional properties. Second, the Parma group from the beginning has distinguished between 'strictly' and 'broadly' congruent mirror neurons (Gallese, Fadiga, Fogassi, & Rizzolatti, 1996). In the case of

sufficient, because there are well-known cases in which the same neural region underpins more than one functionally distinct activity.[2] If that were the case in multi-modal cells or circuits, I doubt that anyone would speak of mirroring. So what are the cognitive or mental units that mirror neurons or mirror circuits underpin? They are units like "planning to grasp an object," "planning to tear an object," "feeling touch in bodily area X," "feeling pain in bodily area X," "feeling disgust", and so forth.

Now if the 'receiving' mirror events are tokens of the same event types (i.e., they co-instantiate the same event types), then they too will be units like "planning to grasp an object," "planning to tear an object," "feeling touch in bodily area X," and so forth. They will not also be beliefs, judgments, or attributions to the effect that the observed agent is planning to grasp an object, planning to tear an object, feeling touch in bodily area X, and so forth. If they were beliefs, judgments, or attributions of these sorts (in addition to being plannings, feelings, etc.), then, since they are mirroring events, the original endogenous occurrences would also have to be beliefs, judgments, or attributions with the same contents. But nobody has ever proposed that the sending mirror events are, or include, beliefs, judgments, or attributions. These are strong considerations in favor of thesis T1.

The truth of T1 does not spell doom for the idea that mirroring is pivotally involved in mindreading. T1 denies that a mirroring process in itself *constitutes* a mindreading event, but it allows a mirroring process to *cause* or *generate* a mindreading event. Lightning does not constitute thunder, but lightning can certainly cause thunder. There is strong evidence for causal links between mirroring processes and selected mindreading events. This is thesis T2, which is addressed in the next several sections.

4 Mirroring and Intention Attribution

Since mirror systems were initially discovered in the domain of motor planning, one would reasonably expect this to be the domain for which mirror-based mindreading is best supported. Also, because the first proposal of a possible link between mirroring processes and mindreading (Gallese & Goldman, 1998) focused on the motoric domain, one might expect this domain to be favored as a locus of evidence for this connection. This is especially so in light of two recent

---

broad congruence, functional properties are not identical. For present purposes, however, we can ignore this issue. Our approach focuses, for simplicity, on strictly congruent mirror neurons (or their analogue in mirror systems or circuits).

[2] For example, lesions to the fusiform gyrus of the right occipital lobe produce both prosopagnosia and achromatopsia (Bartels & Zeki, 2000). But these two deficits have no interesting functional relationship to one another. It just so happens that the impaired capacities are at least partially co-localized in the fusiform gyrus.

studies concerning mirror-based intention attribution, one pertaining to monkeys and one to humans. I do not agree entirely with the researchers' own interpretations of their findings, but the second paper does provide plausible evidence to endorse their principal conclusion with respect to humans.

Fogassi et al. (2005) studied the discharge of monkey parietal mirror neurons during the viewing of a grasping act that would be followed by one of two subsequent acts: either bringing the object to the mouth or placing it in a container. Different grasping neurons of the viewer coded one or the other of the subsequent acts (or neither). In other words, for most of these grasping neurons the level of discharge was influenced by the subsequent motor act, although the discharge occurred in the observing monkey before the subsequent act began. Fogassi et al. write: "Thus, these [mirror] neurons not only code the observed motor act but also allow the observer to understand the agent's intentions" (2005, p. 662). A similar moral is drawn from an experiment on humans by Iacoboni et al. (2005), namely, that the motor mirror system infers an agent's intention. Now, to attribute an intention is to engage in mindreading. Do such intention attributions occur in virtue of the mirroring processes in and of themselves?

This is possible, but it is not definitely implied by the experimental evidence. There are two rival, comparatively 'deflationary,' interpretations of the main findings that would not warrant this conclusion. The first rival interpretation would say that the parietal mirror neuron activity did not constitute the attribution of an *intention*, only the prediction of an *action*. The prediction of an action – since it is not the attribution of a mental state – would not qualify as mindreading. The second rival interpretation would say that the parietal mirror neuron activity in the observer constituted a simulation, or mimicking, of the agent's intention by the observer,[3] but not an intention attribution. *Possessing* an intention (or 'tokening' an intention, as philosophers would say) should not be confused with attributing such an intention to the agent. Only such an attribution would be a belief or judgment *about* an intention.

Under either of these rival interpretations the reported scanning results do not, in isolation, definitively imply mindreading. The observing monkey or human that underwent mirroring processes during the experiment underwent *some* sort of mental events related to the conditions that discriminated between the different intentions of the agent. But on one interpretation the relevant mirror events in the observer were merely *action* predictions (hence not *mind*-reading events); and on the other interpretation the mirror events were intention

[3] I usually speak of the simulation relation as holding between *processes* rather than *states* (including intentions). However, as I use the term, a process is a series of causally related states; so, as a limiting case, we may consider a state to be a process with a single member. Hence, we may also speak of states, such as intentions, as items that figure in simulation relations.

tokenings – again not mindreading events, this time because they were intentions rather than beliefs or attributions.[4]

If this were as far as the evidence goes, it would not firmly establish intention mindreading. But in fact there is additional evidence in the Iacoboni et al. study with human subjects, which had a special feature that lends support to intention attribution. After being scanned, participants were debriefed about the grasping actions they had witnessed. They all reported that they associated the intention of drinking with the grasping action in the 'during tea' condition and the intention of cleaning up with the grasping in the 'after tea' condition. These verbal reports did not depend on the instructions they had been given, that is, whether or not they had been instructed to pay attention to the agent's intention. So here there is independent evidence of intention attribution (at least in humans). One highly probable scenario, then, is that human observers both mirrored the agent's intention by undergoing a matching intention themselves, and, in addition, these intentions were the causal bases for intention attributions to the agent. On this interpretation there is no suggestion that the observers' intentions *constituted* attributions; but it is agreed that intention attributions occurred.

Admittedly, the participants' reports during the debriefing session do not show that their intention ascriptions were *caused* by mirroring processes. But that is a reasonable inference, fully consistent with all findings. Even more open is the question of where their intention ascriptions occurred, whether in a motor mirroring area or elsewhere. Certainly the observers' (mirroring) intentions occurred in a motor mirroring area. But whether the attributions also occurred there is undetermined. However, an act of mindreading produced by a mirroring process need not have a mirroring process as its substrate. An attribution does not have to be *part* of a mirroring process; it only has to be caused by such a process. So the Iacoboni et al. study does provide support, if not conclusive support, for thesis T2.

5 Mirror-Based Attribution of Emotion

Support for thesis T2 finds additional fertile ground in the mindreading of emotion (commonly labeled emotion 'recognition' in the studies to be reviewed). The best way to assemble the relevant evidence is to conjoin two sorts of studies: (1) studies of normal participants that establish the existence of mirroring processes for emotions, and (2) neuropsychological studies of emotion-specific brain damage showing that such damage is accompanied by

---

[4] De Vignemont and Haggard (in press) make a strong case for the claim that the best candidate for what is shared in a pair of mirroring events is an 'intention in action.' If this is right, it argues against the intention-*prediction* interpretation of the Iacoboni et al. (2005) imaging results *per se*.

selective impairment in attributing the same emotion. The two types of studies together yield convincing evidence that the substrate underpinning experience of an emotion is causally implicated in normal attribution of that emotion to an observed other. Failure to (fully) mirror an emotion in oneself while observing its expression in another prevents one from reliably attributing that emotion to the other.

The best example of this two-fold pattern of evidence pertains to disgust. Wicker et al. (2003) conducted an fMRI study of normal participants who were scanned both during the experience of disgust and during the observation of disgust-expressive faces. Participants viewed movies of individuals smelling the contents of a glass (disgusting, pleasant, or neutral) and spontaneously expressing the respective emotions. Then the same participants inhaled disgusting or pleasant odorants through a mask. The left anterior insula and the right anterior cingulate cortex were preferentially activated both during the experience evoked by inhaling disgusting odorants and during the observation of disgust-expressive faces. This establishes the existence of a mirroring process. However, the Wicker et al. study did not feature any emotion recognition tasks; so the study did not address the question of disgust attribution.

There are lesion studies, however, that contain relevant evidence about disgust attribution. Calder, Keane, Manes, Antoun, and Young (2000) studied patient NK who suffered insula and basal ganglia damage. In questionnaire responses NK showed himself to be selectively impaired in the experience of disgust (as contrasted with fear or anger). NK also showed significant and selective impairment in disgust recognition or attribution, which was established in two modalities: visual and auditory. Similarly, Adolphs, Tranel, and Damasio (2003) had a patient B who suffered extensive damage to the anterior insula (among other regions) and was able to recognize the six basic emotions *except disgust* when shown dynamic displays of facial expressions or told stories about actions. Apparently, an inability to mirror disgust because of damage to the anterior insula prevented these patients from attributing disgust, though their ability to attribute other basic emotions remained intact. It is a reasonable inference that when normal individuals recognize disgust when viewing the facial expression of disgust, this recognition is causally based on the production in the viewer of a (mirrored) experience of disgust.[5]

Analogous findings have been made in the case of fear, though here the results are a bit more qualified (Adolphs, Tranel, Damasio, & Damasio, 1994; Sprengelmeyer et al., 1999; Goldman & Sripada, 2005; Goldman, 2006,

---

[5] It is assumed in all of these studies that the participant not only ‘recognizes’ the emotion in the sense of classifying or categorizing it, but also views the emotion as occurring *in the observed target* (whose facial expression is shown or depicted). This implies that the participant is not merely categorizing the emotion but also attributing it *to* the target. If the categorization results from the mirroring process – which includes the observation of the target – it is hardly open to question that the attribution also results from the mirroring process. Thanks to F. de Vignemont for emphasizing this point.

pp. 115–116, pp. 119–124; Keysers & Gazzola, 2006). Turning to the emotion of anger, Lawrence, Calder, McGowan, and Grasby (2002) reported selective anger recognition impairment as a result of 'damage' to one of its 'substrates.' Previous studies indicated that the neurotransmitter dopamine is involved in the experience of anger. Lawrence et al. therefore hypothesized that a temporary, drug-induced suppression of the dopamine system would also result in impairment of the recognition of angry faces while sparing recognition of other emotions. This is indeed what they found, though this has not been replicated in other studies.

Another finding in the emotion category concerns the secondary emotion guilt. According to the 1991 revised psychopathy checklist (PCL-R), psychopaths lack remorse or guilt. Blair et al. (1995) examined the ability of psychopaths and nonpsychopathic controls to attribute emotions to others, using a story understanding task. Responses of psychopaths and controls to happiness, sadness, and embarrassment stories did not significantly differ. But psychopaths were significantly less likely than controls to attribute guilt to others. This is indirect evidence, once again, that possessing the substrate of an emotion is critical to accurate attribution of that emotion to others, implicating a mirroring process as critical to normal attribution.

6 Pain and Touch

Continuing with thesis T2, is there evidence that mirroring plays a causal role in the (third-person) attribution of sensations like touch or pain? We start with touch. Keysers et al. (2004) showed that large extents of the secondary somatosensory cortex that respond to a subject's own legs being touched also respond to the sight of someone else's legs being touched. This is a clear demonstration of empathy for touch (at least in a minimal sense of 'empathy'). However, there have not been tests to determine if observation-mediated somatosensory activity also causes attributions or judgments to the effect that another is undergoing such sensations.

Subsequent experiments do provide dramatic support for the mirroring-of-touch phenomenon, and even show that mirroring events can rise above the threshold of consciousness. Blakemore, Bristow, Bird, Frith, and Ward (2005) described a subject **C** for whom the observation of another person being touched is experienced as tactile stimulation on the equivalent part of **C**'s own body. They call this vision-touch synaesthesia. fMRI experiments also reveal that, in **C**, the mirror system for touch (in both SI and SII) is hyperactive, above the threshold for conscious tactile perception. Banissy and Ward (2007) followed up on this study and confirmed that synaesthetic touch feels like real touch. However, neither of these studies specifically addressed the question of whether synaesthetic touch leads the subject to attribute the felt touch to the observed person, which would be interpersonal mindreading. Their findings are entirely consistent with this claim, but their experimental manipulations did not specifically address the question.

There is more evidence for mirroring-based pain attribution. Mirror cells for pain were initially discovered serendipitously by Hutchison, Davis, Lozano, Tasker, and Dostrovsky (1999) while preparing a neurological patient for cingulotomy. More recently, Singer et al. (2004), Jackson, Meltzoff, and Decety (2004), and Morrison, Lloyd, de Pelligrino, and Roberts (2004) all reported pain resonance or mirroring. All three of these reports were restricted to the affective portion of the pain system, but subsequent transcranial magnetic stimulation (TMS) studies by Avenanti, Bueti, Galati, and Aglioti (2005; Avenanti, Paluello, Bufalari, & Aglioti, 2006) highlighted the sensorimotor side of empathy for pain.

On the question of whether mirrored pain can cause pain attribution to others, results from both Jackson et al. (2004) and Avenanti et al. (2005) are especially pertinent. Jackson et al. had subjects watch depictions of hands and feet in painful or neutral conditions and were asked to rate the intensity of pain they thought the target was feeling. This intensity rating is a third-person attribution task. There was a strong correlation between the ratings (attributions) of pain intensity and the level of activity within the posterior ACC (a crucial component of the affective portion of the pain network). This confirms the idea that a mirror-induced feeling can serve as the causal basis of third-person pain attribution.

Avenanti et al. (2005; for a review, see Singer & Frith, 2005) found that there is sharing of pain between self and others not only in the affective portion of the pain system but also in the fine-grained somatomotor representations. When a participant experiences pain, motor evoked potentials (MEPs) elicited by TMS indicate a marked reduction of corticospinal excitability. Avenanti and colleagues found a similar reduction of corticospinal excitability when participants saw someone else receiving a painful stimulus, for example, when participants watched a video showing a sharp needle being pushed into someone's hand. No change in excitability occurred when they saw a Q-tip pressing the hand or a needle being pushed into a tomato. The neural effects were quite precise. Corticospinal excitability measured from hand muscles was not affected by seeing a needle being thrust into someone's foot. Thus, there appears to be a pain resonance system that extracts basic sensory qualities of another person's painful experience and maps these onto the observers' own sensorimotor system in a somatotopically organized manner. Avenanti et al. also analyzed subjective judgments about the sensory and affective qualities of the pain ascribed to the model during needle penetration. These judgments were obtained by means of the McGill Pain Questionnaire (MPQ) and visual analogue scales, one for pain intensity and one for pain unpleasantness. The amplitude changes of MEPs recorded from the FDI muscle (the first dorsal interosseus) were negatively correlated with sensory aspects of the pain purportedly felt by the model during the 'Needle in FDI' condition, both for the Sensory scale of MPQ and for pain intensity on the visual analogue scale. Thus, judgments of sensory pain to the model seemed to be based on the mirroring process in the sensorimotor pain system. Finally, in a follow-up study, Avenanti et al. (2006) again found a significant reduction in amplitudes of MEPs correlated with the intensity of the pain being attributed to the model, and no MEPs modulation contingent upon different task instructions was found. In particular, specific sensorimotor neural responses did not depend on observers being explicitly asked to mentally simulate sensory qualities of others' sensations.

To sum up, there is adequate evidence in the case of pain to conclude that mirroring states or processes are often causally responsible for third-person mental attributions of the mirrored state, in further support of thesis T2.

7 The Limits of Mirror-Based Mindreading

There is clear evidence, then, that mirroring plays a causal role in certain types of mindreading. How wide a range of mindreading is open to a mirroring explanation? At present the range appears fairly narrow, because the types of mental activity known to participate in mirroring – namely motoric activity, sensation, and emotion – appear to be circumscribed, although their ramifications for other phenomena are quite pervasive. Is it possible, then, that massive amounts of other mindreading are also based on mirroring? Or are there principled reasons to think that other types of mindreading differ from mirror-based mindreading?

Thesis T3 denies that there is a universal connection between mindreading and mirroring. A salient reason for being dubious of a universal connection comes from extensive fMRI studies of 'theory of mind' that identify different brain regions associated with desire and belief attribution, perhaps a dedicated mentalizing network. These regions are disjoint both from the well-known motoric mirror areas and from the areas involving pain and emotion that are cited above as loci of mirroring-based mindreading. The so-called 'theory of mind' regions are sometimes called 'cortical midline structures', and consist in the medial frontal cortex (MFC, perhaps subsuming the anterior cingulate cortex), the temporo-parietal junction, the superior temporal sulcus, and the temporal poles. Because these structures are strongly associated with mentalizing – at least certain types of mentalizing – there is a neuroanatomical challenge to the thesis that *all* mindreading is the product of mirroring. This is the 'argument from neuroanatomy' for T3.

There are two challenges to this neuroanatomy argument for doubting the universality of mirroring in mindreading. The first challenge comes from the study of a stroke patient, *GT*, with extensive damage to the medial frontal lobes bilaterally, including regions identified as critical for 'theory of mind.' Bird, Castelli, Malik, Frith, and Husain (2004) carried out a thorough assessment of *GT*'s cognitive profile and found no significant impairment in 'theory of mind' tasks. They concluded that the extensive medial frontal regions destroyed by her stroke are not necessary for mindreading.

It is not clear, however, that medial prefrontal cortex should have been identified in the first place as the region dedicated to (high-level) mentalizing. Saxe and Wexler (2005) argue that the critical region is RTPJ (right temporo-parietal junction). If they are correct, then *GT* has no substantial bearing on the challenge to a universal role for mirroring in mindreading.

A more dramatic response to the neuroanatomy argument comes from the recent discovery of mirror neurons in the medial frontal lobe by Iacoboni's group (Mukamel, Ekstrom, Kaplan, Iacoboni, & Fried, 2007). Using single-cell recordings in humans, they found mirror cells for grasping and for facial emotional expressions in the medial frontal cortex, the sites being the pre-SMA/SMA proper complex, the dorsal sector of ACC, and the ventral sector of ACC. Iacoboni (personal communication) suggests that the findings may show that even the higher forms of mindreading are based on some mechanism of neural mirroring. Obviously, confirmation of the latter theory would undermine the argument from neuroanatomy. It remains to be seen, however, exactly which types of mindreading, if any, might be subserved by this new group of mirror cells.

Putting aside the argument from neuroanatomy, let us consider a second type of argument for the non-universality of mirroring-based mindreading: a theoretical argument that I will call the 'argument from error.' This argument says that some forms of mindreading are susceptible to a form of error to which mirror-based mindreading is not susceptible.[6] Therefore, not all mindreading is mirror-based. Let us spell this out.

Mirror-based mindreading is comparatively immune from error. In the first stage of mirror-based mindreading, the observer sees a behavioral or expressive sign in the agent that produces a matching mirror event in him. True, there might be a misfire here if the sign does not genuinely manifest a mental event of which it is typical. But this is not the kind of error I am thinking of in the case of 'other' forms of mindreading. In the second stage of mirror-based mindreading, the mindreader classifies the mental event 'received' from the agent and attributes it to the agent. If the classification process is normal, as well as the mirror-matching, the resulting act of mindreading will be accurate.

There are other types of mindreading, however, that are susceptible to a different kind of error. In particular, mindreading is prone to 'egocentric' errors, largely from failures of perspective taking. Children are especially prone to this kind of error but it is also found in adults. One form of perspective-taking failure

---

[6] Saxe (2005) uses a somewhat analogous argument from error to criticize the general simulation theory of mindreading. Here an argument from error is being used to resist the claim that all mindreading takes a specific simulationist form, namely mirroring-based mindreading. Many errors associated with non-mirror-based mindreading are readily accommodated by a second form of simulation, discussed below in section 8. More generally, see Goldman (2006, Chap. 7).

is the failure to inhibit self-perspective (for a review see Goldman, 2006, pp. 164–175). There is no place for this kind of error in mirror-based mindreading. Thus, there must be a kind of mindreading that does not fit the mirroring mold.[7]

Notice that my point does not rest on the claim that mirror-based mindreading leaves no room at all for error. Recent evidence suggests that mirroring does not always guarantee matching, because it can be modulated by other information or preferences. Singer et al. (2006) found that empathic responses to pain are modulated by learned preferences. Participants played an economic game in which two confederates played fairly or unfairly, and participants then underwent functional imaging while observing the confederates receiving pain stimuli. Participants of both sexes exhibited mirrored pain responses, but in males the responses were significantly reduced when observing an unfair person receiving pain. If these mirror responses also generated pain attributions of varying levels, the indicated modulation would tend to produce errors. This is one way that mirror-based mindreading is open to error, but it is quite different from patterns of error found in other cases of mindreading.

A third argument for the non-universality of mirror-based mindreading is more straightforward than the first two. It is the simple point that a great deal of mindreading is initiated by imagination and according to our definition of mirroring processes, imagination-driven events do not qualify as mirroring processes. Thus, if a person attempts to determine somebody else's mental state, not by observing their behavior or their facial or postural expression, but by learning about their situation from an informant's description, this act of mindreading will not involve a mirroring process. It may proceed by inference, by imagination, or by 'putting oneself in the target's shoes,' but none of these qualify as a mirroring process.

A fourth argument pertains to the types of mental states known to possess mirror properties. Most of these states are not states with propositional contents, like beliefs or desires; or if they do have propositional contents, these contents are of a bodily sort, pertaining to bodily location or bodily movement. Thus, states of pain and touch have mirror properties and the mirroring extends to their felt bodily locations. Intentions to act have mirror properties and these

---

[7] It might be replied that mirror-based mindreading *is* susceptible to egocentric error. F. de Vignemont (personal communication) suggests that if I myself have a terrible back pain and I see you carrying a heavy box, I would feel pain and ascribe this feeling to you. This might be an error because you are perfectly fine with a box that heavy; you are not in pain. Is not this an egocentric error? No doubt, it is an egocentric error. The question is whether it is a case of mirroring, at least a *pure* case of mirroring. It is not a case in which I see you exhibiting a behavioral or expressive manifestation of pain. And it is questionable whether the perceived heaviness of the box is a 'sign' of pain comparable to a knife or needle penetrating a body. It might be a case of inference- or imagination-caused pain rather than mirror-produced pain Admittedly, the case puts pressure on our definition of mirroring, but this is not a problem only for me. It is a problem for anyone seeking to be precise about what counts as mirroring. In any case, there are other arguments offered here in favor of thesis T3. It does not rest exclusively on the lesser-liability-to-error argument.

intentions have contents concerning the types of effectors used (hand, foot, mouth) and the types of actions intended (coded in rich motoric terms). But there is no evidence that beliefs, for example, have mirror properties, especially beliefs with abstract contents. Observing another person grasp or manipulate an object with his hands elicits in the observer a covert intention to grasp or manipulate an object. But observing someone else who is reflecting on the problem of global warming does not elicit a similar thought in one's own mind (except by sheer coincidence). Beliefs and other reflective states do not elicit matching contentful states by a mirroring process; nor do desires that go unexpressed in a distinctive motoric signature. Thus, there is a large class of mental states that are not mirrored. Since they surely are the targets of mindreading, they must be read in a different fashion. This establishes thesis T3.

8 High-Level Simulation-Based Mindreading

I turn finally to thesis T4. Many writers equate mirroring with simulation. So if a given mental state cannot be read by a mirroring process, it cannot be read by simulation. I take a different view (Goldman, 2006). Simulation and mirroring are not equivalent; mirroring is just one species of simulation. Hence, if a type of mental state is not readable by mirroring, it is still possible it can be read by simulating, just a different form of simulating. It is also possible, of course, that it can be read by theorizing, and I do not wish to deny that some acts of mindreading, partly or wholly, consist of theorizing (Goldman, 2006, 43–46). Here I shall focus on the second form of simulational mindreading.

The basic idea of simulation of this second kind is to 're-enact' or 're-create' a scenario in one's mind that differs from what one currently experiences in an endogenous fashion. It is to imagine a scenario, not merely in the sense of 'supposing' that it has occurred or will occur, but to imagine being immersed in, or witnessing, the scenario. In other words, it involves engaging in mental 'pretense' in which one tries to construct the scenario as one would experience or undergo it if it were currently happening. This is what philosopher-simulationists had in mind originally by 'simulation' (Gordon, 1986; Heal, 1986; Goldman, 1989, 2006; Currie & Ravenscroft, 2002), not mirroring, which is a more recent entrant onto the scene (Gallese & Goldman, 1998). Mirroring features an automatic re-creation in an observer's mind of an episode that initially occurs in another's mind. In 'enactment simulation,' by contrast, one *attempts* to create such a matching event without currently observing another person who undergoes it. One tries to construct the event with the help of experience or knowledge that, it is hoped, will facilitate the construction. Successful re-enactment or re-creation is more problematic than accurate mirroring. Re-enactment must typically be guided by knowledge stored in memory, the quality of which is quite variable. However, any attempt at re-enactment can be called 'simulation' whether or not there is successful, accurate matching (Goldman, 2006, p. 38).

Enactment simulation as sketched here approximates the notion of simulation evoked by the neuroscientists Buckner and Carroll (2007), who discuss it under the heading of 'self-projection.' They conceive self-projection as the mental exploration and construction of alternative perspectives to one's current actual perspective, including perspectives on one's own future ('prospection'), one's own past (autobiographical memory), the viewpoint of others (theory of mind), and navigation. Buckner and Carroll refer to imagining an alternative perspective as 'simulation.' They also argue that all these forms of self-projection involve a shared neural network involving frontal and medial temporo-parietal lobe systems that are traditionally linked to planning and episodic memory.

Buckner and Carroll cite a variety of evidence in support of their view, beginning with the fact that among the deficits created by frontal lobe lesions are deficits in planning and structuring events in an appropriate temporal sequence. Patients with frontal lesions often perform normally in well-established routines and can show high intellectual function, but when confronted with challenging situations and new environments, reveal an inability to plan. They are unable to order sequences temporally, plan actions on tasks requiring foresight, and adjust behaviors flexibly as rules change. Mesulum (2002) noted that the prefrontal cortex might have a pivotal role in the ability to "transpose the effective reference point [of perception] from the self to other, from here to there, and from now to then."

Other evidence concerns the medial temporal lobe, damage to which often causes amnesia. A lesser-studied aspect of the amnesic syndrome is the inability to conceive the personal future. In his seminal description of amnesia in Korsakoff's syndrome, Talland (1965) noted that his amnesic patients could say little about their future plans. The same was true of the amnesic patient HM. Similarly, Klein, Loffus, & Kihlstrom (2002) observed that their amnesic patient DB, when questioned about his future, either confabulated or did not know what he would be doing. Although DB had general knowledge of the future – he knew there was a threat of weather changes – he lacked the capacity to consider himself in the future.

A propos of theory of mind, Buckner and Carroll draw on Gallagher and Frith's (2003) account of the role of frontopolar cortex. They suggest that the paracingulate cortex, the anterior-most portion of the frontal midline, is recruited in executive components of simulating others' perspectives. This region is contiguous with but distinct from those involved in episodic remembering. Gallagher and Frith also conclude that this region helps to "determine [another's] mental state, such as a belief, that is decoupled from reality, and to handle simultaneously these two perspectives on the world." Obviously, this is the kind of ability crucial in solving false-belief tasks in mindreading.

If Buckner and Carroll are right that a (substantial sector) of mentalizing activities are simulations that conform to the foregoing description, and if they are right that such activities take place (roughly) in the brain systems they identify, then it appears that these are not *mirroring* activities. Nonetheless, they are *simulation* activities, in the sense intended Thus, a substantial chunk of mindreading is simulationist in character without being the product of mirroring. In *Simulating Minds* (Goldman, 2006) I distinguish two types of simulation for mindreading: 'low-level' and 'high-level'. Low-level simulation features mirroring and high-level simulation does not. *Simulating Minds* does not try to pinpoint precisely all the brain regions associated with high-level mindreading, and that is not essential here either. What is interesting about Buckner and Carroll's contribution is that it identifies a certain network or circuit of brain regions that accomplish a certain general type of task (adopting an alternative perspective), which is instantiated in other domains as well as mindreading. This tends to substantiate thesis T4.

9 Interactions between Cortical Midline Structures and Mirror Systems?

Uddin, Iacoboni, Lange, and Keenan (2007) propose a unifying model to account for data on self and social cognition by sketching links between cortical midline structures (CMS) and the (motor) mirror-neuron system (MNS). The former is taken to consist of the medial prefrontal cortex, the anterior cingulate cortex and the precuneus, and the latter is composed of the inferior frontal cortex and the rostral part of the inferior parietal lobule. They argue that a right-lateralized frontoparietal network that overlaps with mirror-neuron areas seems to be involved with self-recognition and social understanding. Because both MNS and CMS are involved in self-other representations, it seems only natural, they propose, that the two systems interact. One pathway by which this might occur is a direct connection between the precuneus (which they regard as a major node of the CMS) and the inferior parietal lobule (the posterior component of the MNS). Also there are direct connections between mesial frontal areas and the inferior frontal gyrus. Thus, the anterior and posterior nodes of the CMS and MNS are in direct communication.

However, it seems that MNS and CMS perform quite different functions vis-à-vis self-understanding. Both the self-face and the self-body activate the right frontoparietal network (Uddin, Kaplan, Molnar-szakacs, Zaidel, & Iacoboni, 2005; Sigiura et al., 2006). So the right-lateralized system, associated with the mirror-neuron system, seems to be related to representations of the *physical* self rather than the *mental* self. CMS structures, on the other hand, seem to be more involved in internal aspects of representing self and others, including mentalizing, as Uddin et al. (2007) themselves concede.

Uddin et al. (2007) propose a division of labor in which the CMS might support 'evaluative simulation' in the same way that the MNS supports 'motor simulation.' This division of labor between the two networks would yield specializations for two related processes that are crucial to navigating the social world: understanding physical actions of intentional agents and understanding the attitudes of others. It is unclear, however, exactly what they mean by 'evaluative simulation.' Not all the mentalizing work done by the CMS involves evaluation in any straightforward sense. Attributing beliefs to other people (including false beliefs) is a principal mentalizing activity executed by portions of the CMS. But there is nothing 'evaluative' (as opposed to 'descriptive') about a belief attribution; nor are beliefs themselves evaluative states. It is also unclear what Uddin et al. (2007) mean by 'simulation' in this context; they offer no explanation of this (somewhat slippery) notion. However, it appears that we agree on two important points: that simulation plays a central role in different sectors of mentalizing and that mirror-neuron systems perform only a portion, albeit a very fundamental portion, of the mentalizing work that the human mind undertakes.

10 Conclusions

Mirroring *per se* does not constitute mindreading. Nonetheless, there is evidence of mirroring-based mindreading in several domains, including action intention, emotion, and pain. Mirroring-based mindreading is what I call 'low-level' mindreading. There are also many reasons, however, to doubt that all mindreading is based on mirroring. How does this bear on the simulation theory of mindreading? Mirroring is one kind of simulational process but not the only one. Attempting to take another person's perspective, or put oneself in their shoes, is another type of simulational process, and this kind of process is extensively used in mindreading. Thus, simulation figures importantly in 'high-level' as well as 'low-level' mindreading.

Acknowledgments The author thanks Vittorio Gallese, Marco Iacoboni, and Frederique de Vignemont for detailed comments that resulted in many helpful changes in the manuscript. Other helpful comments were due to Holly M. Smith.

References

Adolphs, R., Tranel, D., & Damasio, A. R. (2003). Dissociable neural systems for recognizing emotions. *Brain and Cognition*, *52*, 61–69.

Adolphs, R., Tranel, D., Damasio, H., & Damasio, A. (1994). Impaired recognition of emotion in facial expressions following bilateral damage to the amygdala. *Nature*, *372*, 669–672.

Avenanti, A., Bueti, D., Galati, G., & Aglioti, S. M. (2005). Transcranial magnetic stimulation highlights the sensorimotor side of empathy for pain. *Nature Neuroscience*, *8*, 955–960.

Avenanti, A., Paluello, I. M., Bufalari, I., & Aglioti, S. M. (2006). Stimulus-driven modulation of motor-evoked potentials during observation of others' pain. *NeuroImage, 32*, 316–324.

Bartels, A., & Zeki, S. (2000). The architecture of the colour centre in the human visual brain: New results and a review. *European Journal of Neuroscience, 12*, 172–193.

Bird, C. M., Castelli, F., Malik, O., Frith, U., & Husain, M. (2004). The impact of extensive medial frontal lobe damage on 'Theory of Mind' and cognition. *Brain, 127*, 914–928.

Banissy, M. J., & Ward, J. (2007). Mirror-touch synaesthesia is linked with empathy. *Nature Neuroscience, 10*, 815–816.

Blair, R. J. R., Sellars, C., Strickland, I., Clark, F., Williams, A. O., Smith, M., et al. (1995). Emotion attributions in the psychopath. *Personality and Individual Differences, 19*, 431–437.

Blakemore, S.-J., Bristow, D., Bird, G., Frith, C., & Ward, J. (2005). Somatosensory activations during the observation of touch and a case of vision-touch synaesthesia. *Brain, 128*, 1571–1583.

Buckner, R. L., & Carroll, D. C. (2007). Self-projection and the brain. *Trends in Cognitive Sciences, 11*(2), 49–57.

Calder, A. J., Keane, J., Manes, F., Antoun, N., & Young, A. W. (2000). Impaired recognition and experience of disgust following brain injury. *Nature Reviews Neuroscience, 3*, 1077–1078.

Currie, G., & Ravenscroft, I. (2002). *Recreative Minds, Imagination in Philosophy and Psychology*. Oxford: Oxford University Press.

Csibra, G. (2007). Action mirroring and action interpretation: An alternative account. In P. Haggard, Y. Rosetti, & M. Kawato (Eds.), *Sensorimotor foundations of higher cognition: Attention and performance XXII*. Oxford: Oxford University Press.

Decety, J., & Chaminade, T. (2005). The neurophysiology of imitation and intersubjectivity. In S. Hurley & N. Chater (Eds.), *Perspectives on imitation: From neuroscience to social science* (pp. 119–140). Cambridge, MA: MIT Press.

De Vignemont, F., & Haggard, P. (in press). Action observation and execution: What is shared? *Social Neuroscience*.

Fogassi, L., Ferrari, P. F., Gesierich, B., Rozzi, S., Chersi, F., & Rizzolatti, G. (2005). Parietal lobe: from action organization to intention understanding. *Science, 308*, 662–667.

Gallagher, H. L., & Frith, C. D. (2003). Functional imaging of 'theory of mind'. *Trends in Cognitive Sciences, 7*, 77–83.

Gallese, V. (2005). "Being-like-me": Self-other identity, mirror neurons, and empathy. In S. Hurley & N. Chater (Ed.), *Perspectives on imitation*, (Vol. 1, pp. 101–118). Cambridge, MA: MIT Press.

Gallese, V., Fadiga, L., Fogassi, L., & Rizzolatti, G. (1996). Action recognition in the premotor cortex. *Brain, 119*, 593–609.

Gallese, V., & Goldman, A. I. (1998). Mirror neurons and the simulation theory of mind-reading. *Trends in Cognitive Sciences, 2*, 493–501.

Gallese, V., Keysers, C., & Rizzolatti, G. (2004). A unifying view of the basis of social cognition. *Trends in Cognitive Sciences, 8*, 396–403.

Goldman, A. I. (1989). Interpretation psychologized. *Mind and Language, 4*, 161–185.

Goldman, A. I. (2006). *Simulating minds: The philosophy, psychology, and neuroscience of mindreading*. New York: Oxford University Press.

Goldman, A. I., & Sripada, C. R. (2005). Simulationist models of face-based emotion recognition. *Cognition, 94*, 193–213.

Gordon, R. M. (1986). Folk psychology as simulation. *Mind and Language, 1*, 158–171.

Heal, J. (1986). Replication and functionalism. In J. Butterfield (Ed.), *Language, mind and logic*. Cambridge: Cambridge University Press.

Hutchison, W. D., Davis, K. D., Lozano, A. M., Tasker, R. R., & Dostrovsky, J. O. (1999). Pain-related neurons in the human cingulate cortex. *Nature Neuroscience, 2* (5), 403–405.

Iacoboni, M., Woods, R. P., Brass, M., Bekkering, H., Mazziota., J. C. et al. (1999). Cortical mechanisms of human imitation. *Science*, *286*, 2526–2528.
Iacoboni, M. (2005). Understanding others: imitation, language, and empathy. In S. Hurley & N. Chater, eds., *Perspectives on Imitation: From Neuroscience to Social Science* (pp. 76–100). Cambridge, MA: MIT Press.
Iacoboni, M., Molnar-Szakacs, I., Gallese, V., Buccino, G., Mazziotta, J. C., & Rizzolatti, G. (2005). Grasping the intentions of others with one's own mirror neuron system. *PLoS Biology*, *3*, 529–535.
Jackson, P. L., Meltzoff, A. N., & Decety, J. (2004). How do we perceive the pain of others? A window into the neural processes involved in empathy. *NeuroImage*, *24*, 771–779.
Keysers, C., & Gazzola, V. (2006). Towards a unifying neural theory of social cognition. *Progress in Brain Research*, *156*, 383–406.
Keysers, C., Wicker, B., Gazzola, V., Anton, J -L., Fogassi, L., & Gallese, V. (2004). A touching sight: SII/PV activation during the observation of touch. *Neuron*, *42*, 335–346.
Klein, S. B., Loffus, J., & Kihlstrom, J.F. (2002). Memory and temporal experience: the effect of episodic memory loss on an amnesic patient's ability to remember the past and imagine the future. *Social Cognition*, *20*, 353–379.
Lawrence, A. D., Calder, A. J., McGowan, S. M., & Grasby, P. M. (2002). Selective disruption of the recognition of facial expressions of anger. *NeuroReport*, *13* (6), 881–884.
Mesulum, M. M. (2002). The human frontal lobes: transcending the default mode through contingent encoding. In D. T. Stuss & R. T. Knight (Eds.), *Principles of Frontal Lobe Function* (pp. 8–30). Oxford: Oxford University Press.
Morrison, I., Lloyd, D., de Pelligrino, G., & Roberts, N. (2004). Vicarious responses to pain in anterior cingulate cortex. Is empathy a multisensory issue? *Cognitive Affective Behavioral Neuroscience*, *4*, 270–278.
Mukamel, R., Ekstrom, A. D., Kaplan, J., Iacoboni, M., & Fried, I. (2007). Mirror properties of single cells in human medial frontal cortex. *Social Neuroscience Abstracts*.
Rizzolatti, G. (2005). The mirror neuron system and imitation. In S. Hurley & N. Chater (Ed.), *Perspectives on imitation* (Vol. 1, pp. 55–76). Cambridge, MA: MIT Press.
Rizzolatti, G., & Craighero, L. (2004). The mirror-neuron system. *Annual Review of Neuroscience*, *27*, 169–192.
Rizzolatti, G., Fogassi, L., & Gallese, V. (2004). Cortical mechanisms subserving object grasping, action understanding, and imitation. In M. Gazzaniga (Ed.), *The cognitive neurosciences III* (pp. 427–440). Cambridge, MA: MIT Press.
Saxe, R. (2005). Against simulation: The argument from error. *Trends in Cognitive Sciences*, *9*, 174–179.
Saxe, R., & Wexler, A. (2005). Making sense of another mind: The role of the right temporoparietal junction. *Neuropsychologia*, *43*, 1391–1399.
Sugiura, M., Sassa, Y., Jeong, H., Miura, N., Akitsaki., Horie, K., Sato, S., & Kawashima, R. (2006). Multiple brain networks for visual self-recognition with different sensitivity for motion and body part. *Neuroimage*, *32*, 1905–1917.
Singer, T., & Frith, C. (2005). The painful side of empathy. *Nature Neuroscience*, *8*, 845–846.
Singer, T., Seymour, B., O'Doherty, J., Kaube, H., Dolan, R., & Frith, C. (2004). Empathy for pain involves the affective but not sensory components of pain. *Science*, *303*, 1157–1162.
Singer, T., Seymour, B., O'Doherty, J. Stephan, K. E., Dolan, R. J., & Frith, C. D. (2006). Empathic neural responses are modulated by the perceived fairness of others. *Nature*, *439*, 466–469.
Sprengelmeyer, R., Young, A. W., Schroeder, U., Grossenbacher, P. G., Federlein, J., Buttner et al. (1999). Knowing no fear. *Proceedings of the Royal Society, series B: Biology, 266*,: 2451–2456.

Talland, G. A. (1965). *Deranged Memory: A Psychonomic Study of the Amnesic Syndrome*. New York: Academic Press.

Uddin, L. Q., Kaplan, J.T., Molnar-Szakacs, I., Zaidel, E., & Iacoboni, M. (2005). Self-face recognition activates a frontoparietal 'mirror' network in the right hemisphere: an event-related fMRI study. *Neuroimage, 25*, 926–935.

Uddin, L. Q., Iacoboni, M., Lange, C., & Keenan, J. P. (2007). The self and social cognition: the role of cortical midline structures and mirror neurons. *Trends in Cognitive Sciences, 11* (4), 153–157.

Wicker, B., Keysers, C., Plailly, J., Royet, J -P., Gallese, V., & Rizzolatti, G. (2003). Both of us disgusted in *my* insula: The common neural basis of seeing and feeling disgust. *Neuron, 40*, 655–664.

Does the Mirror Neuron System and Its Impairment Explain Human Imitation and Autism?

Victoria Southgate, György Gergely, and Gergely Csibra

Abstract The proposal that the understanding and imitation of observed actions are made possible through the 'mirror neuron system' (Rizzolatti, G., Fogassi, L., & Gallese, V., 2001, Neurophysiological mechanisms underlying the understanding of action. *Nature Review Neuroscience, 2*, 661–670) has led to much speculation that a dysfunctional mirror system may be at the root of the social deficits characteristic of autism (e.g. Ramachandran, V. S., & Oberman, L. M., 2006, Broken mirrors: A theory of autism. *Scientific American*, November 2006). This chapter will critically examine the hypothesis that those with ASD may be in possession of a 'broken' mirror neuron system. We propose that the deficits seen in imitation in individuals with ASD reflect not a dysfunctional MNS, but a lack of sensitivity to those cues that would help them identify *what* to imitate. In doing this, we will also argue that imitation in typically developing children cannot be explained by appealing to a direct-matching mechanism, and that the process by which young children imitate involves a far more complex yet effortless analysis of the communication of those who they learn from.

Keywords Imitation · Emulation · Mirror neuron system · Autism

1 The 'Dysfunctional Mirror Neuron System' Hypothesis of Autism

Action understanding and imitation are both proposed to be subserved by a direct-matching mechanism of the mirror neuron system (MNS, Rizzolatti & Craighero, 2004), and so any dysfunction of this system should be expected to manifest in impairments in either or both of these capacities. There is, in fact, a long history of reports of deficient imitative abilities in individuals with autism (for reviews see Rogers & Pennington, 1991; Smith & Bryson, 1994; Williams, Whiten, & Singh, 2004). Over a large number of studies, children with autism

V. Southgate
Birkbeck College, University of London, London, UK
e-mail: v.southgate@bbk.ac.uk

J.A. Pineda (ed.), *Mirror Neuron Systems*, DOI: 10.1007/978-1-59745-479-7_15,

have consistently been reported to imitate less than typically developing children on a variety of imitation tasks. These difficulties include imitation of object-directed actions (e.g. Whiten & Brown, 1999), pantomimed actions (Rogers et al., 1996) and gestures (Roeyers et al., 1998).

The hypothesized link between the MNS and action imitation has led a number of authors to propose that the deficient imitative abilities of individuals with autism spectrum disorder (ASD) may result from a dysfunctional MNS (Oberman et al., 2005; Rizzolatti & Fabbri Destro, 2007; Williams et al., 2006). In recent years, several papers have appeared reporting evidence supporting this hypothesis. Two of these studies explored the activation of the autistic brain during action observation without requiring any imitative response from participants. Oberman and colleagues (2005) measured the suppression of the mu rhythm, a component of the electroencephalogram (EEG) proposed to reflect MNS activation (Pineda, 2005) because it is suppressed both when we execute actions ourselves and when we observe others executing actions. They found that, although individuals with autism exhibited normal mu-suppression over sensorimotor cortex during the execution of a hand movement, there was no corresponding suppression of the mu-wave when they observed someone else performing a hand movement. In another study, Théoret et al. (2006) used transcranial magnetic stimulation (TMS) to induce motor-evoked potentials (MEPs) while participants watched videos of finger movements, either from their own perspective (the hand appeared in an orientation that was consistent with the subject's own orientation) or from the perspective of another individual so that it appeared upside down to the observer. The authors reported that, for control subjects, MEPs were facilitated during observation of both self- and other-presented views of the hand, but for individuals with autism the facilitation was only apparent during observation of the 'other' hand orientation.

Further studies explored the activity in the MNS during observation for imitation, and during imitation itself, by functional magnetic resonance imaging (fMRI) (Dapretto et al., 2006; Williams et al., 2006) or by magnetoencephalography (MEG) (Nishitani, Avikainen, & Hari, 2004). The fMRI studies reported activation differences during imitation between individuals with ASD and matched controls in areas comprising the mirror neuron system. Dapretto and colleagues reported a difference in activation of the *pars opercularis* of the inferior frontal gyrus (IFG) during imitation of facial expressions in children with autism and those without autism, with higher activation in those without autism. Williams et al. (2006) found activation differences in the posterior parietal area (PPA), but no effects in the IFG. (In fact, unlike the study on which their study was based [Iacoboni et al., 1999], they reported no activation of the IFG even in control subjects during imitation.) Crucially, however, neither of these studies reported any significant difference between the behavioural imitative capacity of the participants with autism and those without, suggesting that whatever the role of the IFG or the PPA, it is not crucial to the ability to be able to accurately imitate another person. A similar dissociation between behavioural and neural responses was found by Nishitani et al. (2004), who reported that the delayed activation of the IFG in Asperger syndrome was not reflected in their overt imitative response of facial expressions, which were executed with the same latency as by control subjects.

Together, these findings support the view that the functioning of the MNS, just like the performance in some imitation tasks, is atypical in autism. However, the 'dysfunctional mirror system' hypothesis puts forward the more specific claim that a dysfunctional MNS is one of the *causal factors* that lead to imitation deficits and possibly further symptoms of ASD. If proper functioning of the mirror neuron system is necessary for imitation, and the ability to imitate is essential for the normal development of social cognition, a dysfunctional mirror neuron system would explain many aspects of autism spectrum disorder.

2 Arguments Against the 'Broken Mirror' Hypothesis

However, we believe that there are good theoretical and empirical reasons to question the dysfunctional MNS hypothesis of autism. In this section, we argue that the nature of the links between the MNS, imitation and autism does not indicate a simple causal relationship among them. We show first that the primary purported function of the mirror neuron system, action understanding in terms of goals (Rizzolatti & Craighero, 2004), appears to be intact in autism. Second, we review evidence demonstrating that, although imitation is atypical in autism, the imitative *abilities* of the individuals with the disorder do not seem to be affected. Third, we challenge the idea that imitation in general is supported by the human mirror system and that variation in imitative performance would map onto variation in motor mirroring processes.

2.1 Action Understanding Is Not Impaired in Autism

Although, compared with imitation, the domain of action interpretation and understanding in ASD has received comparatively little attention, what evidence does exist suggests that those with ASD are not impaired at action understanding.

In a classic study of infants' understanding of the goals and intentions behind actions, Meltzoff (1995) presented typically developing 18-month-old infants with demonstrations of actions which failed to achieve their goals. In a subsequent imitation phase, infants did not imitate the actions that the adult actually did (the failed attempt), but instead performed the action that the adult had been trying to achieve. The same paradigm has recently been employed for use in children with autism (Aldridge, Stone, Sweeney, & Bower, 2000; Carpenter, Pennington, & Rogers, 2001). These studies found that autistic children, like typically developing infants, performed the action that the demonstrator had intended to perform, suggesting that they understood what the demonstrator had intended (but failed) to achieve, rather than interpreting her actions in terms of their actual observable outcomes.

Results from studies that asked observers to describe the patterns of actions of animated shapes further suggest that individuals with ASD are not impaired at action understanding. When asked to provide a verbal description of a short animation in which two triangles and a circle move around each other, adult participants readily attribute an elaborated plot to the scene, for example, describing one triangle as an aggressor who wants to stop the other shapes from getting into his house (Heider & Simmel, 1944). Although individuals with ASD rarely provide mentalistic descriptions of the triangles' behaviour (e.g. that 'one triangle is *tricking* the other triangle') (Abell, Happe, & Frith, 2000; Klin, 2000), they nonetheless provide descriptions in terms of goal-directed actions (e.g. that 'one triangle is *chasing* the other triangle') (Castelli, Frith, Happe, & Frith, 2002) and are also able to distinguish mechanical launching events from intentional reactions (Bowler & Thommen, 2000). Young children with autism also perform as well as typically developing children on an 'unfulfilled intentions' version of this study, in which a circle is depicted rolling up and down a hill, getting closer and closer to a target object. Both typically developing children and those with ASD provide descriptions of the scene in terms of the intentions of the circle to reach the target (Castelli, 2005).

A recent study by Sebanz, Knoblich, Stumpf, and Prinz (2005) investigated the hypothesis that individuals with autism are impaired at representing the actions of others. When instructed to press either the left or the right key depending on whether the stimulus is green or red, respectively (the relevant dimension), typical adults are slower to respond when the colour is presented on an incompatible background (a finger pointing left when the colour is red) than when it is compatible (a finger pointing right when the colour presented is red). If this task is carried out as a go-nogo task, in which subjects have to press a button only if the colour is red, the interference of the incompatible stimulus disappears. However, if the task is distributed between two individuals such that one person's task is to respond if the colour is red and that of the other is to respond if the colour is green, the effect of the irrelevant stimulus returns. Now, if the participant's 'colour' appears in the context of a finger pointing towards the other participant, his response is slower than if he were doing the task alone. Because the task at the individual level is identical between the joint and the individual conditions, the slower reaction time in the joint condition likely arises because the participants are representing the other person's requested response (Sebanz, Knoblich, & Prinz, 2003). Individuals with ASD show the same effect when performing the task with another individual, suggesting that they also represent the actions of others (Sebanz et al., 2005).

2.2 *Imitative Abilities Are Not Impaired in ASD*

Many studies have explored the imitative performance of individuals with autism, but their findings are somewhat contradictory. Early research consistently reported deficits in imitation. For example, Rogers et al. (2003) employed a battery of imitation tasks including manual (e.g. clapping hands), object-directed (e.g. using elbow to touch box) and oral (stick out tongue) actions and found that in all three domains toddlers with ASD imitated less than age-matched controls. However, the majority of tasks reporting imitation deficits in autistic children appear to involve non-object-directed actions (Hamilton, in press). For example, Rogers et al. (1996) report significantly less imitation in older children with autism than in control children on a series of pantomime and meaningless actions.

Nevertheless, recent investigations into the question of imitation in ASD suggest that there may not be such a deficit in the imitative *abilities* of individuals with autism. In a well-known test of 'goal-directed' imitation, Bekkering and colleagues showed 4- to 6-year-old typically developing children actions in which an adult moved her hand to touch dots on the table in front of her. The hand movements were either ipsilateral or contralateral to the adult's body. In a control condition, no dots were present on the table, but the demonstrator performed the same actions to touch the table. When children were asked to copy the demonstrator, they tended to ignore the type of action that was performed (i.e. whether it was ipsilateral or contralateral) and simply performed the touching of the dots with whichever hand was closest. However, when the dots were not present, children tended to imitate the type of action demonstrated. The failure to use the correct hand when the dots are present has been interpreted as the child giving priority to the adult's goal (i.e. touching the correct dots), whereas when the dots are absent, the way in which the action is carried out is itself interpreted as the goal (Bekkering, Wohlschlaeger, & Gattis, 2000). Recently, Hamilton, Brindly, and Frith (2007) reported that children with autism performed in the same way as controls, on this task. This suggests that (1) they had no imitation impairment (their level of imitation was comparable to that of the control children) and (2) they interpret others' actions in terms of goals, and this induces the same kinds of imitative errors as it does in typically developing children.

A number of further studies support this view. For example, the neuroimaging studies on imitation in autism that found abnormal activation of the mirror neuron system (described in Section 1) reported no differences between the abilities of the ASD individuals and the control participants in imitating either facial expressions (Dapretto et al., 2006; Nishitani et al., 2004) or meaningless finger movements (Williams et al., 2006). In fact, one study even suggests an enhanced tendency to imitate in individuals with autism. Bird and colleagues have recently employed an 'automatic' imitation paradigm in subjects with ASD (Bird et al., in press). The term 'automatic' imitation is used to describe the phenomenon that people's motor movements are facilitated by observing the same movement in someone else and impaired when they observe a different movement of the same body part (Brass, Bekkering, & Prinz, 2001; Heyes, Bird, Johnson, & Haggard, 2005). This effect appears to be weaker or absent when the observed action is performed by a non-human actor, like a robot (e.g. Press, Bird, Flach, & Heyes, 2005). In the study by Bird and colleagues, young adults with ASD were instructed to open or close their hand whenever the hand they were watching on the screen began to move. Sometimes the observed hand would open and sometimes it would close. When the participants were instructed to open their hand whenever they saw movement, their responses, just like the ones of individuals without ASD, were slower when the observed hand was closing, and faster when it was opening. This demonstrates that people with ASD are also subject to automatic imitation tendencies. Even more interestingly, individuals with ASD displayed higher specificity of automatic imitation to a human actor compared to a robot than the control subjects. Since, like the mirror neurons of monkeys, the human mirror system is considered to be tuned to biological actions only (Tai, Scherfler, Brooks, Sawamoto, & Castiello, 2004), the especially attenuated effect of the robot hand action compared with the human hand action in individuals with autism would be unlikely if they possessed a dysfunctional mirror neuron system.

Many authors make a distinction between imitation as an automatic process and imitation as a cognitively mediated mechanism for social learning (e.g. Byrne, 2005) or between emulation and mimicry (Hamilton, in press). Automatic imitation or mimicry is evident in the involuntary and unconscious matching of posture, gestures and prosody between individuals (Chartrand & Bargh, 1999), and probably serves to facilitate social functions (Decety & Chaminade, 2003) by generating empathy or mutual identification (Byrne, 2005). Automatic imitation also manifests as response facilitation on a number of experimental tasks (Brass, Bekkering, Wohlschlager, & Prinz, 2000). On the other hand, imitation that serves learning seems to involve more complex processes, like identifying which elements are relevant to the accomplishment of the observed skill and which are incidental or idiosyncratic to the demonstrator. Some authors have suggested that it is only the former, more automatic type of imitation that is likely to be subserved by a process of direct matching (Byrne, 2005). Perhaps then, individuals with ASD might be expected to display a deficit of automatic imitation, but not necessarily of voluntary imitation.

In fact, a recent paper by McIntosh, Reichmann-Decker, Winkielman, and Willbarger (2006) did find a difference between automatic and voluntary imitation in autism. Participants viewed pictures of faces with happy or angry expressions while the activation of their facial muscles was measured. The authors found that, unlike in control subjects, individuals with ASD showed no automatic activation of the muscles associated with performing the expression they observed. Nevertheless, all participants performed well on the task of voluntary imitation, in which they were asked to copy the expression that they saw. However, the idea of a specific deficit in automatic imitation in autism seems to be incompatible with other findings. For example, the imitation impairment observed in the automatic imitation paradigm is higher in those with autism than in controls (Bird et al., submitted). Furthermore, the high instance of echolalia and echopraxia, the excessive vocal and motor imitation of what you hear or see (Fay & Hatch, 1965; Lord, Rutter, & LeCouteur, 1994), could be seen as incompatible with an automatic imitation deficit (Griffin,

2006). Children with autism are also able to recognize when they themselves are being imitated (Tiegerman & Primavera, 1994; Field, Field, Sanders, & Nadel, 2001), which, according to Byrne (2005), is precisely the kind of ability that would be expected to be subserved by the MNS.

In sum, many studies, including some by the proponents of the 'dysfunctional MNS' hypothesis (Dapretto et al., 2006; Williams et al., 2006), suggest that individuals with autism are capable of voluntary imitation. As for automatic imitation, the picture is less clear. Some symptoms of autism, as well as some experimental studies, suggest that automatic imitation is not absent, but even enhanced in ASD. Other findings show a lack of automatic imitation of facial expressions in autism. We find the fact that the conflicting reports on automatic imitation in autism come from studies investigating the imitation of different kinds of actions (i.e. hand actions vs. facial expressions) informative. We will return to this potentially interesting distinction later in the chapter.

2.3 The Mirror Neuron System and Human Imitation

Having reviewed literature showing that neither the mirror neuron system nor imitation is specifically dysfunctional in autism, we now turn to the question of whether it is plausible to assume that human imitation is based on a direct-matching mechanism implemented in the mirror neuron system. Imitation, as is evident from a perusal of the many studies on infant imitation, appears to go beyond direct motor matching. Infants are not blind imitators, as evidenced by the selective nature of their action reproduction. By examining what human infants imitate, we can evaluate whether or not the predictions made by the direct-matching hypothesis are actually borne out.

The hypothesis that the MNS implements the basic neural mechanism that enables the direct transformation of perceptual information into motor commands that lead to imitation is an attractive proposal because it offers a plausible solution to the problem of how an appropriate mapping is created between the body of the demonstrator and the imitator (Nehavniv & Dautenhahn, 2002). According to Rizzolatti and Craighero (2004), two types of newly acquired behaviours are based on imitation learning, and therefore subserved by the mirror neuron system. The first type of learning is 'substitution', in which a pre-existing motor pattern is substituted for a newly observed motor pattern that is better suited to the task. The second type of learning is the acquisition of a new motor sequence, which involves the decomposition of an observed motor pattern into elementary motor acts already in the observer's repertoire. These 'elementary' motor acts have been proposed to activate the corresponding motor representations in mirror areas.

These proposed functions of the mirror neuron system make several predictions concerning the imitative performance of naïve observers. First, Rizzolatti and Craighero (2004) suggest that 'substitution', as a form of imitation learning, occurs when the new action is recognized as being better suited to fulfil the goal of the action. This predicts that a less efficient means of achieving an outcome should never be substituted for a more efficient means. Second, the decomposition-recomposition model proposed by Rizzolatti and Craighero (2004) predicts that imitation should result in actions that bear a high degree of motor resemblance to that of the demonstrator. Finally, the direct-matching hypothesis makes no explicit claims concerning how an observer decides *what* to imitate. However, a number of researchers have highlighted the 'goal-directed' nature of imitation (e.g. Bekkering et al., 2000), which has been proposed to be driven by the MNS (Wohlschlager & Bekkering, 2002). Below we present evidence from studies of imitation in infancy that are difficult to reconcile with these predictions.

2.3.1 Action Substitution and Efficiency

Rizzolatti and Craighero (2004) proposed that imitation learning may take the form of substitution of an old motor pattern for a new behaviour (that already exists in the motor repertoire of the individual), should the new motor behaviour provide a more adequate means to the goal. This proposal presupposes that the observer has some way to evaluate whether the observed action is a better way of fulfilling the task than the action that the observer was previously using.

In a classic study of infant imitation, Meltzoff (1988) asked whether infants of 14 months would imitate a novel act 1 week after seeing it. The novel act that the infant watched was an experimenter using his forehead to illuminate a lightbox. Meltzoff found that the infants who had witnessed the head-touch action also used their own heads to illuminate the box, while infants in a control group who had not witnessed the head-touch action never performed this new action. Clearly, using one's head to press a box is a less 'appropriate' or efficient means than simply using one's hands. However, in a replication of this study, Gergely, Bekkering, and Kiraly (2002) found that most infants used their hands to press the box *before* they imitated the head-touch, demonstrating that they were aware of the availability and efficiency of this action (Gergely & Csibra, 2005, 2006). The fact that they went on and imitated the less efficient means is at odds with the idea that imitative learning is used for substituting less efficient actions with more efficient ones. Further evidence confirmed that 14-month-old infants do know that the inefficient head action is not the most appropriate action for the task. In the Gergely et al. version of the task, infants saw the head action carried out by either someone whose hands were visibly free to perform the task or someone whose hands were covered by a blanket and were not available (Gergely et al., 2002). In this study, the 14-month-olds who were in the condition where the demonstrator's hands were occupied did not perform the head action themselves, whereas those infants, who saw the demonstrator with their hands free, did imitate the head action. This suggests that infants were aware that the hands were more appropriate effectors to fulfil the task.

Other studies confirm that young children imitate unnecessary or inefficient actions, when they are demonstrated by an adult. For example, Nagell, Olguin, and Tomasello (1993) found that while chimpanzees would not imitate a less efficient goal-directed action, young children would copy whatever action was demonstrated to them, irrespective of its physical efficiency. In another study by Horner and Whiten (2006), 3-year-old children imitated an action that they could clearly see was causally unrelated to the goal, and appeared to prefer to model the demonstrator at the expense of efficiency. Children imitate inefficient and unnecessary actions even when their attention is specifically drawn to the fact that some observed actions may be unnecessary (Lyons et al., 2007, The hidden cost of imitation, unpublished manuscript), and this is especially true when the causally unnecessary action is socially cued by the experimenter (Brugger et al., 2007). The fact that children imitate these inefficient actions even in the absence of the experimenter (Gergely et al., in preparation; Horner & Whiten, 2005) suggests that they are not performing them in order to fulfil any social function or simply to please the experimenter.

2.3.2 Fidelity of Imitation

Actions are hierarchically organized (Jeannerod, 2006), and there are different levels on which one can construe and imitate an observed action (Byrne & Russon, 1998; Csibra, 2007). It is not clear how a direct-matching mechanism alone would enable the interpretation and re-enactment of observed actions at different levels, since the level on which one interprets an action appears to depend on a number of factors, such as the presence or absence of an object (Wohlschlager & Bekkering, 2002). Indeed, it is debatable whether it is ever possible for an observer to imitate someone else's actions completely faithfully since her body will never have the capacity to reconstruct exactly what has been done by the demonstrator (Csibra, 2007).

According to Rizzolatti and Craighero (2004), the mechanism by which new motor acts are incorporated into the motor repertoire for reproduction involves a process of decomposing the observed action into known motor acts, and then recomposing these broken-down components into a new behaviour that the individual can then perform. Since this process does not allow for action interpretation at different levels, it should result in a high fidelity action reproduction. The observers' ability to reconstruct the action faithfully will obviously depend on their motor capacity, and it would be expected that faithful motor reproduction may not be entirely possible in infancy. However, major deviations from the demonstration would not be compatible with MNS-driven imitation learning. It is therefore difficult to explain why, in the various versions of the Meltzoff (1988) study, infants rarely copy exactly what the demonstrator has done (a head-touch using the forehead), but illuminate the lightbox using their mouth, cheek, chin or ears (G. Gergely's observations). This kind of imitation fits with an interpretation of the demonstrator's action at a higher level of the action hierarchy ('use the head to contact lightbox'), but it is difficult to see how direct matching would allow for such an interpretation of the action (Csibra, 2007).

2.3.3 Goal-Directed Imitation

According to Iacoboni (2005), the identification of the goal of an observed action by mirror neurons in the premotor cortex dictates what is imitated by the observer. Bekkering and colleagues argue that the object-directed (used synonymously with 'goal-directed') nature of human imitation provides support for the view that imitation is subserved by the MNS. This conclusion was based on the analogy between the finding that mirror neurons in monkeys do not fire unless the observed action is object-directed, and the fact that the presence of an object facilitates imitation in humans. This analogy led Wohlschlager and Bekkering (2002) to suggest that mirror neuron activity and imitation are mediated by the same system of direct matching. The presence of an explicit goal during imitation indeed does result in higher levels of activity in mirror neuron areas in humans (Koski, Wohlschlager, Bekkering, Woods, & Dubeau, 2002).

Notwithstanding the difficulty (arising from the proposal that direct matching is the mechanism by which actions are understood) of how it would be possible to identify the goal of an action that is not yet in one's motor repertoire, there is evidence that imitation in young children and infants is also 'goal-directed'. Bekkering, Wohlschlager, and Gattis (2000) showed that 3- to 5-year-old children tended to ignore the particular hand used by the demonstrator when her action was directed towards a target object, but imitated using the correct hand when no 'goal' object was present. Recently, Carpenter, Call, and Tomasello (2005) have reported a similar finding in 12- and 18-month-old infants. Infants in one condition were shown a toy mouse either hopping or sliding towards a toy house, whereas in another condition infants just saw the demonstrator performing the particular action (hopping or sliding) but with no house present. In the first group, both 12- and 18-month-olds tended to emulate the action of putting the mouse in the house, but neglected to imitate the particular manner in which this was done. However, when no house was present, infants tended to imitate the particular action (hopping or sliding) significantly more.

In a recent study, however, we found evidence that infant imitation is not always goal-directed, but is dependent on a sophisticated interpretation of the communicative intent of the demonstrator. Gergely and Csibra (2006) have proposed that imitation in human children is facilitated and modulated by the presence of ostensive communication cues (e.g. eye contact), which trigger in the recipient the assumption that the demonstrator is going to demonstrate some *new and relevant* information for them. On this basis, we hypothesized that infants would select action elements for imitation on the basis of their communicative novelty rather than relying exclusively on a hierarchical analysis of goals. We presented 18-month-old infants with a variation of the Carpenter et al. (2005) paradigm. Half of the infants received the exact same demonstration as infants in the house-present version of that study, while the other half were first told and shown that the toy animal lived in the house (thus rendering the house aspect of the demonstration 'old') before they were shown the same demonstration as the other group (the animal either hopping or sliding into the house). Infants in this second group imitated the details of the action (hopping or sliding) significantly more than infants in the group where all the information was new, and interestingly, in a non-negligible number of trials, infants even neglected to put the animal in the house at all, in favour of imitating the style of the action modelled by the demonstrator (Southgate, Chevallier, & Csibra, under review). This finding suggests that imitation is more than identifying and re-enacting the goal of the demonstrator.

3 An Alternative Hypothesis for the Connection Between ASD and Imitation

The above study highlights one of the shortfalls of the direct-matching theory of imitation and imitative learning. Although direct matching may offer a plausible solution to the correspondence problem (Heyes & Bird, in press), it cannot tell the observer *what* to imitate. Since much evidence shows that human imitation is a selective process, additional mechanisms are required to account for imitation. In this section, we propose that *communication* plays a key role in human imitation, and this also provides an explanation for the apparent imitation deficit, as well as atypical patterns of MNS activation, in autism.

3.1 Imitation and Communication

The term 'imitation' refers to the phenomenon of the reproduction of some behaviour of an individual by another individual. However, some theorists think about imitation not only as a phenomenon but also as a special mechanism that underlies the reproduction of observed behaviours. Others, like Heyes and Bird (in press), proposed that there is no special mechanism of imitation but it is achieved by simple associative mechanisms. While we agree that imitation is not accomplished by specifically dedicated mechanisms and that action reproduction is sometimes based on associative mechanisms, we think that most instances of imitation are actually achieved by the process of *emulation* (Csibra, 2007).

The term 'emulation' in this context refers to a mechanism that reproduces an action from its description. An action can be described at many levels of precision and resolution, and action descriptions created for the purpose of reproduction usually involve a reference to the outcome of the action. For example, the action to be reproduced in the task that demonstrated 'goal-directed imitation' in children (Bekkering et al., 2000) can be described as 'touch the point', 'touch the point with your right hand', 'touch the point with your right finger' or 'touch the point with the palmar surface of your right finger', each successive description specifying the end of the action with more and more precision. The temporal or sequential aspect of an action can also be specified at various levels of precision, describing details of the subgoals (e.g. path of the hand movement) through which the end goal should be achieved. The process of emulation generates the action simply by feeding these goal specifications into the observer's own motor system, which will achieve them in its own way that will not necessarily match the details of the observed behaviour. Thus, the fidelity of emulative action reproduction depends on what level of precision is chosen to interpret and reconstruct the observed action.

The crucial question that an imitator is confronted with when observing a model is *what* to imitate, i.e. what level of action interpretation and what precision of emulative action reproduction she is expected to perform. Studies on 'goal-directed imitation' demonstrate that children, including children with autism, tend to interpret and emulate the to-be-imitated actions at a higher level of goal descriptions when these are supported by factors like the availability of target objects. Such behaviour is adaptive in many social learning situations, because the minute motor details of action execution are normally irrelevant with respect to goal achievement. However, human cultural practices and norms tend to be *opaque* in the sense that often neither their purpose nor the causal relations between the performed actions and their useful outcome are evident for a naïve observer (Gergely & Csibra, 2006). For example, humans engage in tool making for which, to an observer, there may appear no immediate goal at the time of construction and perform rituals that do not reveal how they are supposed to work. If much of human culture consists of such cognitively opaque practices, it would make little sense for imitation to be driven solely by the identification of goals, as in many cases there will be no obvious goal, but there may nevertheless be important actions worthy of imitation and learning. To cope with the problem of cognitive opacity, Gergely & Csibra (2005, 2006) have recently proposed that humans evolved a suite of adaptations to ensure that cultural knowledge is efficiently transferred across generations.

The key element of this proposal is that communication from the model towards the observer can help to identify the relevant aspect of the modelled behaviour to be learnt, hence what level of emulation the imitator is expected to perform. The necessary adaptations for such a communication system involve many elements, of which the most important in this context is that the observer has to be sensitive to the cues that signal the model's intention to communicate. Human infants show an early, or even innate, sensitivity to such *ostensive cues*, which trigger an expectation of relevant content to be communicated by the source of these cues (Csibra & Gergely, 2006). In the context of action demonstration, these cues will tell children that the model's behaviour will reveal some relevant information for them to be learnt and reproduced.

The implication of this theory is that an observer needs to be sensitive to, and to correctly interpret, the ostensive communication cues that accompany demonstrations in order to benefit from the model's pedagogical efforts.

Without this one would not be able to select the relevant aspects of another person's behaviour that are important to attend to and reproduce. We propose that individuals with autism lack this sensitivity to ostensive cues and the expectation of relevant information to be manifested, which could account for the reported impairments in imitation and a host of other phenomena.

3.2 Understanding Communicative Intent Is Impaired in Autism

From their earliest utterances, human infants appear to abide by Gricean maxims (Grice, 1975) and tailor what they say to their partner's knowledge state (Greenfield & Smith, 1976). However, even high-functioning adults with autism do not find this an easy task, often neglecting to include details that meet the intended recipient's communicative needs (Bruner & Feldman, 1993). Unlike typically developing children, who begin to use pointing around 9–12 months of age, children with autism rarely point (Baron-Cohen, 1989). The sophisticated communicative understanding underlying the pointing behaviour of 12-month-old infants, discussed by Tomasello and colleagues (Tomasello, Carpenter, & Liszkowski, 2007), suggests that the absence of pointing in young children with autism may arise from a lack of understanding of communication.

On the comprehension side, children with autism also seem to be unresponsive to the communicative cues produced by others. For example, in response to others' points, children with autism often look at the hand that is pointing rather than the object being pointed at, suggesting that they lack the understanding that pointing is a communicative, referential act. Similarly, children with autism fail to use gaze to locate a hidden object (Leekam, Lopez, & Moore, 2000), a task that is trivially easy for typically developing children (Behne, Carpenter, & Tomasello, 2005). Furthermore, children with autism do not show the typical reactions to a number of ostensive cues that would enable them to learn from others. Typically developing newborns show a strong preference for faces (Johnson, Dziurawiec, Ellis, & Morton, 1991) and, within the face, for direct gaze over averted gaze (Farroni, Csibra, Simion, & Johnson, 2002). Human infants' attention automatically shifts to the direction of others' perceived gaze shifts (the gaze-cueing effect that is well known in adults, see Driver et al., 1999), but only if the movement of the eyes is preceded by direct gaze (Farroni, Mansfield, Lai, & Johnson, 2003). Autistic children on the other hand do not preferentially attend to faces (Osterling & Dawson, 1994) or the eyes (Klin et al., 2002), and the gaze-cueing effect appears to be absent, or at least divergent (Johnson et al., 2005; Senju, Tojo, Dairoku, & Hasegawa, 2004). While direct gaze cues attention better than a non-social cue in typically developing children, the same differential effect is not present in children with autism (Senju et al., 2004), and it does not facilitate face detection (Senju et al., 2003). Typically developing neonates show a preference for infant-directed speech (Cooper & Aslin, 1990), a pattern of exaggerated prosody that may serve to elicit infants' attention (Fernald & Simon, 1984), and is proposed to serve as an auditory ostensive stimulus for infants (Csibra & Gergely, 2006). Human neonates also prefer to listen to speech rather than a non-speech analogue (Vouloumanos & Werker, 2007). Children with autism, however, do not prefer infant-directed speech (Kuhl, Coffey-Corina, Padden, & Dawson, 2005) and do not even show a preference for their mothers voice over the noise of a busy canteen (Klin, 1991). Finally, a failure to respond to the sound of their own name appears to be one of the earliest observable indications of autism (Nadig et al., 2007; Werner, Dawson, Osterling, & Nuhad, 2000).

If it is the case that imitation is driven by a sensitivity to the communicative intent of the demonstrator, and an impairment in understanding communicative intent inflicts individuals with autism (Sabbagh, 1999), then imitation should be an expected impairment in those with ASD. This impairment may result in inappropriate imitation, but may manifest as either too little imitation (as is often reported) or reproducing unnecessary aspects of another person's behaviour. For example, in studies where there is no instruction to imitate, researchers find that individuals with ASD either fail to imitate or imitate less than typically developing children (e.g. Brown & Whiten, 1999). Since parents rarely explicitly instruct their children to 'do as I do', this may explain the lack of spontaneous imitation in children with autism in their everyday lives. However, as it is evident from recent studies, when instructed to imitate, children with autism imitate as well as non-affected children (Dapretto et al., 2006; Hamilton et al., 2007). Nevertheless, even instruction to imitate would not always result in the same level of imitative competence in individuals with ASD as in typical individuals. This is because telling someone to 'do as I do' does not specify on what level he is supposed to emulate the demonstration (Bird et al., in press). As we have argued above, a sensitivity to the subtle cues in the communication of the other individual is necessary to extract from the demonstration the relevant level of action reproduction.

Gergely and colleagues' study on infants' selective imitation (discussed above) could illustrate how communication cues and expectation of relevance drive children's reproduction of observed actions (Gergely et al., 2002). If infants expect relevant information, they should be sensitive to the relative amount of information inherent in various elements of the demonstration. The amount of information is dependent on the conditional probability of an event (the less probable it is, the more information it carries). Touching the lightbox with the forehead is an unlikely event when the model's hands are free, and thus will be judged as the most informative part, and the most likely content, of the demonstration. In contrast, touching the lightbox with the head is not as unlikely when someone's hands are occupied. In this case, the information that the box can be lit up will be a more informative element of the demonstration than the mode of pressing the box. Thus, 14-month-old infants interpreted the action to be reproduced at the level of the overall goal (lighting up the box) in the hands-occupied condition, but went down to the level of effector (lighting up the box by head-touch) in the hands-free condition.

A further version of this study confirmed that the selective imitation effect in this task was dependent on the ostensive communication cues that could induce the expectation of high relevance in infants. In this version, infants observed the same demonstration, but now the model performed the head-touch actions without ever looking at the infant or emitting any communicative cues. In this situation, infants were less likely to reproduce the head-touch action and, more importantly, the selectivity of imitation disappeared: they touched the box with their head in the hands-occupied condition as often as in the hands-free condition (Kiraly, Csibra, & Gergely, 2004). Recently, a version of this study was also run on a sample of children with autism (Somogyi et al., 2006). Although they received as many ostensive communicative cues as children in the original version of this study (Gergely et al., 2002), they behaved in a similar way to typically developing infants in the non-ostensive condition. Like in a number of studies reported here, the children with autism imitated the head-touch action at a level comparable with typically developing children (around 70%) in the hands-free condition, but crucially, the level of imitation was equivalent in both the hands-free and the hands-occupied conditions in autistic children. This suggests that children with autism are unable to modulate the interpretation of the demonstration on the basis of communicative cues and do not expect that the model's manifestation will reveal relevant information for them. It is thus not the ability to copy an action, but the ability to extract the relevant level of emulation from a communicative demonstration that may be impaired in autism.

This conclusion is consistent with other symptoms of the disorder. If one of the core deficits of autism is the lack of sensitivity to communicative intent, any aspect of cultural knowledge that is transmitted via communication could be expected to be impaired in autism. For example, knowledge about social norms and conventions is learnt exclusively from other individuals of our species, and so may be impaired in autism. In fact, there are reports that this is indeed the case (e.g. Loth, 2007).

3.3 The Mirror Neuron System in Autism

We have argued that, while the ability to imitate may not be especially impaired in autism, the atypical pattern of their imitative behaviour can be derived from their insensitivity to communicative cues, and communicative intent. In this sense, imitation impairment may not be a core deficit, but rather a symptom of autism, not a cause but a consequence of the disorder. This proposal leaves open the question of how to explain the findings of differential functioning of the mirror neuron system in autism, which we reviewed at the beginning of this chapter.

First, when considering mirror neuron system activation in response to the observation of hand actions, the findings are not easy to interpret. Oberman and colleagues (2005) did not find mu rhythm suppression in individuals with autism in response to the observation of hand actions. What is surprising, however, is that the MNS of their control subjects responded to the observation of non-transitive, non-object-directed hand actions. Other studies on mu rhythm suppression (Muthukumaraswamy, Johnson, & McNair, 2004) and neuromagnetic activation of the mirror neuron system (Nishitani & Hari, 2000) reported hardly any MNS activation modulation unless the hand approached or manipulated an object. The findings of Théoret et al. (2006) are also difficult to interpret. They recorded no motor activation when their subjects with ASD saw a hand from their own point of view (as if it was their own hand), but found normal activation when they saw a hand facing the subject (as if it was someone else's hand). This is an unexpected finding and would more likely suggest an impairment in self-body image than in social mirroring. Hobson and Meyer (2005) have, in fact, reported that children with autism, unlike typically developing controls, did not use their own body to indicate to an experimenter where to place a sticker, instead pointing to the experimenter's body, which suggests an impairment of self-body image. Finally, the study by Williams et al. (2006) failed to find inferior frontal activation in adolescent individuals with autism during the imitation or observation of finger movements. However, they did not find such activation in their control subjects either. The authors reported that some parietal regions, considered to be a part of the MNS, are less activated in ASD than in the control subjects during imitation. However, this was also true during action execution, while there was no difference found during action observation. It is thus not clear why this activation difference should be considered to reflect the mirror neuron system. This lack of activation during action execution was also found in a recent study by Cattaneo and colleagues. They used electromyography to record the activity of the mouth-opening mylohyoid muscle when subjects had to grasp an object either for eating or for placing in a container. They found that while the mylohyoid muscle in typically developing children showed activity from the point where the subject began to reach for the object, the same muscle in a group of children with autism was only activated later as they brought the object to their mouth. Typically developing children also showed this anticipatory muscle activity when watching someone else reach for an object they were going to eat, but children with autism did not (Cattaneo et al., 2007). Rather than pointing to a specific impairment of the MNS, this result suggests that children with autism have impairments in their ability to sequence their own actions. In fact, difficulties in motor planning in children with autism are well documented (e.g. Hughes, 1996). As mirroring is, by definition, dependent on one's own action capabilities, any impairment in the production of actions would be expected to lead to impairments in mirroring, but not because there is anything impaired in the mirroring mechanism.

The evidence for atypical MNS activation in autism during observation or imitation of facial actions is also ambiguous. Dapretto and colleagues (2006), using a task of emotional expression imitation or observation, reported that individuals with autism, unlike typically developing children, showed no activation of the inferior frontal gyrus (IFG). Others, however, have failed to replicate these results. For example, Ashwin, Baron-Cohen, Wheelwright, O'Riordan, and Bullmore (2007) found no activation difference in the IFG during an emotional face perception task between a group of individuals with ASD and control subjects. Even if the reduced MNS activation to facial expressions in autism were reliable, it does not necessarily indicate a causal role of motor mirroring in understanding emotional expressions. An alternative explanation for this effect could be that it is the result, rather than the basis, of emotion understanding. Facial expressions are inherently communicative (Fridlund, 1994), and so it is not surprising that autism, whose primary symptoms include disordered communication, affects the comprehension of emotional expressions as well. It is well known that individuals with autism do not spend as much time looking at faces, and especially looking at the eye regions, as non-affected people (Gliga & Csibra, 2007), and their relative inexperience with faces could also contribute to the failure, or slower speed, of recognizing facial expressions. Such an impairment would also make it less likely that individuals with ASD would generate the appropriate facial expression in response to the perception of an emotional signal, which could explain the reduced activity of, for example, appropriate facial muscles during observation of emotional expressions (McIntosh et al., 2006). However, this may be the result, rather than a cause, of impaired understanding of emotions.

Translating this account into neurological terms, it is possible that the lack of IFG activation observed in children with autism results from dysfunction of the amygdala, which has reciprocal connections with IFG. An abnormal amygdala has been given a central role in autism (Schultz, Romanski, & Tsatsanis, 2000) and has been proposed to underlie difficulties in emotion recognition (Baron-Cohen et al., 2000). Indeed, numerous recent studies have reported anatomical abnormalities in the amygdalae of individuals with autism (e.g. Aylward et al., 1999; Munson et al., 2006). Functional studies also found reduced amygdala activations in autism. When asked to judge the emotion conveyed by the eyes, non-affected individuals activated their amygdala whereas the ASD subjects did not, and performed significantly worse on the task than the controls (Baron-Cohen et al., 1999). Ashwin et al. (2007), who did not find activation difference in the MNS, also reported reduced amygdala activation in people with ASD in a facial expression perception task. Even Williams et al. (2006), who studied observation and imitation of hand movements, found significantly reduced amygdala activation in their ASD subjects.

One possibility then is that the additional IFG activation evident in control subjects (and lacking in ASD subjects) in the Dapretto et al. (2006) study stems from an automatic process of emotion identification. Hypoactivation of the amygdala, known to be involved in emotion recognition, could lead to differences in the activation of the mirror neuron system by virtue of the connections between the two, but this would not necessarily imply that the mirror neuron system itself was dysfunctional. Whether or not this was a plausible explanation for this result is difficult to judge because Dapretto et al. (2006) reported in their results that there was no difference in amygdala activation between the groups, but their data reveals significantly lower amygdala activation in the ASD group compared with the typically developing children (Supplementary Table 3).

The amygdala is not the only brain region that is affected in autism. Consistent with the proposal that autism is a primarily communicative disorder, ASD individuals also show differences in activation of the medial frontal cortex, an area proposed to play an important role in recognizing the communicative intent of others (Amodio & Frith, 2006; Kampe, Frith, & Frith, 2003). For example, Wang and colleagues found that whereas typically developing children showed increased activation of the medial prefrontal cortex when interpreting the intended meaning of an ironic scenario, children with autism did not, and were less accurate than controls in detecting the communicative intent behind the remark (Wang, Lee, Sigman, & Dapretto, 2006). Facial emotional expressions also fail to activate medial prefrontal areas in autism (Ashwin et al., 2007).

While we propose that, on the cognitive level, the imitation deficit in autism could be satisfactorily explained by the communicative impairment that is undoubtedly implicated in the disorder, we are not committed to any particular theory about the neural bases of this impairment. Nevertheless, we think that the available evidence on the MNS dysfunction in autism is too ambiguous to support the 'broken mirror' hypothesis, and alternative proposals, like the ones that emphasize the role of the amygdala or the medial prefrontal cortices, or even impairments in the individual's own motor capabilities, are not less compatible with recent findings in neuroimaging.

4 Conclusions

We have argued that the numerous studies reporting intact imitative abilities in individuals with autism, both voluntary (e.g. Dapretto et al., 2006; Hamilton et al., 2007; Somogyi et al., 2006) and automatic (Bird et al., in press), do not fit with the view that their mirror neuron system is impaired. We have presented evidence from infants and children which suggests that imitation goes far beyond a process of direct matching, entailing a sophisticated analysis of the communicative intent of the demonstrator. While it is an open question whether a direct-matching mechanism enables or facilitates imitation, it cannot account for the selective nature of imitation reported here. Since individuals with autism can imitate (even if they need to be told to do so), and their imitation impairments appear to arise at the level of selecting when and what to imitate, it seems unlikely that a dysfunctional mirror neuron system underlies the social difficulties they face.

Nevertheless, atypical patterns of activation in the MNS of individuals with autism are to be expected as the result of other deficits (like impaired communication or motor sequencing) that would have consequences for normal MNS functioning, and could explain the controversial findings reported in the literature. Individuals with ASD are less interested and less engaged in social interactions, are less willing to cooperate or communicate with others and have less experience with dealing with social stimuli. It is thus not surprising that their mirror neuron system, whose main function probably involves making social interactions smooth (Csibra, 2007), does not show the same patterns of activation as the mirror neuron system of unaffected individuals.

Acknowledgements We thank Coralie Chevallier, Uta Frith, Teodora Gliga, Tobias Grossmann, Eva Loth, Olivier Morin, Atsushi Senju, Dan Sperber and John S. Watson for their valuable comments on this paper. This work was supported by the UK Medical Research Council (Programme Grant #G9715587) and a Nest-Path Cooperative Research Grant (EDICI) from the European Commission.

References

Abell, F., Happe, F., & Frith, U. (2000). Do triangles play tricks? Attribution of mental states to animated shapes in normal and abnormal development? *Journal of Cognitive Development, 15*, 1–20.

Aldridge, M. A., Stone, K. R., Sweeney, M. H., & Bower, T. G. R. (2000). Preverbal children with autism understand the intentions of others. *Developmental Science, 3*(3), 294–301.

Amodio, D. M., & Frith, C. D. (2006). Meeting of minds: The medial frontal cortex and social cognition. *Nature Reviews Neuroscience, 7*, 268–277.

Ashwin, C., Baron-Cohen, S., Wheelwright, S., O'Riordan, M., & Bullmore, E. T. (2007). Differential activation of the amygdala and the 'social brain' during fearful face-processing in Asperger Syndrome. *Neuropsychologica, 45*(1), 2–14.

Aylward, E. H., Minshew, N. J., Goldstein, G., Honeycutt, N. A., Augustine, A. M., Yates, K. O., et al. (1999). MRI volumes of amygdala and hippocampus in non-mentally retarded autistic adolescents and adults. *Neurology, 53*, 2145–2150.

Baron-Cohen, S. (1989). Perceptual role taking and protodeclarative pointing in autism. *British Journal of Developmental Psychology, 7*(2), 113–127.

Baron-Cohen, S., Ring, H. A., Bullmore, E. T., Wheelwright, S., Ashwin, C., & Williams, S. C. R. (2000). The amygdala theory of autism. *Neuroscience and Biobehavioural Reviews, 24*, 355–364.

Baron-Cohen, S., Ring, H. A., Wheelwright, S., Bullmore, E. T., Brammer, M. J., Simmons, A., et al. (1999). Social intelligence in the normal and autistic brain: an fMRI study. *European Journal of Neuroscience, 11*(6), 1891–1898.

Baron-Cohen, S., Leslie, A. M., & Frith, U. (1986). Mechanical, behavioural and Intentional understanding of picture stories in autistic children. *British Journal of Developmental Psychology, 4*, 113–125.

Behne, T., Carpenter, M., & Tomasello, M. (2005). One-year-olds comprehend the communicative intentions behind gestures in a hiding game. *Developmental Science, 8*, 492–499.

Bekkering, H., Wohlschlager, A., & Gattis, M. (2000). Imitation of Gestures in Children is Goal-directed. *The Quarterly Journal of Experimental Psychology, 53A* (1), 153–164.

Bird, G., Leighton, J., Press, C., & Heyes, C. (in press). Intact automatic imitation of human and robot actions in autism spectrum disorders. *Proceedings of the Royal Society of London: B.*

Bowler, D. M., & Thommen, E. (2000). Attribution of mechanical and social causality to animated displays by children with autism. *Autism, 4*(2), 147–171.

Brass, M., Bekkering, H., & Prinz, W. (2001). Movement observation affects movement execution in a simple response task. *Acta Psychologica, 106*(1), 3–22.

Brass, M., Bekkering, H., Wohlschlager, A., & Prinz, W. (2000). Compatibility between observed and executed finger movements: Comparing symbolic, spatial, and imitative cues. *Brain and Cognition, 44*(2), 124–143.

Bruner, J., & Feldman, C. (1993). Theories of mind and the problem of autism. In S. Baron-Cohen et al. (Eds.), *Understanding other minds: Perspectives from autism* (pp. 267–291). Oxford: Oxford University Press.

Byrne, R. W. (2005). Social cognition, imitation, imitation, imitation. *Current Biology, 15*(13), R498–R500.

Byrne, R. W., & Russon, A. E. (1998). Learning by imitation: A hierarchical approach. *Behavioral and Brain Sciences, 21*, 667–684.

Carpenter, M., Call, J., & Tomasello, M. (2005). Twelve- and 18-month-olds copy actions in terms of goals. *Developmental Science, 8*, F13–F20.

Carpenter, M., Pennington, B. F., & Rogers, S. J. (2001). Understanding of others' intentions in children with autism. *Journal of Autism and Developmental Disorders, 31*(6), 589–599.

Castelli, F. (2005). The Valley task: Understanding intention from goal-directed motion in typical development and autism. *British Journal of Developmental Psychology, 24*, 655–668.

Castelli, F., Frith, C., Happe, F., & Frith, U. (2002). Autism, Asperger syndrome, and brain mechanisms for the attribution of mental states to animated shapes. *Brain, 125*, 1839–1849.

Cattaneo, L., Fabbri-Destro, M., Boria, S., Pieraccini, C., Monti, A., Cossu, G., et al. (2007). Impairment of action chains in autism and its possible role in intention understanding. *Proceedings of the National Academy of Sciences, 104*(45), 17825–17830.

Chartrand, T. L., & Bargh, J. A. (1999). The chameleon effect: The perception-behavior link and social interaction. *Journal of Personality and Social Psychology, 76*, 893–910.

Cooper, R. P., & Aslin, R. N. (1990). Preference for infant-directed speech in the first month after birth. *Child Development, 61*(5), 1584–1595.

Csibra, G. (2007). Action mirroring and action interpretation: An alternative account. In: P. Haggard, Y. Rosetti, & M. Kawato (Eds.), *Sensorimotor foundations of higher cognition. Attention and performance XX11* (pp. 427–451). Oxford: Oxford University Press.

Csibra, G., & Gergely, G. (2006). Social learning and social cognition: The case for pedagogy. In Y. Munakata & M. H. Johnson (Eds.), *Processes of change in brain and cognitive development. Attention and performance XXI* (pp. 249–274). Oxford: Oxford University Press.

Dapretto, M., Davies, M. S., Pfeifer, J. H., Scott, A. A., Sigman, M., Brookheimer, S. Y., et al. (2006). Understanding emotions in others: Mirror neuron dysfunction in children with autism spectrum disorders. *Nature Neuroscience, 9*, 28–30.

Decety, J., & Chaminade, T. (2003). When the self represents the other: A new cognitive neuroscience view on psychological identification. *Consciousness and Cognition, 12*, 577–596.

Driver, J., Davis, G., Ricciardelli, P., Kidd, P., Maxwell, E., & Baron-Cohen, S. (1999). Gaze perception triggers reflexive visuospatial orienting. *Visual Cognition, 6*, 509–540.

Farroni, T., Csibra, G., Simion, F., & Johnson, M. H. (2002). Eye contact detection in humans from birth. *Proceedings of the National Academy of Sciences of the United States of America, 99*, 9602–9605.

Farroni, T., Mansfield, E. M., Lai, C., & Johnson, M. H. (2003). Infants perceiving and acting on the eyes: Tests of an evolutionary hypothesis. *Journal of Experimental Child Psychology, 85*, 199–212.

Fay, W. H., & Hatch, V. R. (1965). Symptomatic echopraxia. *American Journal of Occupational Therapy, 19*, 189–191.

Fernald, A., & Simon, T. (1984). Expanded intonation contours in mother's speech to newborns. *Developmental Psychology, 20*, 104–113.

Field, T., Field, T., Sanders, C., & Nadel, J. (2001). Children with autism display more social behaviours after repeated imitation sessions. *Autism*, *5*(3), 317–323.

Fridlund, A. J. (1994). *Human facial expression: An evolutionary view*. San Diego, CA: Academic Press.

Gergely, G., & Csibra, G. (2006). Sylvia's recipe: The role of imitation and pedagogy in the transmission of human culture. In: N. J. Enfield & S. C. Levinson (Eds.), *Roots of human sociality: Culture, cognition, and human interaction* (pp. 229–255). Oxford: Berg Publishers.

Gergely, G., & Csibra, G. (2005). The social construction of the cultural mind: Imitative learning as a mechanism of human pedagogy. *Interaction Studies*, *6*, 463–481.

Gergely, G., Bekkering, H., & Kiraly, I. (2002). Rational imitation in preverbal infants. *Nature*, 415, 755.

Gliga, T., & Csibra, G. (2007). Seeing the face through the eyes: A developmental perspective on face expertise. *Progress in Brain Research*, *164*, 323–339.

Greenfield, P. M., & Smith, J. H. (1976). *The structure of communication in early language development*. New York: Academic Press.

Grice, P. (1975). Logic and conversation. In P. Cole & J. Morgan (Eds.), *Syntax and semantics, Vol. 3* (pp. 41–58). New York: Academic Press.

Griffin, R. (2006). Social learning in the non-social: Imitation, intentions, and autism. *Developmental Science*, *5*(1), 30–32.

Hamilton, A. F. (in press). Emulation and mimicry for social interaction: A theoretical approach to imitation in autism. *Quarterly Journal of Experimental Psychology*.

Hamilton, A. F., Brindley, R. M., & Frith, U. (2007). Imitation and action understanding in autistic spectrum disorders: How valid is the hypothesis of a deficit in the mirror neuron system? *Neuropsychologia*, *45*(8), 1859–1868.

Heider, F., & Simmel, M. (1944). An experimental study of apparent behavior. *American Journal of Psychology*, *57*, 1377–1259.

Heyes, C. M., & Bird, G. (in press). Mirroring, association and the correspondence problem. In P. Haggard, Y. Rossetti, & M. Kawato (Eds.). *Sensorimotor foundations of higher cognition. Attention and performance XX11*. Oxford: Oxford University Press.

Heyes, C. M., Bird, G., Johnson, H., & Haggard, P. (2005). Experience modulates automatic imitation. *Cognitive Brain Research*, *22*, 233–240.

Hobson, R. P., & Meyer, J. A. (2005). Foundations for self and other: A study in autism. *Developmental Science*, *8*(6), 481–491.

Horner, V., & Whiten, A. (2005). Causal knowledge and imitation/emulation switching in chimpanzees (Pan troglodytes) and children (Homo sapiens). *Animal Cognition*, *8*, 164–181.

Hughes, C. (1996). Planning problems in autism at the level of motor control. Journal of *Autism and Developmental Disorders*, *26*, 99–107.

Iacoboni, M. (2005). Understanding others: Imitation, language, empathy. In S. Hurley & N. Chater (Eds.), *Perspectives on imitation: From cognitive neuroscience to social science* (Vol. 1, pp. 77–99). Cambridge, MA: MIT Press.

Iacoboni, M., Woods, R. P., Brass, M., Bekkering, H., Mazziotta, J. C., & Rizzolatti, G. (1999). Cortical mechanisms of human imitation. *Science*, *286*, 2526–2528.

Jeannerod, M. (2006). *Motor cognition: What actions tell the self*. Oxford: Oxford University Press.

Johnson, M. H., Griffin, R., Csibra, G., Halit, H., Farroni, T., de Haan, M., et al. (2005). The emergence of the social brain network: Evidence from typical and atypical development. *Development and Psychopathology*, *17*, 599–619.

Johnson, M. H., Dziurawiec, S., Ellis, H. D., & Morton, J. (1991). Newborns' preferential tracking of face-like stimuli and its subsequent decline. *Cognition*, *40*, 1–19.

Kampe, K. K. W., Frith, C. D., & Frith, U. (2003). "Hey John": Signals conveying communicative intention toward the self activate brain regions associated with "mentalizing," regardless of modality. *Journal of Neuroscience*, 23, 5258–5263.

Kiraly, I., Csibra, G., & Gergely, G. (2004). The role of communicative-referential cues in observational learning during the second year. Poster presented at the 14th Biennial International Conference on Infant Studies, May 2004, Chicago, IL, USA.

Klin, A. (2000). Attributing social meaning to ambiguous visual stimuli in higher-functioning autism and asperger syndrome: The social attribution task. *Journal of Child Psychology and Psychiatry*, *41*(7), 831–846.

Klin, A. (1991). Young autistic children's listening preferences in regard to speech: A possible characterization of the symptom of social withdrawal. *Journal of Autism and Developmental Disorders*, *21*(1), 29–42.

Klin, A., Jones, W., Schultz, R., Volkmar, F., & Cohen, D. (2002). Visual fixation patterns during viewing of naturalistic social situations as predictors of social competence in individuals with autism. *Arch Gen Psychology*, *59*, 809–816.

Koski, L., Wohlschlager, A., Bekkering, H., Woods, R. P., & Dubeau, M. C. (2002). Modulation of motor and premotor activity during imitation of target-directed actions. *Cerebral Cortex*, *12*, 847–855.

Kuhl, P. K., Coffey-Corina, S., Padden, D., & Dawson, G. (2005). Links between social and linguistic processing of speech in preschool children with autism: Behavioural and electrophysiological measures. *Developmental Science*, *8*(1), F1.

Leekam, S. R., Lopez, B., & Moore, C. (2000). Attention and joint attention in preschool children with autism. *Developmental Psychology*, *26*(2), 261–273.

Lord, C., Rutter, M., & Le Couteur, A. (1994). Autism diagnostic interview – Revised: A revised version of a diagnostic interview for caregivers of individuals with possible pervasive developmental disorders. *Journal of Autism and Developmental Disorders*, *24*, 659–685.

Loth, E. (in press). Abnormalities in 'cultural knowledge' in autism spectrum disorders: A link between behavior and cognition? McGregor, M. Nunez, K. R. Williams, & J. C. Gomez (Eds.), *An integrated view of autism: Perspectives from neurocognitive, clinical and intervention research*. Oxford: Blackwell.

McIntosh, D. N., Reichmann-Decker, A., Winkielman, P., & Willbarger, J. L. (2006). When the social mirror breaks: Deficits in automatic, but not voluntary, mimicry of emotional facial expressions in autism. *Developmental Science*, *9*(3), 295–302.

Meltzoff, A. N. (1995). Understanding the intentions of others: Re-enactment of intended acts by 18-month-old children. *Developmental Psychology*, *31*, 1–16.

Meltzoff, A. N. (1988). Infant Imitation After a 1-Week Delay: Long-term memory for novel acts and multiple stimuli. *Developmental Psychology*, *24*(4), 470–476.

Muthukumaraswamy, S. D., Johnson, B. W., & McNair, N. A. (2004). Mu rhythm modulation during observation of an object-directed grasp. *Cognitive Brain Research*, *19*, 195–201.

Munson, J., Dawson, G., Abbott, R., Faja, S., Webb, S. J., Friedman, S. D., et al. (2006). Amygdalar volume and behavioral development in autism. *Archives of General Psychiatry*, *63*(6), 686–693.

Nadig, A. S., Ozonoff, S., Young, G. S., Rozga, A., Sigman, A., & Rogers, S. J. (2007). A prospective study of response to name in infants at risk for autism. *Archives of Pediatrics and Adolescent Medicine*, *161*(4), 378–383.

Nagell, K., Olguin, R., & Tomasello, M. (1993). Processes of social learning in the imitative learning of chimpanzees and human children. *Journal of Comparative Psychology*, *107*, 174–186.

Nehaniv, C. L., & Dautenhahn, K. (2002). The correspondence problem. In K. Dautenhahn & C. L. Nehaniv (Eds.), *Imitation in animals and artifacts*. Cambridge, MA: MIT Press.

Nishitani, N., & Hari, R. (2000). Temporal dynamics of cortical representation for action. *Proceedings of the National Academy of Sciences*, *97*, 913–918.

Nishitani, N., Avikainen, S., & Hari, R. (2004). Abnormal imitation-related cortical activation sequences in Asperger's syndrome. *Annals of Neurology*, *55*, 558–562.

Oberman, L. M., Hubbard, E. M., McCleery, J. P., Altschuler, E. L., Ramachandran, V. S., & Pineda, J. A. (2005). EEG evidence for mirror neuron dysfunction in autism spectrum disorders. *Cognitive Brain Research, 24*, 190–198.

Osterling, J., & Dawson, G. (1994). Early recognition of children with autism: A study of first birthday home videotapes. *Journal of Autism and Developmental Disorders, 24*(3), 247–257.

Pineda, J. (2005). The functional significance of mu rhythms. *Brain Research Reviews, 50*, 57–68.

Press, C., Bird, G., Flach, R., & Heyes, C. M. (2005). Robotic movement elicits automatic imitation. *Cognitive Brain Research, 25*, 632–640.

Rizzolatti, G., & Fabbri Destro, M. (2007). Understanding actions and the intentions of others: The basic neural mechanism. *European Review, 15*, 209–222.

Rizzolatti, G., & Craighero, L. (2004). The mirror-neuron system. *Annual Review of Neuroscience, 27*, 169–192.

Royeurs, H., Van Oost, P., & Bothuyne, S. (1998). Immediate imitation and joint attention in young children with autism. *Developmental Psychopathology, 10*, 441–450.

Rogers, S. J., Hepburn, S., Stackhouse, T., & Wehner, E. A. (2003). Imitation performance in toddlers with autism and those with other developmental disorders. *Journal of Child Psychology and Psychiatry, 44*(5), 763–781.

Rogers, S. J., Bennetto, L., McEvoy, R., & Pennington, B. F. (1996). Imitation and pantomime in high-functioning adolescents with autism spectrum disorders. *Child Development, 67*, 2060–2073.

Rogers, S. J., & Pennington, B. F. (1991). A theoretical approach to the deficits in infantile autism. *Development and Psychopathology, 3*, 137–162.

Sabbagh, A. (1999). Communicative intentions and language: Evidence from right-Hemisphere damage and autism. *Brain and Language, 70*, 29–69.

Schultz, R., Romanski, L. M., & Tsatsanis, K. D. (2000). Neurofunctional models of autistic disorder and Asperger syndrome. In A. Klin, F. R. Volkmar, & S. S. Sparrow (Eds.), *Asperger syndrome* (pp. 172–209). New York: Guilford Press.

Sebanz, N., Knoblich, G., Stumpf, L., & Prinz, W. (2005). Far from action-blind: Representation of others' actions in individuals with autism. *Cognitive Neuropsychology, 22*(3), 433–454.

Sebanz, N., Knoblich, G., & Prinz, W. (2003). Representing others' actions: Just like one's own? *Cognition, 88*, B11–B21.

Senju, A., Tojo, Y., Dairoku, H., & Hasegawa, T. (2004). Reflexive orienting in response to eye gaze and an arrow in children with and without autism. *Journal of Child Psychology and Psychiatry, 45*(3), 445–458.

Senju, A., Yaguchi, K., Tojo, Y., & Hasegawa, T. (2003). Eye contact does not facilitate detection in children with autism. *Cognition, 89*, B43–B51.

Smith, I. M., & Bryson, S. E. (1994). Imitation and action in autism: A critical review. *Psychological Bulletin, 116*, 259–273.

Somogyi, E., Kiraly, I., Gergely, G., de Portzamparc, V., & Nadel, J. (2006). The development of intentional imitation – A comparative study. Poster presented at the International Conference on Infant Studies, June 2006, Kyoto, Japan.

Southgate, V., Chevallier, C., & Csibra, G. (under review). Sensitivity to commnicative relevance tells young children what to imitate.

Tai, Y. F., Scherfler, C., Brooks, D. J., Sawamoto, N., & Castiello, U. (2004). The human premotor cortex is 'mirror' only for biological actions. *Current Biology, 14*, 117–120.

Theoret, H., Halligan, E., Kobayashi, M., Fregni, F., Tager-Flusberg, H., & Pascual-Leone, A. (2006). Impaired motor facilitation during action observation in individuals with autism spectrum disorder. *Current Biology, 15*(3), R84–R85.

Tiegerman, E., & Primavera, L. (1984). Imitating the autistic child: Facilitating communicative gaze behavior. *Journal of Autism and Developmental Disorders, 11*, 427–438.

Tomasello, M., Carpenter, M., & Liszkowski, U. (2007). A new look at infant pointing. *Child Development*, 78(3), 705–722.
Vouloumanos, A., & Werker, J. F. (2007). Listening to language at birth: evidence for a bias for speech in neonates. *Developmental Science*, *10*(2), 159–171.
Wang, A. T., Lee, S. S., Sigman, M., & Dapretto, M. (2006). Neural basis of irony comprehension in children with autism: the role of prosody and context. *Brain*, *129*, 932–943.
Werner, E., Dawson, G., Osterling, J., & Nuhad, D. (2000). Recognition of autism spectrum disorder before one year of age: A retrospective study based on home videotapes. *Journal of Autism and Developmental Disorders*, *30*(2), 157–162.
Whiten, A., & Brown, J. (1999). Imitation and the reading of other minds: Perspectives from the study of autism, normal children and non-human primates. In S. Braten (Ed.), Intersubjective communication and emotion in early ontogeny. Cambridge: Cambridge University Press.
Williams, J. H. G., Waiter, G. D., Gilchrist, A., Perrett, D. I., Murray, A. D., & Whiten, A. (2006). Neural mechanisms of imitation and 'mirror neuron' functioning in autistic spectrum disorder. *Neuropsychologia*, *44*, 610–621.
Williams, J. H. G., Whiten, A., & Singh, T. (2004). A systematic review of action imitation in autistic spectrum disorder. *Journal of Autism and Developmental Disorders*, *34*(3), 285–299.
Wohlschlager, A., & Bekkering, H. (2002). Is human imitation based on a mirror-neuron system? *Experimental Brain Research*, *143*, 335–341.

Neural Simulation and Social Cognition

Shaun Gallagher

Abstract This article reviews the claim that mirror neurons are simulating neurons and the basis of an implicit simulation theory in regard to how we understand other persons. I claim that the equation of mirror system activation with an implicit simulation is unjustified, and I offer an alternative interpretation of the scientific data. The alternative considers mirror system activation as underlying part of an enactive perception in the social context.

Keywords Simulation · Mirror Neurons · Social Cognition · Embodiment · Enactive Perception

1 Introduction

Like many important discoveries the discovery of mirror neurons gains more and more significance as scientists work out its implications in a variety of areas. Unfortunately this can lead to an overenthusiastic explanatory over-extension. Not unlike the enthusiasm for computational models forty years ago, when some people thought that eventually everything could be explained in computational terms, it sometimes seems as if mirror neurons are involved in all aspects of cognition and human behavior. This has even been exploited in the popular press. Thus, in a recent article in the *New York Times*, mirror neurons are said to explain not only how we are capable of understanding another person's actions, but also language, empathy, "how children learn, why people respond to certain types of sports, dance, music and art, why watching media violence maybe harmful and why many men like pornography" (Blakeslee, 2006). Without denigrating the importance of mirror neurons, I suggest that we may have to wait until a lot more science is done before endorsing many of these claims. Perhaps the most established of these claims, however, is the first

---

S. Gallagher
Philosophy and Cognitive Sciences, University of Central Florida, Orlando, FL, USA
e-mail: gallaghr@mail.ucf.edu

J.A. Pineda (ed.), *Mirror Neuron Systems*, DOI: 10.1007/978-1-59745-479-7_16,

one: that mirror neurons must be part of an explanation of how we are capable of understanding another person's actions. I'm not a skeptic on this point. Indeed, I'm an enthusiast. But I want to be a careful enthusiast in this regard, and I suspect that how we should understand this claim is closely tied both to the fine details of how mirror neurons function, and to the way that we interpret concepts such as perception, simulation, and understanding.

One interpretation of how mirror neurons can fit into an explanation of how we understand the actions of others has been gaining in prominence amongst both neuroscientists and philosophers. Indeed, I think it is safe to say that this is the dominant view of most of the important researchers in this area. Simply stated, the view is that the function of mirror neurons (or the mirror system) is to simulate the actions of others. In this paper I want to offer a critical discussion of this claim and propose an alternative interpretation.

2 Explicit and Implicit Versions of Simulation Theory

Claims that mirror neurons function to simulate the actions of others and that this simulation is what provides our understanding of the other person's actions have been made by many researchers. In the *New York Times* article cited above, for example, Sandra Blakeslee quotes Giacomo Rizzolatti: "Mirror neurons allow us to grasp the minds of others not through conceptual reasoning but through direct simulation. By feeling, not by thinking." She also quotes Marco Iacoboni: "When you see me perform an action - such as picking up a baseball - you automatically simulate the action in your own brain." I start with these two quotations, not only because these two neuroscientists have done important work on mirror neurons but also because they already indicate an important distinction in the claims that are made about mirror neurons and simulation. Rizzolatti suggests that mirror neurons allow us to simulate the *minds* of other; Iacoboni suggests that they allow us to simulate the *actions* of others.[1] Rizzolatti's claim, of course, is a much stronger and controversial one and it ties directly to the dominant theory-of-mind (ToM) approaches to social cognition that speak of 'mind-reading' or 'mentalizing.' Both claims, however, are closely tied to one particular version of ToM: simulation theory.

Simulation theory (ST) is usually contrasted with 'theory theory' (TT). The latter suggests that we 'mind-read' the other person's mind by taking a theoretical stance. We infer the other's beliefs, desires, and intentions based on a theory (a folk psychology) of how people experience, act and behave. In contrast, ST argues that

[1] Elsewhere, however, Iacoboni does invoke mindreading: "Mirror neurons suggest that we pretend to be in another person's mental shoes. In fact, with mirror neurons we do not have to pretend, we practically are in another person's mind. ... You either simulate with mirror neurons, or the mental states of others are completely precluded to you" (quoted in Than 2005).

instead of a theory, we use our own mind as a model to simulate what must be going on in the other's mind. Given what we can see of the other's overt behavior, we consider what it would be like to be in their situation (we put ourselves 'in their shoes') and we create corresponding simulations or imaginary enactments of 'as if' beliefs, desires, or intentional states in our own minds. We then project those mental states onto the other person as an explanation of their behavior or as a way to predict their behavior. Both TT and ST hold to the mentalizing supposition, that is, that to understand other persons we need to figure out their mental states – we need to be engaged in some form of mind-reading.

Simulation theorists, however, are not in full agreement on every issue. Some claim that the simulation in question is explicit and involves the exercise of conscious imagination and deliberative inference (e.g., Goldman, 2005a); some insist that the simulation, although explicit, is non-inferential in nature (Gordon, 2005), and finally there are those who argue that the simulation, rather than being explicit and conscious, is implicit and subpersonal. The latter position is the one taken up by those who claim that mirror neurons function to simulate the actions or minds of others. Vittorio Gallese is perhaps the best representative of this view.

> Whenever we are looking at someone performing an action, beside the activation of various visual areas, there is a concurrent activation of the motor circuits that are recruited when we ourselves perform that action. ... Our motor system becomes active *as if* we were executing the very same action that we are observing. ... Action observation implies action simulation ... our motor system starts to covertly simulate the actions of the observed agent (Gallese, 2001, 37).

To understand what simulation means in this implicit version of ST, we need to look first at the original explicit version, and here the best representative is the philosopher, Alvin Goldman.

According to the explicit version of ST, simulation involves conscious or introspective mental states in which I imagine (simulate) myself in the other's situation and use that model to predict the other's mental states. Goldman, for example, argues that simulation is explicit insofar as it involves a conscious introspective use of the imagination to conceptually manipulate propositional attitudes (beliefs, desires). "When a mindreader tries to predict or retrodict someone else's mental state by simulation, she uses pretense or imagination to put herself in the target's 'shoes' and generate the target state" (Goldman, 2005a). According to Goldman, simulation involves three steps.

> First, the attributor creates in herself pretend states intended to match those of the target. In other words, the attributor attempts to put herself in the target's 'mental shoes.' The second step is to feed these initial pretend states [e.g., beliefs] into some mechanism of the attributor's own psychology ... and allow that mechanism to operate on the pretend states so as to generate one or more new states [e.g., decisions]. Third, the attributor assigns the output state to the target ... [e.g., we infer or project the decision to the other's mind]. (Goldman, 2005b, 80–81.)

One problem that appears in the very first step can motivate us to consider the role of mirror neurons even in explicit simulation models: "the attributor creates in herself pretend states intended to match those of the target." It seems that the simulator already has some sense of what is going on with the other person. Where does that knowledge come from and why is not that already the very thing we are trying to explain. Hybrid theorists who combine TT and ST suggest that folk psychology provides, not a sense of what is going on with the other person, but some general rules about how people think and behave in a certain situations, and that this is what the simulationist can use to initiate the first step (e.g., Currie & Ravenscroft, 2002). In contrast, Goldman appeals to sub-personal mirror resonance processes, although he then faces the problem of how to translate these processes into a conceptual grasp of propositional attitudes (see Goldman, 2006).

It is important to understand that the claim of explicit simulation theorists is not that we use this kind of simulation process occasionally. Rather, it is a more universal claim. On the explicit version of ST, simulation is not only explicit but pervasive. That is, we use it all the time, or at least it is the default way of understanding others. Goldman (2002: 7–8) thinks this is a moderate claim.

> The strongest form of ST would say that all cases of (third-person) mentalization employ simulation. A moderate version would say, for example, that simulation is the *default* method of mentalization ... I am attracted to the moderate version .... Simulation is the primitive, root form of interpersonal mentalization.

This claim to near universality motivates what I have called the simple phenomenological objection.[2] If simulation is both explicit and pervasive, then one should have some awareness of the different steps that one goes through as one consciously simulates the other's mental states. But there is no phenomenological evidence for this. When I encounter another person there is no experiential evidence that I use such conscious (imaginative, introspective) simulation routines. That is, when we consult our own common experience of how we understand others, we do not find such processes. Of course, this is not to say that we never use conscious simulations, but that is a telling point in itself. Confronted with some strange or unaccountable behavior it may be the case that we do try to understand the other person by running a simulation routine. This is, however, the rare case and it tends to stand out in its rarity. That is, I can easily become aware that I am in fact taking this approach and it is all the more apparent when I do this, simply because it tends to be the exception. But this suggests that I do not employ simulation in the usual everyday circumstance.

This simple phenomenological objection can be defeated, however, if ST adopts the idea of implicit simulation. Indeed, the implicit version of ST can be used as an argument against the explicit version of ST: if understanding others is in fact mediated by an implicit and automatic simulation process, then we have little need for the more explicit version. To the extent that implicit ST would explain the phenomenological scarcity of explicit simulation, it would support

[2] See Gallagher (2007) for this and other objections to explicit ST.

the simple phenomenological argument against explicit simulation. This is quite consistent with the claim made by Gallese: "Whenever we face situations in which exposure to others' behavior require a response by us, be it active or simply attentive, we seldom engage ourselves in an explicit, deliberate interpretive act. Our understanding of a situation most of the time is immediate, automatic, and almost reflex like" (2005:102).

Implicit ST appeals to good neuroscientific evidence involving sub-personal activation of mirror neurons, shared representations, or more generally, resonance systems. Let us take a close look at the precise statements of implicit ST. We know that one's motor system reverberates or resonates in one's encounters with others. My motor system is activated when I perceive another person performing an intentional action, for example. Mirror neurons in the premotor cortex, in Broca's area, and in the parietal cortex of the human brain are activated both when the subject engages in specific instrumental actions, and when the subject observes someone else engage in those actions (Rizzolatti et al. 1996; 2000). Also, specific overlapping neural areas (shared representations), in parts of the frontal and parietal cortexes, are activated under the following conditions: (1) when I engage in intentional actions; (2) when I observe some other person engage in that action; and (3) when I imagine myself or another person engage in that action (e.g., Grèzes & Decety, 2001). One claim that can be made by explicit simulation theorists is that these processes underpin (or are the neural correlates) of explicit acts of simulation (Goldman, 2006; Ruby & Decety, 2001). For the implicit simulation theorists, however, these subpersonal processes themselves just are a simulation of the other's intentions. Gallese captures it clearly in his claim that activation of mirror neurons involves "automatic, implicit, and nonreflexive simulation mechanisms ..." (Gallese, 2005, 117; also see Gallese, 2007). Gallese refers to his model as the 'shared manifold hypothesis' and distinguishes between three levels (2001, 45):

- The *phenomenological level* is the one responsible for the sense of similarity ... that we experience anytime we confront ourselves with other human beings. It could be defined also as the *empathic* level ....
- The *functional level* can be characterized in terms of simulation routines, *as if* processes enabling models of others to be created.
- The *subpersonal level* is instantiated as the result of the activity of a series of mirror matching neural circuits.

On this hypothesis, at the explicit, phenomenological level, one is not explicitly simulating; rather one is experiencing an empathic sense of the other person, and this is the result of a simulation process that happens on the subpersonal level.

Implicit ST understood in these or in similar terms is, as we have seen, the growing consensus. Indeed, use of the term 'simulation' is becoming the standard way of referring to mirror system activation. Thus, for example, Marc Jeannerod and Elizabeth Pacherie write:

> As far as the understanding of action is concerned, we regard simulation as the default procedure .... We also believe that simulation is the root form of interpersonal mentalization and that it is best conceived as a hybrid of explicit and implicit processes, with subpersonal neural simulation serving as a basis for explicit mental simulation (Jeannerod and Pacherie, 2004, p. 129; see Jeannerod, 2001; 2003).

Jean Decety and Julie Grèzes (2006, 6), explaining Rizzolatti's position, put it this way:

> By automatically matching the agent's observed action onto its own motor repertoire without executing it, the firing of mirror neurons in the observer brain simulates the agent's observed action and thereby contributes to the understanding of the perceived action.

Goldman (2006) now distinguishes between simulation as a high-level (explicit) mind-reading and simulation as a low-level (implicit) mind-reading where the latter is "simple, primitive, automatic, and largely below the level of consciousness" (p. 113), and the prototype for which is "the mirroring type of simulation process" (p. 147). That mirror neuron activation is a simulation not only of the goal of the observed action but of the intention of the acting individual, and therefore a form of mind-reading, is suggested by research that shows mirror neurons discriminate identical movements according to the intentional action and contexts in which these movements are embedded (Fogassi et al. 2005; Icoboni et al. 2005; Kaplan and Iacoboni, 2007). Neural simulation has also been extended as an explanation of how we grasp emotions and pain in others (Avenanti and Aglioti, 2006; Minio-Paluello, Avenanti, and Aglioti, 2007; Gallese, Eagle, & Migone, 2007). The idea that 'simulator neurons' are responsible for understanding actions, thoughts, and emotions is taken up by Oberman and Ramachandran (2007) who amass evidence that the mirror neuron system as an internal simulation mechanism is dysfunctional in cases of autism.

Thus a growing number of researchers appear to take up the simulation terminology as a generally accepted way to describe mirror system processes. In their research papers they do not (and of course, it is not expected that they should) identify this interpretation as controversial, even if they are familiar with the debates in these areas and have thought about the use of this term. Statements such as "specific perceptual contexts automatically trigger body action simulation in motor areas" (Urgesi et al. 2007, 31) are made in a matter-of–fact way, so that the idea of neural simulation is fast becoming the default or received doctrine (see e.g., Molnar-Szakacs et al. 2006; Newman-Norlund et al. 2007; Uddin et al. 2005).

3 Why Mirror Processes Are Not Simulations

I want to suggest that there are several things wrong with calling mirror system processes simulations. Consider, first, the meaning of 'simulation' as defined by ST. Specifically two aspects are of importance here: (1) simulation involves pretense; (2) simulation has an instrumental character, that is, it is characterized in terms of a mechanism or model that we manipulate or control in order to understand something to which we do not have instrumental access. We find both aspects discussed in the ST literature; indeed, they are ubiquitous and considered essential to the concept of simulation. Dokic and Proust (2002, viii) provide a good example of the instrumental aspect: Simulation means '*using* one's own evaluation and reasoning mechanisms as a model for theirs.' Gordon (2004: 1) locates this instrumentalism at the neuronal level by suggesting that on the 'cognitive-scientific' model, "one's own behavior control system is employed as a *manipulable model* of other such systems. (This is not to say that the 'person' who is simulating is the model; rather, only that *one's brain can be manipulated to model other persons*)." Both the instrumental and the pretense aspects are reflected in Goldman's (2002, 7) explanation: Simulation involves 'pretend states' where, "by pretend state I mean some sort of surrogate state, which is *deliberately adopted* for the sake of the attributor's task . . . In simulating practical reasoning, the attributor *feeds* pretend desires and beliefs into her own practical reasoning system." Adams (2001, 384) indicates that "it is a central feature of ST that one takes perceptual inputs off-line," that is, that simulation involves pretense. Bernier (2002, 34) takes both instrumental and pretense aspects to be essential elements of simulation.

> According to ST, a simulator who runs a simulation of a target would use the resources of her own decision making mechanism, in an 'off-line' mode, and then the mechanism would be fed with the mental states she would have if she was in the target's situation.

The aspect of pretense seems essential for simulation if it is to be distinguished from a theoretical model or a simple practice of reasoning (see Fisher, 2006). Simulation involves the use of the model 'as if' I were in the other person's situation. As Gallese puts it, "our motor system becomes active *as if* we were executing that very same action that we are observing" (2001: 37). Likewise for Gordon (2005: 96) the neurons that respond when I see your intentional action, respond "*as if* I were carrying out the behavior . . ."

For ST, then, the concept of simulation clearly needs to meet these two conditions: it is a process that I control in an instrumental way (in the explicit version it is 'deliberately adopted'), and it involves pretense (I put myself 'as if' in the other person's shoes). It seems clear, however, that neither of these conditions is met by mirror neurons. First, in regard to the instrumental aspect, if simulation is characterized as a process that I (or my brain) instrumentally use(s), manipulate(s), or control(s), then it seems clear that what is happening in the implicit processes of motor resonance is not simulation. We, at the personal level, do not manipulate or control the activated brain areas – in fact, we have no instrumental access to neuronal activation, and we cannot use it as a model. Nor does it make sense to say, *pace* Gordon, that at the subpersonal level the brain itself is *using* a model or methodology, or that one set of neurons makes use of another set of neurons as a model in order to generate an understanding of something else. Indeed, in precisely the intersubjective circumstances that we are considering, these neuronal systems do not take the initiative; they do not activate themselves. Rather, they are activated by the other person's action. The other person *has an effect on us* and *elicits* this activation. It is not us (or our brain) *initiating* a simulation; it is the other who does this to us via a perceptual elicitation.

Second, in regard to pretense, in sub-personal mirror processes there is no pretense. Obviously, as vehicles or mechanisms, neurons either fire or do not fire. They do not pretend to fire. Furthermore, as representations, what these neurons represent or register cannot involve pretense in the way required by ST. Since mirror neurons are activated both when I engage in intentional action and when I see you engage in intentional action, the mirror system is neutral with respect to the agent; no first- or third-person specification is involved (deVignemont, 2004; Gallese, 2005; Hurley, 2005; Jeannerod and Pacherie, 2004). In that case, it is not possible for them to register *my* intentions as pretending to be *your* intentions; there is no 'as if' of the sort required by ST because there is no 'I' or 'you' represented.

One could argue that the instrumental and pretense conditions are not necessary conditions for simulation and that a necessary condition for simulation is something more minimal. Goldman (2006; Goldman & Sripada, 2005), for example, giving some worry to the concept of neural simulation, acknowledges a discrepancy between the ST definition of simulation and the working of subpersonal mirror processes. "Does [the neural simulation] model really fit the pattern of ST? Since the model posits unmediated resonance, it does not fit the usual examples of simulation in which pretend states are created and then operated upon by the attributor's own cognitive equipment (e.g. a decision-making mechanism), yielding an output that gets attributed to the target. . . ." To address this discrepancy Goldman and Sripida propose a generic definition of simulation:

> However, we do not regard the creation of pretend states, or the deployment of cognitive equipment to process such states, as essential to the generic idea of simulation. The general idea of simulation is that the simulating process should be similar, in relevant respects, to the simulated process. Applied to mindreading, a minimally necessary condition is that the state ascribed to the target is ascribed as a result of the attributor's instantiating, undergoing, or experiencing, that very state. In the case of successful simulation, the experienced state matches that of the target. This minimal condition for simulation is satisfied [in the neural model] (Goldman and Sripada, 2005, 208).

Goldman (2006, 131 ff.) realizes, however, that if this is a necessary condition, it is not a sufficient one, since on this generic definition and without something further, it is not clear why I would treat this state as something other than simply my own state, or as representative of someone else's state. What one would have to add to this minimal condition in order to make it a simulation of the other's action, intention, or mental state are precisely the pretense and instrumental conditions that are being eliminated. Another way to say this is that the very notion of simulation implies that it is a personal-level process. Indeed, in part, this is what motivates Goldman to adopt a hybrid simulation model. If implicit mirroring, in the sense of being in a matching state, may be considered a minimal condition for simulation, for the full-fledged simulation that constitutes mindreading (the attribution of mental states to another) one requires higher-level processes of 'classification' (which he characterizes as introspective) and 'projection' (2006, 245 ff.). Although pretense is not required for low-level mirroring, it, or something like it, seems necessary to realize high-level simulation and mindreading (49).

Let me further suggest, however, against ST, that the minimal condition of matching, or any simulation that one can build on this, cannot be the pervasive or default way of attaining an understanding of others. There are many cases of encountering others in which we simply do not adopt, or find ourselves in, a matching state. Furthermore, with respect to implicit ST, if simulation were as automatic as mirror neurons firing, then it would seem that we would not be able to attribute a state different from our own to someone else. But we do this all the time. Consider the example of the snake woman. I see a woman in front of me enthusiastically and gleefully reaching to pick up a snake; at the same time, I am experiencing revulsion and disgust about that very possibility. Her action, which I fully sense and understand from her enthusiastic and gleeful expression to be something that she likes to do, triggers in me precisely the opposite feelings. In this case, neither my neural states nor my motor actions (I maybe retreating with gestures of disgust just as she is advancing toward the snake with gestures of enthusiasm), nor my feelings/cognitions match hers. Yet I understand her actions and emotions (which are completely different from mine), and I do this without even meeting the minimal necessary condition for simulation, that is, matching my state to hers.

In this regard, also consider the difficulties involved if we were interacting with more than one other person, or trying to understand others who are interacting with each other. Is it possible to enter into the same, or what are likely different states, and thereby simulate the neural/motor/mental/emotional states of more than one person at the same time? Or can we alternate quickly enough, going back and forth from one person to the other, if in fact our simulations must be such that we instantiate, undergo, or experience, the states in question? How complicated does it get if there is a small crowd in the room? Would there not be an impossible amount of cognitive work or subpersonal matching required to predict or to understand the interactions of several people if the task involves simulating their mental states, especially if in such interpersonal interactions the actions and intentions of each person are affected by the actions and intentions of the others (Morton, 1995 makes a similar point).[3]

---

[3] The requirement that the simulation has to be concretely similar also raises problems for the instrumental and pretense conditions even for the explicit version of ST. If our simulation has to be concretely similar to the simulated state for it to be considered a simulation, assuming explicit instrumental control of our simulation process, how will we know how to run or control our

Beyond such examples, the scientific research on mirror neurons suggests good reasons to think mirror neuron activation cannot involve a precise match between motor system execution and observed action. Csibra (2005) points out that conservatively, between 21 and 45% of neurons identified as mirror neurons are sensitive to multiple types of action; of those activated by a single type of observed action, that action is not necessarily the same action defined by the motor properties of the neuron; approximately 60% of mirror neurons are 'broadly congruent,' which means there maybe some relation between the observed action(s) and their associated executed action, but not an exact match. Only about one-third of mirror neurons show a one-to-one congruence.[4] Newman-Norlund et al. (2007, 55) suggest that activation of the broadly congruent mirror neurons may represent a complementary action rather than a similar action. In that case they could not be simulations.

Again, in denying that mirror neurons are simulating or specifically creating a match in such cases, I am not denying that mirror neurons may be involved in our interactions with others, possibly contributing to our ability to understand others or to keep track of ongoing intersubjective relations. I am simply suggesting that mirror processes do not constitute simulations in any acceptable use of that term and that furthermore many other kinds of processes are going on in the brain that complicate our explanation of how we understand others. Moreover, I do not want to deny that in some cases we may in fact engage in explicit simulation, although I suggest these are rare cases, and that in some of these cases we maybe engaged in a different kind of simulation from the standard dyadic model described by ST. Matthew Ratcliffe (2007) has proposed that as we engage with another person we may sometimes in fact simulate, not their mental states, but our interaction, for example, in an attempt to predict where our interaction is heading. One can generalize this and suggest that we may also simulate the potential interaction of agents other than ourselves. I say 'potential' interaction, because if it is a current interaction, whether between ourselves and someone else, or between other agents that we can see or overhear, etc., then there would be very little reason to try to simulate what is directly accessible.

simulation unless we already know in some detail what the other's state is like. And how do we come by that knowledge? If the answer is through simulation, then we have an infinite regress. In regard to explicit pretense, Fisher (2006), who models simulation as a reasoning process, rejects this aspect as inconsistent with simulation being concretely similar, for if we simulate a reasoning process, we are really reasoning, and not just pretending to.

[4] Csibra concludes: "With strongly unequal distribution of types of action or types of grip, one could find a relatively high proportion of good match between the [observed action vs executed action] domains even if there were no causal relation between them. Without such a statistical analysis, it remains uncertain whether the cells that satisfy the definition of 'mirror neurons' (i.e., the ones that discharge both with execution and observation of actions) do indeed have 'mirror properties' in the everyday use of this term (i.e., are generally activated by the same action in both domains)" (2005, 3).

4 An Alternative Interpretation

I want to suggest that an alternative interpretation of mirror neuron activation is possible, one that is more parsimonious with the way that we normally come to understand others in our everyday encounters. Rather than simulation, mirror neuron activation can easily be interpreted as part of the neuronal processes that underlie intersubjective *perception*. That is, the articulated neuronal processes that include activation of mirror neurons or shared representations may underpin a non-articulated immediate perception of the other person's intentional actions, rather than a distinct process of simulating their intentions. This claim requires that we conceive of perception as a temporal phenomenon and as an enactive process.

We note that mirror neurons fire 30–100 milliseconds after appropriate visual stimulation (Gallese, personal correspondence). This short amount of time between activation of the visual cortex and activation of the mirror neuron system raises the question of where precisely to draw the line between perceptual processes and something that would count as a sub-personal simulation. Even if it is possible to distinguish at the neuronal level between activation of the visual cortex and activation of the pre-motor cortex, this certainly does not mean that this constitutes a distinction between processes that are purely perceptual and processes that involve something more than perception. I think two issues can be raised in this regard.

The first issue is this: deciding that mirror neurons function as simulations depends on invoking the step-wise model developed at the explicit, conscious or personal level, and looking for that step-wise model at the neuronal level. On various versions of ST, simulation begins with perception and ends with some form of inferential understanding. We first see an action that we need to understand; we then simulate it in our own mind or motor system; and then we infer something about the other's experience. But even if neuronal processes that involve information flow from sensory cortex to pre-motor cortex take some time (as much as 100 milliseconds), it is not clear that we should identify these step-wise processes as perception plus simulation, rather than a temporally extended and enactive perceptual process.

Second, if we think of perception as an enactive process (Hurley, 1998; Noë, 2004; Varela, Thompson, and Rosch, 1991), as involving sensory-motor skills rather than as just sensory input/processing – as an active, skillful, embodied engagement with the world rather than as the passive reception of information from the environment – then it may be more appropriate to think of mirror resonance processes as part of the structure of the perceptual process when it is a perception of another person's actions. Fogassi and Gallese, despite their simulationist interpretation, put this point clearly: "perception, far from being just the final outcome of sensory integration, is the result of sensorimotor coupling" (2002, 27). Mirror activation, on this interpretation, is not the initiation of simulation; it is part of a direct intersubjective perception of what the other is doing.

This interpretation of mirror neuron activation provides a tight fit with an alternative account of intersubjective understanding and interaction which is not framed in terms of ToM or ST. On this account the capacities for human interaction and intersubjective understanding are already operative in infancy in certain embodied practices – practices that are emotional, sensory-motor, perceptual, and nonconceptual. These embodied practices constitute our primary way of understanding others and they continue to do so even after we attain our more sophisticated abilities in this regard (Gallagher, 2001; 2005).

This primary, perceptual sense of others is already implicit in the behavior of the newborn. In neonate imitation, which depends on a sense that the other is of the same sort as oneself (Bermúdez, 1996; Gallagher, 1996; Gallagher & Meltzoff, 1996), infants are able to distinguish between inanimate objects and agents. The fact that they imitate only *human* faces (Johnson, 2000; Johnson et al., 1998; Legerstee, 1991) suggests that infants are able to parse the surrounding environment into those entities that perform human actions and those that do not (Meltzoff and Brooks, 2001). For the infant, the other person's body presents opportunities for action and expressive behavior – opportunities that it can pursue through imitation. There is, in this case, a common bodily intentionality that is shared by the perceiving subject and the perceived other. From early infancy humans (and perhaps some animals) have capabilities for an interaction with others that fall under the heading of what Trevarthen (1979; 1980) calls 'primary intersubjectivity.' On this view the mind of the other person is not something that is entirely hidden away and inaccessible. Rather, in most intersubjective situations we have a direct perceptual access to another person's intentions because their intentions are explicitly expressed in their embodied actions and their expressive behaviors.

The early capabilities that contribute to primary intersubjectivity constitute an immediate, non-mentalistic mode of interaction. Infants, notably without the intervention of theory or simulation, are able to see bodily movement as goal-directed intentional movement, and to perceive other persons as agents. This does not require advanced cognitive abilities or simulation skills; rather, it is a perceptual capacity that is 'fast, automatic, irresistible and highly stimulus-driven' (Scholl and Tremoulet, 2000, 299). Evidence for this early, non-mentalistic interpretation of the intentional actions of others can be found in numerous studies. The infant, even at 9 months, follows the other person's eyes (Senju, Johnson and Csibra, 2007), and starts to perceive various movements of the head, the mouth, the hands, and more general body movements as meaningful, goal-directed movements. Baldwin and colleagues, for example, have shown that infants at 10–11 months are able to parse some kinds of continuous action according to intentional boundaries (Baldwin & Baird, 2001; Baird & Baldwin, 2001). Such perceptions give the infant, by the end of the first year of life, a non-mentalistic, perceptually-based embodied understanding of the intentions and dispositions of other persons (Allison, Puce, and McCarthy, 2000; Baldwin, 1993; Johnson, 2000; Johnson et al. 1998).

Infants do not simply perceive others in an observational mode; they perceive while interacting with them. With the onset of joint attention around 9–14 months (Phillips, Baron-Cohen, and Rutter, 1992), they perceive others interacting with the world. In such interactions, the child looks to the body and the expressive movement of the other to discern the intention of the person or to find the meaning of some object. The child can understand that the other person *wants* food or *intends* to open the door; that the other can *see* him (the child) or is *looking at* the door. This is not taking an intentional stance, that is, treating the other *as if* they had desires or beliefs hidden away in their minds; rather, the intentionality is there to be perceived in the embodied actions of others.

Primary intersubjectivity also includes affective coordination between the gestures and expressions of the infant and those of caregivers with whom they interact. Infants "vocalize and gesture in a way that seems 'tuned' [affectively and temporally] to the vocalizations and gestures of the other person" (Gopnik & Meltzoff, 1997, 131). Infants at 5 to 7 months detect correspondences between visual and auditory information that specify the expression of emotions (Walker, 1982). The perception of emotion in the movement of others, however, does not involve creating a simulation of some inner state. It is a perceptual experience of an embodied comportment (Bertenthal, Proffitt, and Cutting, 1984; Moore, Hobson, and Lee, 1997).

This kind of perception-based understanding, therefore, is not a form of mindreading or mentalizing. In seeing the actions and expressive movements of the other person, one already sees their meaning in the context of the surrounding world; no inference to a hidden set of mental states (beliefs, desires, etc.) is necessary. At the phenomenological level, when I see the other's action or gesture, I see (I *immediately perceive*) the meaning in the action or gesture. I see the joy or I see the anger, or I see the intention in the face or in the posture or in the gesture or action of the other. I see it. No simulation of what is readily apparent is needed, and no simulation of something more than this is required in the majority of contexts.

This is far from the complete story of social cognition. The fuller version includes (1) the role of secondary intersubjectivity: shared attention beginning at around one year of age allows us to understand other people's actions in terms of the pragmatic and social contexts of everyday life (Trevarthen & Hubley, 1978); (2) the expectations associated with culturally defined social roles; and (3) the enriching interpretive processes associated with narrative competency that lead to the more nuanced understandings in adulthood (Gallagher and Hutto, 2007).

In the case of seeing someone reach to pick up a snake, for example, it seems possible that my understanding of the other's enthusiastic action is constituted purely in enactive perceptual processes that would include activation of mirror neurons but that would not involve a matching-state simulation. Indeed, the contrast between her enthusiastic action, which I perceive, and my repulsion, which I feel, may motivate a more explicit interpretive process. That is, depending on circumstances, I may explicitly frame her action in a narrative that makes sense of it (e.g., "this woman is, after all, the curator of the zoo's snake house, she's trained to handle snakes, and does it all the time"), or if there is some puzzling element involved (e.g., this woman is my wife who I know to be afraid of snakes) I may even engage in one of those rare instances of explicit simulation (e.g., assuming I am not in a position to ask her what she's doing, I might try to imagine what she is thinking – perhaps that this is a toy snake etc.).

This alternative, non-simulationist interpretation of the neuroscience of mirror neurons, then, suggests that before we are in a position to theorize, simulate, explain, or predict mental states in others, we are already in a position to interact with and to understand others in terms of their expressions, gestures, and purposive movements, reflecting their intentions and emotions. By way of the capabilities involved in primary intersubjectivity we already have specific perception-based understandings about what others feel, whether they are attending to us or not, how they are acting toward ourselves and others, whether their intentions are friendly or not, and so forth; and we have this without the need to simulate what the other person believes or desires. What Merleau-Ponty (1968) calls an 'intercorporeality,' an intermodal communication between a proprioceptive-kinaesthetic sense of one's own body and the perceived body of the other, is already functioning from birth, and continues to function throughout our social life.[5] In primary intersubjectivity, a common bodily intentionality is shared across the perceiving subject and the perceived other, and the evidence from research on mirror neurons and resonance systems in social neuroscience supports this. If mirror neuron activation is part and parcel of intersubjective perception, we can say that even as young infants, we map the visually perceived actions of others onto our own sensory-motor systems. This is not a simulation of something that lies hidden behind or beneath the action that we grasp perceptually since that action is already expressive of intentions. Intentions are expressed in action, not as an expression of something more than what we can perceive, but simply as what the action means.

References

Adams, F. (2001). Empathy, neural imaging, and the theory versus simulation debate. *Mind and Language*, *16* (4): 368–92.

Allison, T., Puce, Q., and McCarthy, G. (2000). Social perception from visual cues: role of the STS region. *Trends in Cognitive Science*, *4* (7): 267–278.

Avenanti, A., & Aglioti S. M. (2006). The sensorimotor side of empathy for pain. In M. Mancia (ed.), *Psychoanalysis and neuroscience* (pp. 235–256). Milan: Springer.

[5] In the phenomenological tradition, this idea of kinaesthetic coordination was expressed by Husserl as early as 1908 (see Husserl 1973). See Berthoz and Petit (2006).

Baird, J. A. & Baldwin, D. A. (2001). Making sense of human behavior: Action parsing and intentional inference. In B. F. Malle, L. J. Moses, & D. A. Baldwin (Eds.). *Intentions and intentionality: Foundations of social cognition* (pp. 193–206). Cambridge, MA: MIT Press.
Baldwin, D. A. (1993). Infants' ability to consult the speaker for clues to word reference. *Journal of Child Language*, *20*, 395–418.
Baldwin, D. A. & Baird, J. A. (2001). Discerning intentions in dynamic human action. *Trends in Cognitive Science*, *5* (4), 171–78.
Bermúdez, J. L. (1996). The moral significance of birth. *Ethics*, *106*, 378–403.
Bernier, P. (2002). From simulation to theory. In J. Dokic & J. Proust (eds.), *Simulation and knowledge of action* (pp. 33–48). Amsterdam: John Benjamins.
Bertenthal, B. I., Proffitt, D. R. & Cutting, J. E. (1984). Infant sensitivity to figural coherence in biomechanical motions. *Journal of Experimental Child Psychology*, *37*, 213–30.
Berthoz, A. & Petit, J-L. (2006). *Phénoménologie et physiologie de l'action*. Paris: Odile Jacob.
Blakeslee, S. (2006). Cells that read minds. *New York Times*, January 10, 2006. http://www.nytimes.com/2006/01/10/science/10mirr.html. Accessed January 2006; May 2007.
Csibra, G. (2005). Mirror neurons and action observation. Is simulation involved? ESF Interdisciplines. http://www.interdisciplines.org/mirror/papers/.
Currie, G. & Ravenscroft, I. (2002). *Recreative Minds*. Oxford: Oxford University Press.
deVignemont, F. (2004). The co-consciousness hypothesis. *Phenomenology and the Cognitive Sciences*, *3*(1): 97–114.
Decety, J. & Grèzes, J. (2006). The power of simulation: Imagining one's own and other's behavior. *Brain Research*, 1079, 4–14.
Dokic, J. & Proust, J. (2002). Introduction. In J. Dokic and J. Proust (Eds.), *Simulation and knowledge of action* (pp. vii–xxi). Amsterdam: John Benjamins.
Fisher, J. C. (2006). Does simulation theory really involve simulation? *Philosophical Psychology*, *19* (4), 417–432
Fogassi, L., Ferrari, P.F., Gesierich, B., Rozzi, S., Chersi, F. & Rizzolatti, G., (2005). Parietal lobe: From action organization to intention understanding. *Science*, *308*, 662–667.
Fogassi, L. & Gallese, V. (2002). The neural correlates of action understanding in non-human primates. In M. I. Stamenov & V. Gallese (Eds.), Mirror Neurons and the Evolution of Brain and Language (pp. 13–35). Amsterdam: John Benjamins Publ.
Gallagher, S. (1996). The moral significance of primitive self-consciousness. *Ethics*, *107*, 129–140.
Gallagher, S. (2001). The practice of mind: Theory, simulation, or interaction? *Journal of Consciousness Studies*, *8* (5–7), 83–107.
Gallagher, S. (2005). *How the Body Shapes the Mind*. Oxford: Oxford University Press.
Gallagher, S. (2007). Simulation trouble. *Social Neuroscience*, *2* (3–4), 353–365.
Gallagher, S. & Hutto, D. (2008). Primary interaction and narrative practice. In: Zlatev, Racine, Sinha & Itkonen (Eds.). *The Shared Mind: Perspectives on Intersubjectivity* (pp. 17–38). Amsterdam: John Benjamins.
Gallagher, S. & Meltzoff, A. (1996). The earliest sense of self and others: Merleau-Ponty and recent developmental studies. *Philosophical Psychology*, 9: 213–236.
Gallese, V. (2007). Before and below 'theory of mind': embodied simulation and the neural correlates of social cognition. *Philosophical Transactions of the Royal Society, B-Biological Sciences*, *362* (1480), 659–669.
Gallese, V. (2005). 'Being like me': Self-other identity, mirror neurons and empathy, In Hurley, S. & Chater, N. (eds.), *Perspectives on Imitation I* (pp. 101–118). Cambridge, MA: MIT Press.
Gallese, V. (2001). The 'shared manifold' hypothesis: from mirror neurons to empathy'. *Journal of Consciousness Studies, 8*, 33–50.
Gallese, V., Eagle, M.N. & Migone, P. (2007). Intentional attunement: Mirror neurons and the neural underpinnings of interpersonal relations. *Journal of the American Psychoanalytic Association, 55* (1), 131–176.

Goldman, A. I. (2002). Simulation theory and mental concepts. In J. Dokic & J. Proust (Eds.), *Simulation and knowledge of action* (pp. 1–19). Amsterdam: John Benjamins.

Goldman, A. (2005a). Mirror Systems, Social Understanding and Social Cognition. *Interdisciplines.* (http://www.interdisciplines.org/mirror/papers/3).

Goldman, A. (2005b). Imitation, mind reading, and simulation. In Hurley and Chater (Eds.) *Perspectives on Imitation* II (pp. 79–93). Cambridge, MA: MIT Press.

Goldman, A. (2006). *Simulating minds: The philosophy, psychology and neuroscience of mindreading*. Oxford, England: Oxford University Press.

Goldman, A. I. Sripada, C. S. (2005). Simulationist models of face-based emotion recognition. *Cognition*, *94* (2005), 193–213.

Gopnik, A. & Meltzoff, A. (1997). *Words, Thoughts, and Theories.* Cambridge, MA: MIT Press.

Gordon, R. M. (2004). Folk psychology as mental simulation. In N. Zalta (Ed.), *The Stanford Encyclopedia of Philosophy.* (http://plato.stanford.edu/archives/fall2004/entries/folk psych-simulation/).

Gordon, R. M. (2005). Intentional agents like myself. In S. Hurley & N. Chater (eds.), *Perspectives on Imitation* I (pp. 95–106). Cambridge, MA: MIT Press.

Grèzes, J. & Decety, J. (2001). Functional anatomy of execution, mental simulation, and verb generation of actions: A meta-analysis. *Human Brain Mapping* 12, 1–19.

Hurley, S. L. (2005). Active perception and perceiving action: The shared circuits model. In T. Gendler & J. Hawthorne (Eds.), *Perceptual experience*. New York: Oxford University Press.

Hurley, S. L. (1998). *Consciousness in action*. Cambridge, MA: Harvard University Press.

Husserl, E. (1973). *Ding und Raum*. Husserliana 16. The Hague: Martinus Nijhoff.

Iacoboni, M., Molnar-Szakacs, I., Gallese, V., Buccino, G., Mazziotta, J. & Rizzolatti, G. (2005). Grasping the intentions of others with one's own mirror neuron system. *PLoS Biology*, *3* (79), 1–7.

Jeannerod, M. (2003). The mechanism of self-recognition in humans. *Behavioural Brain Research*, *142*: 1–15

Jeannerod, M. (2001) Neural simulation of action: A unifying mechanism for motor cognition. *Neuroimage*, *14*, 103–109.

Jeannerod, M. & Pacherie, E. (2004). Agency, simulation, and self-identification. *Mind and Language*, *19* (2): 113–46.

Johnson, S. C. (2000). The recognition of mentalistic agents in infancy. *Trends in Cognitive Science*, *4*: 22–28.

Johnson, S. et al. (1998). Whose gaze will infants follow? The elicitation of gaze-following in 12-month-old infants. *Developmental Science*, 1: 233–38.

Kaplan, J. T. & Iacoboni, M. (2006). Getting a grip on other minds: Mirror neurons, intention understanding, and cognitive empathy. *Social neuroscience*. 1 (3–4): 175–83.

Legerstee, M. (1991). The role of person and object in eliciting early imitation. *Journal of Experimental Child Psychology*, 51: 423–33.

Meltzoff, A. N. & Brooks, R. (2001). 'Like Me' as a building block for understanding other minds: Bodily acts, attention, and intention. In B. F. Malle, *et al.* (eds.), *Intentions and Intentionality: Foundations of Social Cognition* (pp. 171–191). Cambridge, MA: MIT Press.

Merleau-Ponty, M. (1968). *The Visible and the Invisible*. Evanston: Northwestern University Press.

Minio-Paluello I, Avenanti A, and Aglioti, S. M. (2006). *Social Neuroscience* 1 (3–4): 320–333.

Molnar-Szakacs I., Kaplan J., Greenfield PM, Iacoboni M. (2006). Observing complex action sequences: The role of the fronto-parietal mirror neuron system. *Neuroimage* 33 (3): 923–935.

Moore, D. G., Hobson, R. P. & Lee, A. (1997). Components of person perception: An investigation with autistic, non-autistic retarded and typically developing children and adolescents. *British Journal of Developmental Psychology,* 15: 401–423.

Morton, A. (1996). Folk psychology is not a predictive device. *Mind* 105: 119–37.

Newman-Norlund, RD. Noordzij, ML. Meulenbroek, RGJ, Bekkering, H. (2007). Exploring the brain basis of joint attention: Co-ordination of actions, goals and intentions. *Social Neuroscience* 2 (1): 48–65.

Noë, A. (2004). *Action in Perception*. Cambridge, MA: MIT Press.

Oberman, L. M. & Ramachandran, V. S. (2007). The Simulating Social Mind: The Role of the Mirror Neuron System and Simulation in the Social and Communicative Deficits of Autism Spectrum Disorders. *Psychological Bulletin* 133 (2): 310–327.

Phillips, W., Baron-Cohen, S. & Rutter, M. (1992). The role of eye-contact in the detection of goals: Evidence from normal toddlers, and children with autism or mental handicap. *Development and Psychopathology*, 4: 375–83.

Ratcliffe, M.J. (2006). *Rethinking Commonsense Psychology: A Critique of Folk Psychology, Theory of Mind and Simulation*. Basingstoke: Palgrave Macmillan.

Rizzolatti, G., Fogassi, L. & Gallese V. (2000). Cortical mechanisms subserving object grasping and action recognition: A new view on the cortical motor functions. In M. S. Gazzaniga (ed.), *The New Cognitive Neurosciences* (pp. 539–52). Cambridge, MA: MIT Press.

Rizzolatti, G., Fadiga, L., Gallese V. & Fogassi, L. (1996). Premotor cortex and the recognition of motor actions. *Cognitive Brain Research*, 3: 131–141.

Ruby, P., Decety, J., (2001). Effect of subjective perspective taking during simulation of action: a PET investigation of agency. *Nature Neuroscience* 4, 546–550.

Senju, A, Johnson MH and Csibra G. (2006). The development and neural basis of referential gaze perception. *Social Neuroscience* 1 (3–4): 220–234.

Scholl, B. J. & Tremoulet, P. D. (2000). Perceptual causality and animacy. *Trends in Cognitive Sciences*, 4 (8): 299–309.

Than, K. (2005). Scientists Say Everyone Can Read Minds. *LiveScience* 27 April 2005 (http://www.livescience.com/humanbiology/050427_mind_readers.html)

Trevarthen, C. (1980). The foundations of intersubjectivity: Development of interpersonal and cooperative understanding of infants. In D.R. Olson (ed.), *The Social Foundation of Language and Thought: Essays in Honor of Jerome S. Bruner* (316–341). New York: Norton.

Trevarthen, C. (1979). Communication and cooperation in early infancy: A description of primary intersubjectivity. In M. Bullowa (ed.), *Before Speech*. Cambridge: Cambridge University Press.

Trevarthen, C. & Hubley, P. (1978). Secondary intersubjectivity: Confidence, confiding and acts of meaning in the first year. In A. Lock (ed.), *Action, gesture and symbol:The emergence of language*. San Diego,CA: Academic Press.

Urgesi, C. Candidi, M. Ionta, S., & Aglioti, S.M. (2007). Representation of body identity and body actions in extrastriate body area and ventral premotor cortex. *Nature Neuroscience* 10 (1): 30–31.

Uddin, LQ, Kaplan JT, Molnar-Szakacs I, Zaidel E, and Iacoboni, M. (2005). Self-face recognition activates a frontoparietal "mirror" network in the right hemisphere: an event-related fMRI study. *Neuroimage* 25 (3): 926–935.

Varela, F. J., Thompson, E. & Rosch, E. (1991). *The Embodied Mind: Cognitive Science and Human Experience*. Cambridge: MIT Press.

Walker, A. S. (1982). Intermodal perception of expressive behaviors by human infants. *Journal of Experimental Child Psychology*, *33*, 514–35.

Index

Printed in the United States of America